Springer Hydrogeology

Series Editor

Juan Carlos Santamarta Cerezal, San Cristóbal de la Laguna, Sta. Cruz Tenerife, Spain

The *Springer Hydrogeology* series seeks to publish a broad portfolio of scientific books, aiming at researchers, students, and everyone interested in hydrogeology. The series includes peer-reviewed monographs, edited volumes, textbooks, and conference proceedings. It covers the entire area of hydrogeology including, but not limited to, isotope hydrology, groundwater models, water resources and systems, and related subjects.

More information about this series at https://link.springer.com/bookseries/10174

Alejandro García Gil ·
Eduardo Antonio Garrido Schneider ·
Miguel Mejías Moreno ·
Juan Carlos Santamarta Cerezal

Shallow Geothermal Energy

Theory and Application

Alejandro García Gil
Geological and Mining Institute of Spain (IGME)
Spanish National Research Council (CSIC)
Madrid, Spain

Eduardo Antonio Garrido Schneider
Geological and Mining Institute of Spain (IGME)
Spanish National Research Council (CSIC)
Madrid, Spain

Miguel Mejías Moreno
Geological and Mining Institute of Spain (IGME)
Spanish National Research Council (CSIC)
Madrid, Spain

Juan Carlos Santamarta Cerezal
Department of Agricultural and Environmental Engineering
University of La Laguna (ULL)
Santa Cruz de Tenerife, Spain

ISSN 2364-6454 ISSN 2364-6462 (electronic)
Springer Hydrogeology
ISBN 978-3-030-92257-3 ISBN 978-3-030-92258-0 (eBook)
https://doi.org/10.1007/978-3-030-92258-0

Translation from the Spanish language edition: *Geotermia somera - fundamentos teóricos y aplicación* by Alejandro García Gil, et al., © Instituto Geológico y Minero de España 2020. Published by Instituto Geológico y Minero de España. All Rights Reserved.

This Springer imprint is published by the registered company Springer Nature Switzerland AG
The registered company address is: Gewerbestrasse 11, 6330 Cham, Switzerland

Contents

About the Authors

Alejandro García Gil received the Ph.D. degree in Geology from Zaragoza University in 2015. He subsequently joined the Geological Survey of Spain (IGME-CSIC), where he is currently coordinating the Applied Hydrogeology and Shallow Geothermal Energy Research Group. His principal research interest is shallow geothermal energy–aquifers interaction. He has worked extensively in city scale environmental impacts of groundwater heat pump systems. He has published over 70 papers in peer-reviewed journals and international conferences. During his research career, he has carried out different research stays at the Institute of Environmental Assessment and Water Research (IDAEA-CSIC) in Barcelona, the Department of Environmental Sciences of the University of Basel (Switzerland) and the British Geological Survey (BGS) in Oxford (UK). He received the *Casañal Poza* Prize and the Extraordinary Ph.D. Award from the University of Zaragoza. The International Association of Hydrogeologists (IAH-GE) awarded him the *Alfons Bayó* Prize for young researchers. e-mail: a.garcia@igme.es

Eduardo Antonio Garrido Schneider received the B.Sc. degree in Geological Sciences and a Diploma of Advanced Studies in External Geodynamics from the University of Zaragoza; a Diploma of Groundwater Hydrology from the Polytechnic University of Catalonia; and a Diploma of Environmental Engineering and Management from the School of Industrial Organisation. He joined IGME office in Zaragoza in 1990, participating and coordinating since then numerous projects on groundwater and hydrogeology, most of them in the Ebro basin. He is the author of more than 80 publications in scientific and technical journals, books and contributions to conferences. Between 2014 and 2018, he was a member of the Governing Board and Water Council of the Ebro Hydrographic Confederation. e-mail: e.garrido@igme.es

Miguel Mejías Moreno received his B.Sc. degree in Geological Sciences from the Complutense University of Madrid and his M.Sc. in Hydrogeological Technology from the Polytechnic University of Madrid. He also received a M.Sc. degree in General and Applied Hydrology from CEDEX. He joined the Spanish State's General Administration as a civil servant in 1986. He has held the position of Head of the Applied Hydrogeology Area of the Spanish Geological Survey (IGME) for 11 years and author or co-author of twelve books and more than 100 scientific and technical articles. He is a specialist in low permeability geological formations, regional hydrogeological studies and deep hydrogeology. He is Member of the Scientific Committee of National Parks, Member of the Scientific Council of the Spanish Committee of the UNESCO MAB Programme, Member representing the Ministry of Science and Innovation on the Governing Board and the Water Council of the Guadiana Hydrographic Demarcation and Member of the Tablas de Daimiel National Park Scientific Board. e-mail: m.mejias@igme.es

Juan Carlos Santamarta Cerezal received his Ph.D. degree in Hydraulic and Energy Engineering from the Universidad Politécnica de Madrid (UPM). He received a B.Sc. degree in Forestry Engineer and Mining Engineer from the UPM. He also holds a M.Sc. degree in Water Engineering from the University of Seville. Currently, he is Lecturer at the University of La Laguna. He is a collaborating researcher at the University Institute of Water and Environmental Sciences of the University of Alicante and the Institute for Water Research (IdRA) of the University of Barcelona. He has authored more than 200 scientific publications, including 50 articles, author or co-author of 12 books or scientific monographs, 14 books as scientific editor, 49 book chapters (national and international), 15 scientific-technical reports and 13 articles in specialised journals. He has participated in 50 national and international congresses. He has been the scientific, technical and advisory committee on 11 occasions. He has more than 100 R&D&I activities organised in the fields of water, energy and the environment. At the research level, he has participated in 33 R&D&I projects and contracts, 9 of them international (H2020, INTERREG Atlantic, Erasmus +, National Plan, regional), of which he has been the principal investigator (PI) in 24 of them. He is a scientific advisor for the hydrological plans of the Canary Islands and for the management of water resources in Equatorial Guinea. He has been distinguished with different awards, such as the Teaching Innovation Award (2013), the Canary Islands Research Award in Civil Engineering *Agustín de Betancourt* 2018 and the Award for Excellence in Internationalisation in 2021. e-mail: jcsanta@ull.es

Abbreviations

ACS	Agua Caliente Sanitaria
ASHRAE	American Society of Heating, Refrigerating and Air-Conditioning Engineers
ATES	Aquifer thermal energy storage
BHE	Borehole heat exchanger
BTES	Borehole thermal energy storage
CHE	Confederación Hidrográfica del Ebro
COP	Coefficient of performance
CTE	Código Técnico de Construcción
DHC	District Heating and Cooling Network
DMA	Directiva Marco del Agua
DX	Direct expansion
EAHP	Exhaust air heat pump
EGEC	European Geothermal Energy Council
EGRT	Enhanced geothermal response test
EPA	Environmental Protection Agency (USA)
ETIP	Plataformas Europeas de Tecnología e Innovación
ETT-NB	Energía Térmica Total no Balanceada
FLS	Finite line source
GEI	Gases de Efecto Invernadero
GEOTERZ	Modelo Geotérmico de la ciudad de Zaragoza
GHP	Guarded hot plate
GSHP	Ground source heat pump
GWHP	Groundwater heat pump
HAPs	Hidrocarburos Aromáticos Policíclicos
HDPE	High-density polyethylene
HFCs	Hidrofluorocarburos
HVAC	Heating, ventilation and air conditioning
HWST	Hot water storage tank
ICS	Infinite cylindrical source
IGME	Instituto Geológico y Minero de España

IGSHPA	International Ground Source Heat Pump Association
ILS	Infinite line source
IPCC	Intergovernmental Panel on Climate Change
IRENA	Agencia Internacional de Energías Renovables
IRF	Factor de Relajación Indirecto
LSI	Langelier saturation index
MILS	Moving infinite line source
MUSE	Managing urban shallow geothermal energy
NREAP	Planes de Acción Nacionales en Energía Renovable
PCTS	Perfil Característico de Temperaturas Subterráneas
PFCs	Perfluorocarbonos
RDPH	Reglamento del Dominio Público Hidráulico
RH&C	Renewable heating and cooling
RITE	Regulación para las Instalaciones Térmicas en Edificios
RSI	Ryznar stability index
SBM	Superposition borehole model
SCOP	Seasonal coefficient of performance
SEER	Seasonal energy efficiency ratio
STES	Seasonal thermal energy storage
SUHI	Subsurface urban heat island
TDS	Total dissolved solids
THM	Acoplamiento Termohidromecánico
TRT	Test de Respuesta Térmica
TTT	Test de Trazador Térmico
UHI	Urban heat island
UNFCCC	United Nations Framework Convention on Climate Change
UTA	Unidades Climatizadoras o de Tratamiento de Aire
UTES	Underground thermal energy storage
VER	Volume elemental representative
WGPS	Water-gravel pit storage

Symbols

Q	Heat (J)
m	Mass (kg)
C	Volumetric heat capacity (J m^{-3} K^{-1})
c	Specific heat capacity (J kg^{-1} K^{-1})
T	Absolute temperature (K)
J	Radiated heat flux (J s^{-1})
e	Emissivity (-)
σ	*Stefan–Boltzmann* constant (W m^{-2} K^{-4})
A	Surface area (m^2)
t	Time (s)
E	Energy (J)
P	Power (W)
U	Internal energy (J)
W	Work (J)
p	Pressure (kg m^{-1} s^{-2})
H	Enthalpy (J)
ρ	Density (kg m^{-3})
F	Force (N)
V	Volume (m^3)
η	Efficiency (-)
μ	Mass of a gas molecule (kg)
γ	Specific heat (-)
ϖ	*Boltzmann* constant (J K^{-1})
S	Entropy (J K^{-1})
r	Radius (m)
x	Position vector (m, m, m)
ϕ	Porosity (-)
l	Characteristic length (m)
q	Heat flow (W)
λ	Thermal conductivity (W K^{-1} m^{-1})
∇	Nabla operator (-)

x, y, z	Space coordinates (m)
g	Source/sink term (W m^{-3})
g	Earth's acceleration constant (m s^{-2})
∇^2	Laplacian operator (-)
u	Internal energy per unit mass (J kg^{-1})
α	Thermal diffusivity (m^2 s^{-1})
v	Fluid velocity (m s^{-1})
D	Hydrodynamic thermal dispersion (m)
R	Thermal resistivity (K W^{-1})
ϵ	Thermal expansion (K^{-1})
ε	Axial internal thermal expansion (-)
χ	Saturation (-)
τ	Shear stress (Pa)
μ	Dynamic viscosity (Pa s)
υ	Kinematic viscosity (m^2 s^{-1})
Re	Reynolds number (-)
k	Intrinsic permeability (m^2)
b	Heat transfer coefficient (W m^{-2} K^{-1})
Fo	Fourier number (-)
Pe	Peclet number (-)
K	Hydraulic permeability (m s^{-1})
h	Hydraulic head (m)
Q	Flow rate (m s^{-3})
ω	Medium compressibility (m kg^{-1} s^{-2})
S	Storage coefficient (m^{-1})
J	Radiation of heat (W m^{-2})
Υ	Long-wave radiation (W m^{-2})
G	CO_2 emissions (kg)
ξ	CO_2 emission factor (kg kWh^{-1})
D	Diameter (m)
a	Fit coefficient (-)
Θ	Dimensionless temperature (-)
Z	Depth dimensionless (-)
τ	Temperature change (K)
b	Aquifer (saturated) thickness (m)
φ	Velocity potential (s^{-1})
Ψ	Stream function (m^2 s^{-1})
R	Radius of influence (m)
a	Spacing distance (m)
κ	Heat recovery ratio (-)
ψ	Energy balance ratio (-)
Ex	Exergy (J)
Λ	Energy efficiency (-)
γ	*Euler-Mascheroni* constant (-)
δ	Extraction factor (-)

s Dropdown (m)
T Transmissivity ($m^2\ s^{-1}$)
M Retardation factor (-)

Superscripts

X_{sys} System
X_{ext} Exterior
X_W Work
X_Q Heat
X_C Convection
X_T Total
X_K Kinetics
X_V Potential
X_{ent} Environment
X_P Constant pressure
X_0 Initial
X_F Final
X_A High
X_B Low
X_{ref} Reference
X_{abs} Absolute
X_{rev} Reversible
X_{irrev} Irreversible
X_{Temp} Constant temperature
$X_{x,y,z}$ Coordinates in space
X_f Fluid
X_m Solid mineral
X_w Water
X_{fs} Source-sink
X_V Constant volume
X_e Equivalent
X_D Darcy
X_h Hydrocarbons
X_{ga} Gas
X_{tb} Pipe
X_{con} Heat conduction
X_{alm} Storage
X_{po} Pores
X_{co} Pores interconnected
X_r Real
X_{ef} Effective
X_{int} Internal

X_E	Latent
X_H	Sensible
X_a	Atmosphere
X_{ab}	Absorbed
X_t	Terrain
X_n	Clouds
X_G	Geothermal
X_{RF}	Refrigeration
X_{CF}	Heating
X_{cnd}	Condenser
X_{ev}	Evaporator
X_{ie}	Isentropic
X_{ad}	Adiabatic
X_{el}	Electric
X_{HP}	Heat pump
X_{sav}	Savings
X_{sps}	Substituted production system
X_i	In
X_o	Out
X_g	Grout
X_{th}	Thermal
X_{s}	Borehole
X_{sup}	Superior
X_{ft}	Convective heat resistance
X_{ss}	Steady state
X_I	Exchanger
X_{2D}	Bidimensional
X_{ca}	Capture
X_{re}	Residence
X_{max}	Maximum
X_{cr}	Critical
X_{st}	Stagnation
X_b	Heat front
X_{rq}	Recovered
X_{B}	Background
X_d	Dissipated
X_{TC}	In place
X_R	Reservoir
X_{min}	Minimum
X_{adv}	Advection
X_{TEC}	Technical
X_{suf}	Ground surface

X_{geo} Geological
X_s Specific
X_{fm} Mass flow

Superscripts

X^{adv} Advection
X^{dis} Hydrodynamic dispersion
X^r Reflection
X^* Absorption
X^{emi} Emission
X^T Total
X^{net} Net
X'' Per unit of cross section
X^o Per linear metre
X^{Δ} Triangular circuit
X^T Tensor
X^{rq} Recovered
X' Per mass unit

Chapter 1
Introduction

1.1 Background

Planetary change in climate and ecological functioning associated with the phenomenon of global warming is attributed, with a high degree of acceptance in the scientific community (IPCC 2007, 2011, 2013), although not universally, to the massive emission of greenhouse gases (GHGs) (carbon dioxide, methane, nitrogen oxides, sulphur hexafluoride, HFCs and PFCs) derived from the production, distribution and consumption of energy. This energy is obtained through the use of fossil fuels as the main primary energy source in the production of electricity, heat or locomotion.

Since the Industrial Revolution, the world's population has grown exponentially as has per capita energy consumption (Glassley 2010). In the mid-twentieth century, after the Second World War, in order to meet the growing energy demand, there was a boom in the use of petroleum derivatives and fossil fuels in general, their use spreading across the planet. Since then, the combustion of carbon-rich fossil fuels has increased the concentration of CO_2 and other GHGs in the atmosphere to concentrations never before experienced on this planet. There is ample scientific evidence linking global warming to anthropogenic GHG emissions. Evidence includes instrumental, glacial and sedimentary records. Recent technological advances, especially in the field of satellite remote sensing, have made it possible to obtain ground-scale data on the decrease in infrared radiation over the last 40 years (Brindley and Bantges 2016). This is seen as unequivocal evidence of the anthropogenic origin of GHGs being responsible for the increase in global warming and, therefore, climate change.

Different possible strategies to reduce the concentration of GHGs in the upper atmosphere to sustainable levels include incorporating large amounts of low-carbon resources into the energy sector. The use of primary energy not based on fossil fuels, so-called *low-carbon energy*, has been promoted as a first energy policy for several years.

A. García Gil et al., *Shallow Geothermal Energy*, Springer Hydrogeology,
https://doi.org/10.1007/978-3-030-92258-0_1

The concerns of the scientific, technological and social community lie in understanding the relationship between climate and the chemical composition of the atmosphere and oceans, as well as predicting the impact on climate of different possible energy production scenarios. Global energy demand will double by 2050 due to global economic and population growth, with a large impact from the emerging market economy (Whitesides and Crabtree 2007). In principle, this demand can be met by fossil fuel-based energy resources. However, even to maintain CO_2 levels in the atmosphere at twice pre-industrial levels by 2050 will require intervention in the energy production sector and decarbonisation at a magnitude equal to or greater than current global energy production.

The growing demand for energy can only be met in the future if supply is increased by using low-carbon technologies or through good management and deceleration of energy demand (Narsilio and Aye 2018).

The negotiation of protocols for action on climate change has been carried out at the international level through international conferences such as those held in Rio de Janeiro (Brazil) in 1992 and Kyoto (Japan) in 1997. These international conferences have sought to obtain commitments from participating nations to significantly reduce GHG emissions over the coming decades.

Everything seems to indicate the need to develop technologies capable of using primary energy sources that are not based on fossil fuels, that ideally do not run out and that are environmentally friendly, i.e. they must meet three conditions: (1) low-carbon energy (2) renewable energy and (3) clean energy. Energy sources and resources that meet these three premises are known as *renewable energy*.

The very concept of renewability depends on the rate of replenishment. If this replenishment occurs on the scale of a human lifetime, only then is it considered renewable. Oil and other hydrocarbons renew themselves but only after several hundred thousand years and are, therefore, not considered renewable. The most important renewable energy resources and sources are solar, hydro, wind, geothermal, solid biomass, biogas and biofuels. The contributions of these renewable energy sources and resources still play a minor role in global energy production. They have additional advantages, such as improving industrial profits and balance sheets, contributing to technological development and creating jobs.

There has been a succession of agreements within the framework of the *United Nations Framework Convention on Climate Change (UNFCCC)* to try to reduce carbon dioxide emissions into the atmosphere. At the 2015 convention (COP21), the 2015 *Paris Agreement* was signed, a binding climate agreement among 195 countries, with the aim of strengthening global response to the threat of climate change. A common goal was set to keep the average global temperature increase to less than 2 °C (preferably 1.5 °C) above pre-industrial levels, which entails keeping atmospheric CO_2 concentration below 450 ppm. A target was set to reduce GHG emissions by at least 40% below 1990 levels by 2030 (UNFCCC 2015).

The *Intergovernmental Panel on Climate Change* (IPCC) scientific consortium has identified the global need to phase out the use of fossil fuels in power generation by 20% by 2050 and to phase this use out completely by the end of the century.

We are searching for economically profitable energy generation technologies that meet society's energy demand and do not continue to damage the environment. However, there is also considerable scepticism that GHG emissions will be adequately reduced in time to significantly curb the effects of global warming. Strategies to be considered include reducing energy consumption, using energy sources that do not rely on carbon-intensive fossil fuels, and finding alternatives to atmospheric GHG emissions.

The use of shallow geothermal energy is one of the alternatives that contributes to achieving these goals. Geothermal heat pumps improve energy efficiency in air conditioning, compared to other conventional technologies that are based on the combustion of fossil fuels. They use renewable thermal energy and their GHG emissions are very low and even zero in some specific situations (EPA 1997; Saner et al. 2010). The Kyoto Protocol is being incorporated into European Community legislation and transposed into national legislation. In addition to requiring the energy efficiency of buildings, building regulations require the use of low-carbon technologies in thermal installations. In this context, the use of shallow geothermal heat production systems has important potential.

However, at the moment the whole world continues to consume energy based on fossil resources for social, economic and human development. Currently, 65% of the energy used for air conditioning in buildings comes from power plants based on fossil fuels such as hard coal and lignite, which are the source of the highest GHG emissions on the planet. Therefore, it is crucial to look for alternatives that require more efficient energy consumption for the air conditioning of these spaces.

Although the use of geothermal energy has clear advantages to be the technology with the most potential for decarbonisation of the global thermal sector, it is true that it does not necessarily lead to a drastic reduction of GHGs (Bayer et al. 2012; Blum et al. 2010; Saner et al. 2010). Shallow geothermal systems require a certain amount of electrical energy for operation of the integrated heat pump. If the fraction of electrical energy needed for the operation of a geothermal installation is based on fossil fuels, contribution to the reduction in greenhouse gas emissions is smaller. Disregarding the possible share of fossil fuel-derived electricity can lead to an overestimation of the technology's benefit, which has sometimes been counterproductive and has led to negative responses from the scientific community and industry. Overestimation of the technology's benefit has added to cases of bad practice due to poor design of shallow geothermal installations (Florea et al. 2017; Vienken et al. 2015). A responsible stance, adoption of good practices and efficient management of shallow geothermal resources are required.

In addition to the environmental cause, there are other compelling reasons to promote alternative primary energy sources over fossil fuels. From a geopolitical point of view, it is desirable to reduce dependence on fossil fuels, given the instability of the regions hosting the main hydrocarbon deposits and the obvious reasons for conflicts of interest between nations. From an economic and social point of view, shallow geothermal systems offer new services from borehole drilling companies, so that this activity is currently displacing the water borehole drilling market, especially in countries such as Norway and the UK (Banks 2011). Shallow geothermal activity has a positive impact on employment and the household economy, providing large savings to households in the medium term.

1.2 Shallow Geothermal Energy

1.2.1 Geothermal Energy

Geothermal energy, in the thermodynamic and general sense, refers to the internal energy of geological materials (rocks, groundwater, sediments, magma, etc.) contained between the earth's surface and down to the earth's core itself. Geothermal energy tends to transfer in a solidary way, from higher internal energy to lower internal energy. The process of transferring internal energy from one system to another, where internal energy is conserved, is known as heat. If, during the transfer of internal energy from one body to another, part of this energy is transformed into another type of energy, this transformation process is called work. Therefore, geothermal energy describes the potential ability of a geological system (rocks, groundwater, sediment, etc.) to transfer thermal energy in the form of heat or to perform work.

The study of how internal energy is transferred between systems is one of the fundamental objectives of thermodynamics. The geothermal energy discipline combines thermodynamic and geological knowledge to understand and quantify heat transfer in the subsurface and its possible exploitation as heat (direct use) or as work (electricity generation). Thermodynamics demonstrates how part of the heat flow from one body to another can be transformed into mechanical energy, through work on a steam turbine. It also establishes that the amount of work that can be done is a function of the difference in internal energy of the two systems.

1.2.2 Types and Classification of Geothermal Energy

In the absence of changes in the nature and internal structure of a geological thermodynamic system, the change in internal energy is accompanied by a more or less proportional increase in temperature. In the geological environment there is a great variety of systems at different temperatures. On a planetary scale, the highest internal energy of the planet is found in its core, with temperatures of about 6000 °C (Anzellini et al. 2013), more than a hundred times higher than earth's surface and atmosphere temperatures. The internal energy difference between these extremes induces an energy flow from the core to the earth's surface, known as terrestrial heat flow. The result of this continuous heat flow is an associated decreasing temperature gradient towards the surface. With this conceptual model, it can be roughly understood how geothermal energy is distributed on a planetary level.

In general, in regions where there is a normal geothermal gradient of approximately 25–32 °C km^{-1} (Limberger et al. 2018), boreholes of several kilometers deep are required to reach geological materials presenting temperatures exceeding 175 °C, which are needed for electrical power generation. Therefore, geological materials at a depth of several kilometers could be of economic interest for geothermal energy transformation into electricity through mechanical work. In volcanic areas, where the geothermal gradient can be much higher (up to 200 °C km^{-1}), electrical power can be generated by very shallow boreholes. This primary energy source has enormous potential, and its development has resulted in the production of electricity from geothermal energy on a commercial scale (Narsilio and Aye 2018). Once access to geological materials with sufficient internal energy is gained using deep piped boreholes, steam is recirculated and heated by *hot* rock at a high temperature. It is then conducted to steam turbines to generate work capable of producing economically exploitable electricity (Toth and Bobok, 2017). This type of geothermal exploitation and its study is known as *deep geothermal energy* or *high-temperature* (*enthalpy*) *geothermal* energy (Fig. 1.1).

Lower temperature geological materials with lower geothermal energy are closer to the surface and are therefore more technically and economically accessible. However, their internal energy transferred to another system in the surface would not be able to generate economically exploitable work. They do, however, constitute a thermal reservoir of great interest for internal energy transfer as heat. It should not be forgotten that Western society consumes approximately 50% of the total energy produced for heating and cooling (Sanner et al. 2013). The internal energy exchange of geological materials with surface thermodynamic systems to satisfy a heat demand without its transformation into work is called *direct geothermal use*. The term *direct* refers to the absence of transformation of internal energy into work; a direct use of geothermal energy is realised through the transfer of internal energy, i.e. heat.

Since prehistoric times, direct use of geothermal energy has been carried out in several ways such as in caves, avoiding environmental extreme temperatures, or in the use of natural hot springs. Throughout the history of mankind, attempts have been made to maximise productivity of thermal water reservoirs with existing drilling technology. With the advent of mechanical drilling machines in the nineteenth century until the present day, it has been possible to reach geothermal reservoirs of varying depths. This has made it possible to utilise geothermal energy for direct use in buildings and district heating networks. As drilling and heat pump technologies have been developed, the great potential for transferring thermal energy with the more accessible subsoil (first 400 m of depth) has become increasingly apparent, to a point where it has become economically viable, and even viewed as the most economical option in the long term. Two types of geothermal energy use can be distinguished according to the drilling technology considered which, in turn, is a function of the maximum drilling depth expected. These are deep well drilling technology (technology similar to oil drilling) and shallow well drilling technology (technology similar to groundwater drilling). The drilling cost per linear meter is radically different. In fact, deep drilling is usually only costeffective for oil wells. This is evidenced by the conversion of abandoned oil wells into geothermal wells (Templeton et al. 2014). This type

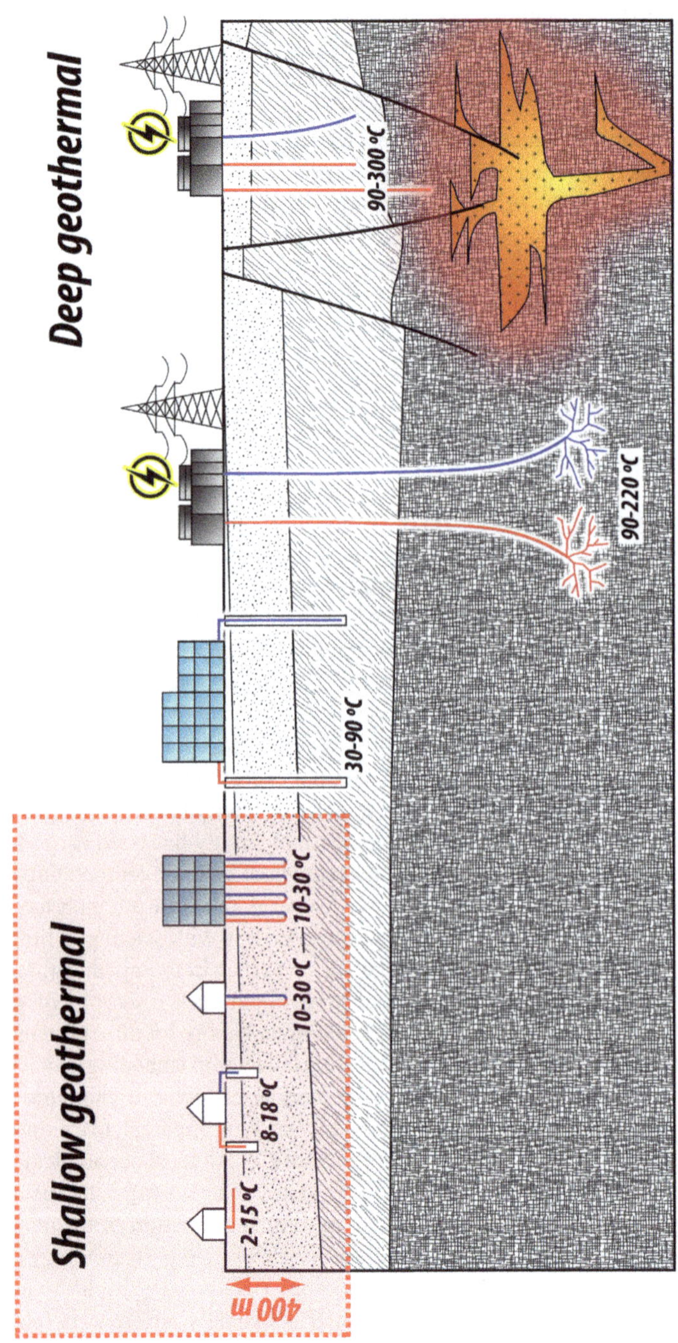

Fig. 1.1 Conceptual diagram of shallow geothermal energy versus medium and deep geothermal energy. Modified from www.alpine-space.eu/projects/greta

of geothermal exploitation and its study is known as *medium geothermal energy* or *medium or low temperature (enthalpy) geothermal energy* (Fig. 1.1). Shallow (arguably conventional) drilling technology, although cheaper, requires considerable initial investment. However, the low primary energy consumption in heat production enhances offsets of the initial investment, resulting in a quick payback. This type of geothermal exploitation and its study is known as *shallow geothermal energy* or *very low temperature (enthalpy) geothermal energy* (Fig. 1.1). Furthermore, it should be noted that while medium and deep geothermal energy (cogeneration) are capable of providing heat for heating, only shallow geothermal energy has potential for cooling, since the subsurface temperature in the shallow domain roughly coincides with the annual average atmospheric temperature of the region. Therefore, the shallower ground (<400 m) can provide heating and cooling, including domestic hot water (DHW), all year round (Banks 2012). Another very significant difference between shallow and deep geothermal energy, apart from its economic efficiency and its use as a heat sink, is its ubiquity. While deep and medium geothermal energy (to some extent) depend on thermal anomalies in the ground, shallow geothermal energy is available everywhere, independent of the geology found. Geology and hydrogeology will condition the design to maximise the efficiency of exploitation but do not question its economic viability. Figure 1.2 represents a summary of the classification of geothermal energy according to depth and type of end use.

Geothermal energy, in general, is a huge and versatile resource capable of helping to meet the world's energy demand and reduce the use of fossil fuels as primary

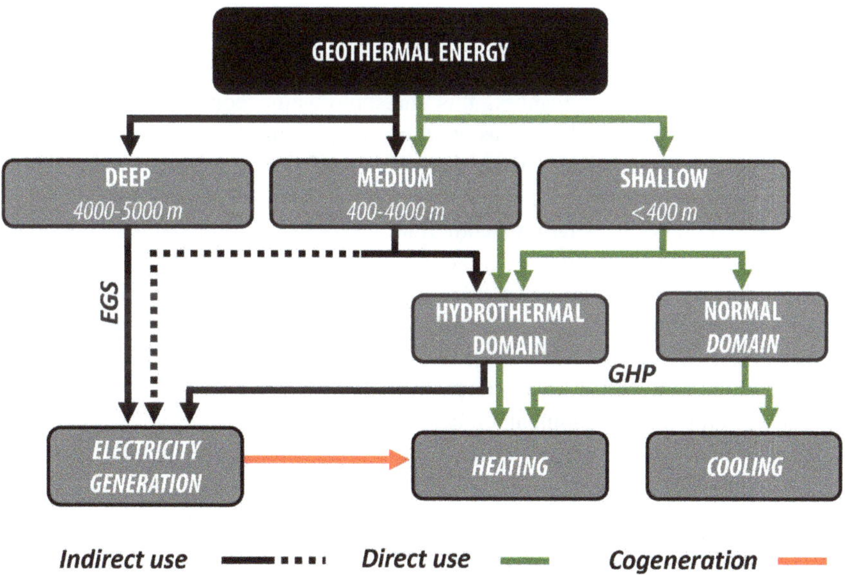

Fig. 1.2 Classification of geothermal energy according to depth and type of end use. EGS: Enhanced geothermal systems. GHP: Geothermal heat pump

energy. Geothermal energy can produce electricity and efficiently meet the needs for air conditioning and DHW generation in residential, commercial and industrial buildings (Glassley 2010; Narsilio and Aye 2018).

1.2.3 Shallow Geothermal Energy

The first 400 m below the ground surface is a unique thermal reservoir for the transfer and storage of thermal energy. All the thermal energy that can be transferred as heat to this thermal reservoir, either by heat dissipation or heat absorption, is called shallow geothermal energy. The shallowest subsurface, together with bodies of surface water and the atmosphere, form the main thermal reservoirs of ambient thermal energy (Fig. 1.3). The set of industrial processes aimed at optimising heat transfer with these environmental reservoirs are called *geothermal* (shallow), *hydrothermal* and *aerothermal.*

Among the ambient thermal reservoirs, the shallow geothermal reservoir is the most efficient in the long term. Nevertheless, this technology requires a significant initial investment in the drilling of the necessary geothermal heat exchangers (Fig. 1.3). The shallow geothermal reservoir is characterised by a constant stable temperature throughout the year and is renewable, thus offering immense potential, not only for heating, ventilation and air conditioning (HVAC) of domestic and commercial buildings, but also for heat production in industrial and other infrastructures.

The threshold value of 400 m depth is conditional on the depth limits of conventional drilling technologies needed to construct geothermal heat exchangers. As drilling technology advances, this threshold value could increase. Shallow geothermal energy is also known as very low temperature (or enthalpy) geothermal energy, since the vast majority of the shallow subsurface of the continental domain is in thermal equilibrium with atmospheric conditions and solar radiation (Oke 1987). Under these conditions, shallow geothermal reservoirs present stable temperatures throughout the year of about two degrees Celsius above the annual mean atmospheric temperature. The shallow geothermal reservoirs are constituted by rocks, unconsolidated sediment and groundwater with relatively similar thermal properties. These properties, which do not vary in large amounts, ensure the cost-effective use of shallow geothermal energy in any part of the territory considered.

The continuous development of heat pump technology, based on the vapour compression cycle, has made its heat transfer efficiency even higher and more competitive. When highly efficient heat pumps are combined with shallow geothermal reservoirs, the resulting technology, known as geothermal heat pumps, becomes the most efficient technology for heating and cooling buildings (EPA 1993). The energy efficiency of ground source heat pumps is 50–70% higher than that of conventional heating systems, and 20–40% better than air-to-air heat pumps used in aerothermal systems (Letcher 2013). The use and promotion of ground source heat pumps is also justified by the decarbonisation of the heating sector as a measure of

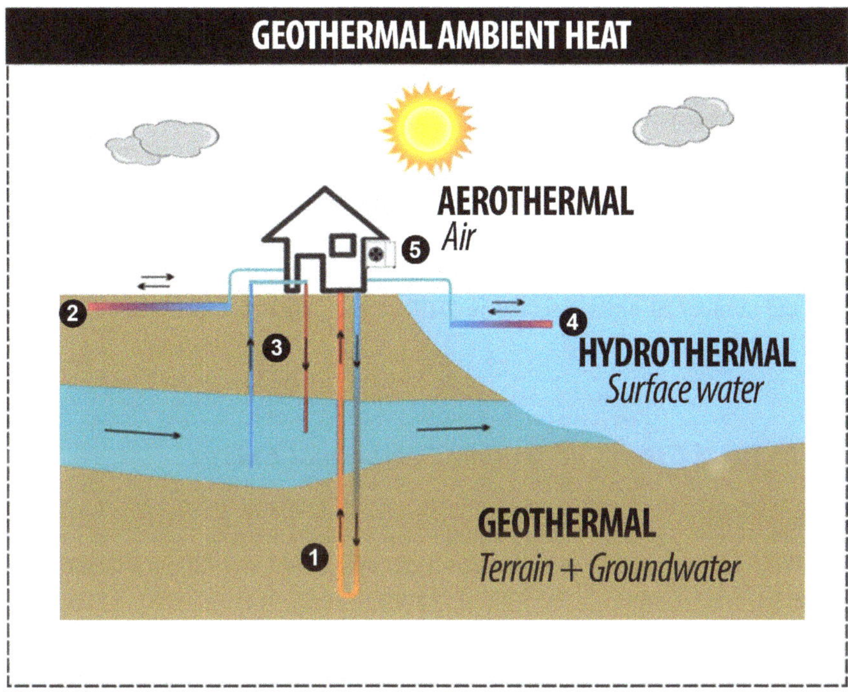

GEOTHERMAL
Geothermal heat pump

Closed loop geothermal Heat exchanger

1. Borehole heat exchanger (BHE)
2. Horizontal ground heat exchanger

Open loop geothermal Heat exchanger

3. Geothermal well doublet

HYDROTHERMAL
Water-to-water heat pump

4. Heat exchanger-serpentine

AEROTHERMAL
Air-to-water heat pump

5. Air-to-air fins heat exchanger

Fig. 1.3 Environmental thermal energy concept. Figure courtesy of Gregor Göetzl (*Geologische Bundesanstalt für Österreich*)

mitigating climate change. In 2008, the use of around 879,000 thermal installations with ground source heat pumps in 19 European countries saved 3.7×10^6 tCO_2eq compared to thermal installations making use of conventional production systems based on the combustion of fossil fuels (Bayer et al. 2012).

GHG emissions and particulate air pollution only occur indirectly and far from urban areas where most of the electrical energy is consumed. Its high efficiency consumes approximately 1 kWh electrical for every 4–8 kWh of thermal heat transfered (Self et al. 2013). Therefore, GHG emissions will depend on the emissions caused during the production of electrical energy (electricity mix), with zero emissions when the electrical energy consumed by the heat pump is produced entirely from renewable energy. Minimal electricity consumption and low GHG emissions are the most important reasons why shallow geothermal energy is in vogue today; it is promoted by governments and is increasingly seen as a promising measure to reduce fossil fuel consumption and mitigate climate change (Stauffer et al. 2013).

For all these reasons, the use of shallow geothermal systems for heating and cooling, including seasonal heat storage, is experiencing significant growth worldwide. Between 2010 and 2015, the total global installed capacity of geothermal heat pumps increased at an annual rate of 13.2%, going up to 50,258 MW thermal of total capacity, with an annual energy use of 326,848 TJ year^{-1}. The equivalent number of installed 12 kW units is approximately 4.19 million, representing in a 52% increase over the number of units installed in 2010 (Lund and Boyd 2015; Lund and Toth 2020).

1.2.4 Brief History of Shallow Geothermal Energy

The first shallow geothermal exchangers had a closed-loop configuration but were installed inside groundwater wells. This happened in the late 1920s in Klamath Falls, Oregon, USA (Chiasson 2016). Iron pipes were introduced inside hydrothermal water wells (65–95 °C), in which a heat transfer fluid was circulated and the heat was exchanged directly (without a heat pump) to heat home spaces, schools and other buildings. This type of rig laid the foundation for the closed-loop geothermal heat exchangers as we know them today.

The first commercial applications of low-temperature heat exchangers for heat pumps are documented from the 1940s in Portland, USA. In this case, by making use of groundwater wells as open-loop geothermal heat exchangers, groundwater was used for heat exchange. The theoretical understanding of the operation of heat exchangers was developed by Adler et al. (1951), providing some of the theoretical foundations for their design. The real revolution in shallow geothermal knowledge came during the oil crisis of the early 1970s. This was a time when the development of alternative energy sources, such as solar, wind and geothermal, was encouraged. At that time, the use of heat pumps for domestic air conditioning became widespread. Heat pumps were designed to utilise ambient heat (Fig. 1.3) by including air (aerothermal), surface water (hydrothermal) and ground/groundwater (geothermal) as heat sources/sinks. Of these heat sources/sinks, ground and groundwater, when available in sufficient quantities with good chemical quality, are the best option because their temperature is very stable throughout the whole year.

Since groundwater is not available in sufficient quantity and quality for thermal use throughout the territory, it was necessary to investigate another technology to exchange heat with the ground without using groundwater. Soils that do not constitute aquifers are also thermally stable throughout the year bellow a given depth (10–15 m). In this case, if an efficient technology could be developed for heat exchange with any terrain, the potential would be immense given that it is ubiquitous. Research and development of closed-loop geothermal heat exchanger technology began simultaneously in Sweden and the United States in the 1970s, allowing shallow geothermal energy exploitation of any terrain.

The great success of heat pump prototypes using geothermal heat exchangers, both open and closed loop, resulted in the emergence of a new market for ground source heat pumps in the late 1980s. At this time, many thermal installations started to install ground source heat pumps to meet the heat demand in many types of buildings such as homes, schools and offices. At the same time, new design tools and manuals for technicians and installers appeared. It was at this time that the main associations emerged: *International Ground Source Heat Pump Association* (IGSHPA) and *American Society of Heating, Refrigerating and Air-Conditioning Engineers* (ASHRAE), and certifications for designers and installers were promoted.

Today, the scientific community is continuously reviewing and improving mathematical methods for thermal analysis of geothermal heat exchangers. During the first decade of this century, research in this field has focused on refining design and installation procedures, particularly related to hybrid systems and sustainable designs (Chiasson 2016). The industry also to improve drilling technology through automation and by improving the technology of the heat pumps themselves, with intelligent controls and variable speed compressors.

Although shallow geothermal energy is the most energy and environmentally efficient technology, it has not been implemented on a massive scale in society. The high initial cost of building geothermal heat exchangers is the biggest barrier, although costs are quickly amortised. These systems require a significant initial financial investment. Various economic models and incentives have been used to overcome this barrier, including leasing the geothermal installation to third parties, offering advantageous credit line programmes by administrations and developing district systems, among others.

References

Adler FT, Ingersoll AC, Ingersoll LR, Plass HJ (1951) Theory of earth heat exchangers for the heat pump. ASHVE Trans 57:167–188

Anzellini S, Dewaele A, Mezouar M, Loubeyre P, Morard G (2013) Melting of iron at earth's inner core boundary based on fast X-ray diffraction. Science 340(6131):464. https://doi.org/10.1126/science.1233514

Banks D (2011) The application of analytical solutions to the thermal plume from a well doublet ground source heating or cooling scheme. Q J Eng Geol Hydrogeol 44(2):191–197. https://doi.org/10.1144/1470-9236/09-028

Banks D (2012) An introduction to thermogeology: ground source heating and cooling. Wiley

Bayer P, Saner D, Bolay S, Rybach L, Blum P (2012) Greenhouse gas emission savings of ground source heat pump systems in Europe: a review. Renew Sustain Energy Rev 16(2):1256–1267. https://doi.org/10.1016/j.rser.2011.09.027

Blum P, Campillo G, Münch W, Kölbel T (2010) CO_2 savings of ground source heat pump systems—a regional analysis. Renew Energy 35(1):122–127. https://doi.org/10.1016/j.renene.2009.03.034

Brindley HE, Bantges RJ (2016) The spectral signature of recent climate change. Curr Clim Change Rep 2(3):112–126. https://doi.org/10.1007/s40641-016-0039-5

Chiasson AD (2016) Geothermal heat pump and heat engine systems: theory and practice. Wiley

EPA (1993) Space conditioning: the next frontier—the potential of advanced residential space conditioning technologies for reducing pollution and saving consumers money. Environmental Protection Agency, U.S

EPA (1997) Manual on environmental issues related to geothermal heat pump systems. In: Agency USEP (ed) National Service Center for Environmental Publications, p 98

Florea LJ, Hart D, Tinjum J, Choi C (2017) Potential Impacts to groundwater from ground-coupled geothermal heat pumps in district scale. Groundwater 55(1):8–9. https://doi.org/10.1111/gwat.12484

Glassley W (2010) Geothermal energy: renewable energy and the environment. CRC Press, FL

IPCC (2007) Climate change 2007: the physical science basis. In: Contribution of working group I to the fourth assessment report of the intergovernmental panel on climate change. Cambridge University Press, UK

IPCC (2011) Special report on renewable energy sources and climate change mitigation. Cambridge University Press, UK

IPCC (2013) Climate change 2013: the physical science basis. In: Contribution of working group I to the fifth assessment report of the intergovernmental panel on climate change. Cambridge University Press, Cambridge

Letcher TM (2013) Future energy: improved, sustainable and clean options for our planet. Elsevier Science

Limberger J et al (2018) Geothermal energy in deep aquifers: a global assessment of the resource base for direct heat utilization. Renew Sustain Energy Rev 82:961–975. https://doi.org/10.1016/j.rser.2017.09.084

Lund JW, Boyd TL (2015) Direct utilization of geothermal energy 2015 worldwide review. In: World geothermal congress 2015. Melbourne, Australia, p 31

Lund JW, Toth AN (2020) Direct utilization of geothermal energy 2020 worldwide review. In: World geothermal congress 2020. International Geothermal Association (IGA) Reykjavik, Iceland, p 39

Narsilio GA, Aye L (2018) Shallow geothermal energy: an emerging technology. In: Sharma A, Shukla A, Aye L (eds) Low carbon energy supply. Springer, Singapore, p 441

Oke TR (1987) Boundary layer climates. Routledge

Saner D et al (2010) Is it only CO_2 that matters? A life cycle perspective on shallow geothermal systems. Renew Sustain Energy Rev 14(7):1798–1813. https://doi.org/10.1016/j.rser.2010.04.002

Sanner B et al (2013) Strategic research and innovation agenda for renewable heating and cooling. RHC-Platform, Brussels

Self SJ, Reddy BV, Rosen MA (2013) Geothermal heat pump systems: status review and comparison with other heating options. Appl Energy 101:341–348. https://doi.org/10.1016/j.apenergy.2012.01.048

Stauffer F, Bayer P, Blum P, Giraldo NM, Kinzelbach W (2013) Thermal use of shallow groundwater. Taylor & Francis

Templeton JD, Ghoreishi-Madiseh SA, Hassani F, Al-Khawaja MJ (2014) Abandoned petroleum wells as sustainable sources of geothermal energy. Energy 70:366–373. https://doi.org/10.1016/j.energy.2014.04.006

Toth A, Bobok E (2017) Flow and heat transfer in geothermal systems. In: Basic equations for describing and modelling geothermal phenomena and technologies. Flow and heat transfer in geothermal systems. Elsevier, Oxford, 380pp. https://doi.org/10.1016/B978-0-12-800277-3.00001-3

UNFCCC (2015) Adoption of the Paris agreement. Proposal by the president (draft decision). In: Change, FCOC (ed) United Nations Office, Geneva, Switzerland, p 32

Vienken T, Schelenz S, Rink K, Dietrich P (2015) Sustainable intensive thermal use of the shallow subsurface—a critical view on the status quo. Groundwater 53(3):356–361. https://doi.org/10.1111/gwat.12206

Whitesides GM, Crabtree GW (2007) Don't forget long-term fundamental research in energy. Science 315(5813):796–798. https://doi.org/10.1126/science.1140362

Chapter 2
Theoretical Background

2.1 Thermodynamic Principles

The word thermodynamics originates from the Greek words *therme* (heat) and *dynamis* (force) and was intended to represent efforts to convert heat into mechanical energy. Today the meaning is much broader, encompassing all aspects of energy and its transformations, including production, cooling and relationships between the properties of matter.

Thermodynamics is one of the four main branches of physics, along with mechanics, electricity and magnetism. Its aim is to study the processes of transformation of work into heat and vice versa. This transformation is of vital importance for shallow geothermal energy. The relatively low temperatures of the shallow underground (<30 °C) require mechanical work to generate a heat source and a heat sink, with a significant temperature difference, in order to transfer heat quickly and efficiently between the subsurface and the heated/cooled space. Therefore, in order to understand shallow geothermal installations, it is necessary to establish the equivalence between work and heat. In deep geothermal energy, the reverse process is usually of greater interest, i.e., determining the conditions under which work can be obtained from thermal energy. The main objective of deep geothermal systems is the transformation of primary energy, extracted from the geological media in the form of thermal energy, into mechanical work and its subsequent transformation into electrical energy.

In general, thermodynamics studies the relationship between the energy of a system and the ways in which this energy is transferred through heat and work. Therefore, the main goal of thermodynamics is the development of universal laws governing the use and conservation of energy on a macroscopic scale.

Thermodynamic studies are based on the definition of thermodynamic systems, consisting of matter contained in a region of space bounded by a closed surface called an enclosure, which separates the system from its surroundings. Systems can be open or closed, depending on whet the enclosure allows the exchange of matter with the surroundings. The system will be diathermic or adiabatic if the enclosure does or

A. García Gil et al., *Shallow Geothermal Energy*, Springer Hydrogeology,
https://doi.org/10.1007/978-3-030-92258-0_2

does not allow heat conduction with the surrounding environment. If the enclosure does not allow any kind of interaction with the system it is considered to be isolated and, therefore, there is no transfer of matter or energy.

Systems subjected to an external mechanical process that causes work is called a mechanical action and, if the physical process produces an exchange of heat, it is called a heat transfer process. Finally, when an exchange of matter takes place in an open system, it is called a mass or matter transfer process.

In closed diathermic systems, the process of heat transfer without mass transfer is called *conduction.* In open diathermic systems, the process of energy transfer by mass transfer is called *convection.* In most cases, a mass transfer is associated simultaneously with a mechanical action and a heat transfer.

The state of a thermodynamic system is characterised by its physical properties described by a set of state variables. Variables can be intensive—when their value remains unchanged when dividing the system into two or more subsystems (e.g., temperature, pressure, density, specific heat, etc.) or they can be extensive—when dividing the system into subsystems their value is reduced proportionally (e.g., mass, volume, enthalpy, entropy, etc.).

The system is considered to have reached thermodynamic equilibrium when all its state variables have constant values. When the environment interacts with the system through mechanical, thermal or chemical processes (exchange of matter), the state of the system is modified until a new state of equilibrium is reached.

2.1.1 Concept of Energy

Energy is a property of matter that gives it the ability to do work. Depending on the nature of the work to be done, matter will possess a certain amount of energy of that nature. Determining the absolute value of the energy of a system can be difficult to quantify, whereas a change in energy is relatively easy to calculate.

Energy manifests itself in macroscopic and microscopic forms. Macroscopic forms of energy are those where a system possesses, as a whole, a quantity of energy with respect to an external frame of reference (kinetic, potential energy, etc.). Microscopic forms are those related to the molecular structure of a system independent of an external frame of reference. The sum of all microscopic forms of energy of a system defines its *internal energy* of the system. The internal energy of a system depends on its composition, structure (phase) and environmental variables (temperature, pressure, magnetic field, electric field, etc.).

There are two forms of microscopic internal energy: internal kinetic energy and internal potential energy. Both refer to the kinetic and potential energy of the atomic or molecular particles that make up the system under study at the microscopic scale. The kinetic energy of the particles determines the temperature of the system, and the potential energy of the particles determines the type of chemical bond or interaction between particles in the system. The internal kinetic energy of a system determines its thermal energy. Consequently, thermal energy is a component of the internal energy of a system. However, thermal energy and internal potential energy are closely related, and their transformation is common in many thermodynamic processes. If the amount of thermal energy transferred between two systems is called heat, the heat transferred to a system can change the internal kinetic energy of the system and consequently change the potential energy of the system's particles. If the heat transferred only affects the internal kinetic energy (thermal energy), the system will only change its temperature, and the heat will be called sensible heat. If the transferred heat affects the internal potential energy, the system will only change its chemical bonds and type of interactions between particles, without changing its temperature (constant temperature), and the heat will be called latent heat.

In the first case, the thermal energy transferred modifies the vibration energy (solids) or movement (gases and liquids) of the particles that compose it, and in the second case it modifies the potential energy of chemical bonds that determine its structure. If, during a heat transfer process within a system, a phase change occurs, the heat transferred will be latent heat until the phase change ends. During the process, the temperature will be constant regardless of the heat transferred. Energy can be transferred between different systems in three ways: heat, work or mass flow.

2.1.2 Temperature and Heat

The microscopic definition of temperature is based on the kinetic theory, which attributes the temperature of a solid, liquid or gaseous body to the movement and interactions of its constituent atoms or molecules. Bodies, in their microscopic structure, are made up of a large number of atoms or molecules in continuous movement (fluids) or vibration (solids) and temperature is the effect of the kinetic energy of the molecules. The kinetic energy of the particles defines the absolute temperature of a body; when the particles are at rest their absolute temperature is zero Kelvin. This temperature is −273.15 °C. The macroscopic definition of temperature is limited to defining it as a state variable of a thermodynamic system that indicates thermal equilibrium when two systems are identical in magnitude.

The transfer of energy caused by a temperature difference between two systems is called heat flow or heat transfer, and the amount of energy transferred is called heat. The heat transferred from one system to other will tend to change the temperature in both systems until a thermal equilibrium is reached. An equilibrium will be reached only if both systems are at the same temperature. The heat transfer will occur from the higher temperature system to the lower temperature system. The heat required to raise the temperature of one gram of water by one Kelvin is one calorie (cal), one of the reference units of heat. However, the unit of reference for all forms of energy in the International System of Units (SI) is the joule (J), and one calorie is equivalent to 4.186 J of mechanical energy.

The heat Q (J) required to change the temperature ΔT (K) of a system of a given mass m (kg) is given by

$$Q = mc\Delta T \tag{2.1}$$

where c (J kg^{-1} K^{-1}) is the specific heat capacity of the system under consideration. Each material, depending on its internal structure and physical state will require a different amount of heat for the same temperature increase, which is called the specific heat of the material.

2.1.3 Heat Transfer Mechanisms

There are three fundamental mechanisms of heat transfer (thermal energy): conduction, convection and radiation. Heat conduction refers to the transfer of heat between two closed diathermic bodies in physical contact. In heat transfer by convection, there is a transfer of mass between the two systems. In nature, when a region of a system transfers heat, thermal expansion or contraction occurs, changing the density of the system and inducing buoyant mass movements known as free convection. When the mass transfer is due to external mechanical work, it is called forced convection or advection. The radiation mechanism is the transfer of heat by electromagnetic waves. Every system emits energy in the form of electromagnetic radiation. The radiated heat flow J (W) is given by:

$$J = \frac{dQ}{dt} = Ae\sigma T^4 \tag{2.2}$$

where A (m^2) is the surface area of the body emitting the radiation, T (K) is the absolute temperature, e (−) is the emissivity, and σ (W m^{-2} K^{-4}) is the Stefan-Boltzmann constant (5.670400×10^{-8} W m^{-2} K^{-4}). The emissivity is a dimensionless value between 0 and 1, representing a correction factor for the ideality of the opaque black radiating surface.

2.1.4 First Law of Thermodynamics

The first law of thermodynamics states the following:

For any thermodynamic system, there is an extensive scalar state function called energy (E_{sys}). When the system is isolated, the energy is conserved.

Therefore, we can speak of a law of conservation of energy over time (t), mathematically expressed for an isolated system as:

$$\frac{dE_{sys}}{dt} = 0 \tag{2.3}$$

If the thermodynamic system interacts with the environment (open and diathermic system), the time evolution of the energy will depend on the power of the external processes acting on the system. The variation of the energy of the external process per unit time is understood as the power of the external process acting on the thermodynamic system. Four external powers are differentiated (among many other possible ones):

- Power P_{ext}: process by which forces external to the system modify the kinetic energy of its centre of mass, both translational and rotational. The applied forces do not deform the system enclosure.
- Power P_W: process by which mechanical work done on the system by the environment produces a deformation of the system enclosure. During the deformation of the system there is no change in its kinetic energy.
- Power P_{con}: The process by which heat is exchanged between the environment and the system by conduction.
- Power P_C: The process by which an exchange of matter takes place between the environment and the system by means of convection.

A thermodynamic system interacting with its environment (open system) will be given by:

$$\frac{dE_{sys}}{dt} = \frac{dE_{ext}}{dt} + \frac{dE_W}{dt} + \frac{dE_{Con}}{dt} + \frac{dE_C}{dt} \tag{2.4}$$

And expressed in power:

$$P_{sys} = P_{ext} + P_W + P_{Con} + P_C \tag{2.5}$$

The total energy of a system can be understood as the sum of its external and internal energies. The first law expanded to the two mechanical conservation laws requires that the total energy (E_T) of a system is the sum of its kinetic energy (E_K), its potential energy (E_V) and internal energy (U):

$$E_T = E_K + E_V + U \tag{2.6}$$

The *internal energy* is an intrinsic property of the matter constituting the system. The internal energy of a thermodynamic system is defined as the sum of the kinetic and potential energies of all the particles within the enclosure. In a non-inertial system and ignoring the potential fields to which the system may be subjected, $E_T = U$. Therefore, the internal energy is a conservative quantity and, consequently, any change in the energy of the system (dU_{sys}) leads to another of the same magnitude in the environment (dU_{etn}) so that the total energy in the system universe remains constant. That is to say:

$$dU_{sys} + dU_{ent} = 0 \tag{2.7}$$

In a closed non-inertial system and ignoring the external potential fields, it is known that its internal energy can be modified by heat transfer (Q) or by mechanical work (W):

$$\Delta U = Q - W \tag{2.8}$$

Ordering the terms:

$$Q = \Delta U + W \tag{2.9}$$

Either of these equations is the first law of thermodynamics.

It follows from the first law of thermodynamics that if we transfer energy in the form of heat to a system, part of the heat transferred can be transformed into mechanical work. The rest can be used to modify its internal energy, by modifying the energy of the constituent particles, variation of the state of aggregation, etc. It may happen that the system does not do any work in the process, so the heat transferred will be equal to the increase in internal energy ($Q = \Delta U$). According to the first law of thermodynamics, it may happen that (from a theoretical point of view) all heat is transformed into mechanical work ($Q = W$) in an isothermal process.

It is experimentally proven that the internal energy does not depend on the path of the thermodynamic process. It only depends on the initial and final state, unlike the Q and W, which do depend on the trajectory. Therefore, the ΔU in a system during a thermodynamic process depends on the initial and final states. The direct consequence is that a cyclic process in which the initial internal energy is returned to the initial ($\Delta U = 0$) implies that the heat transferred to the system will have been transformed into work during the process ($Q = W$).

Enthalpy H (J) is a variable that comprehensively characterises the amount of energy contained in a system. Enthalpy not only represents the total thermodynamic energy of that system but also includes all forms of energy (e.g., chemical, nuclear, elastic) as well as the energy being transferred by the system to its surroundings. One possible enthalpy equation per unit of mass would be:

$$\Delta H = \Delta U + \Delta\left(\sum_{i=0}^{n} U_i^V\right) + \Delta\left(\frac{p}{\rho}\right) \tag{2.10}$$

where ΔU (J kg^{-1}) is the variation of internal energy, the sum of U^V (J kg^{-1}) represents the n potential energies considered, p (kg m^{-1} s^{-2}) is the pressure, and ρ (kg m^{-3}) is the density. The last term of Eq. (2.10) represents the energy that the system has to maintain its volume under high pressure conditions (high depth in geothermal energy). Under a condition of equilibrium, the environment exerts the same pressure on the system, so it is an amount of energy recoverable to do work and is included in the enthalpy term. The increments in Eq. (2.10) represent the need to use a reference enthalpy since it is not possible to measure it directly, only its variations. When shallow geothermal energy is considered, only the first term of

Eq. (2.10) is significant. It is much larger than the rest, which can be considered negligible. In the environments in which shallow geothermal systems operate, there are no significant changes in temperature, pressure, density or volume, nor are there any phase changes (e.g., liquid–gas), which occur in deep geothermal energy and are important components of the enthalpy change, especially due to increases in enthalpy. $\Delta(p/\rho)$. Therefore, it can be assumed that in shallow geothermal energy:

$$\Delta H \cong \Delta U \tag{2.11}$$

If we consider that the net sensible heat transferred to a system will increase the temperature proportionally as a function of its heat capacity (Eq. 2.1), we can relate both quantities by means of the relationship:

$$\Delta U = c\Delta T \tag{2.12}$$

There are four specific types of thermodynamic processes that acquire their own name. Processes without heat transfer are adiabatic, constant volume isochoric, constant pressure isobaric and constant temperature isothermal.

The internal energy is a quantity that depends on the $Q(T)$ and $W(V)$ and, therefore, on the T and V. Work is defined in mechanics as the scalar product of a force F (kg m s^{-2}) by the displacement dx (m) exerted in moving an object.

$$dW = F \cdot dx \tag{2.13}$$

If we consider a system in the form of a cylinder, the work will be given by the force exerted by the area (A) at one end, i.e., the pressure (p), by the displacement (dx) in the direction of the axis of symmetry of the cylinder.

$$dW = pAdx = pdV \tag{2.14}$$

Taking into account the first law of thermodynamics and substituting at constant pressure for a closed system:

$$Q = \Delta U + W \tag{2.15}$$

$$dQ_P = dU + pdV \tag{2.16}$$

And integrating:

$$Q_P = \int_1^2 dU + \int_1^2 pdV \tag{2.17}$$

$$Q_P = (U_2 + pV_2) - (U_1 + pV_1) = H_2 - H_1 \tag{2.18}$$

where $(U + pV)$ is called *enthalpy* (H), is a state function and represents the heat input to the isobaric thermodynamic process.

In a controlled process, the U and p of a system are composite functions:

$$U = U(T(t), V(t)) \tag{2.19}$$

$$p = p(T(t), V(t)) \tag{2.20}$$

Called equations of state, they describe the state of aggregation of matter by means of a mathematical relationship between T, p, V, U and other associated state functions.

The first law of thermodynamics is formally expressed as:

$$\frac{dU}{dt} = \frac{dQ}{dt} + \frac{dW}{dt} \tag{2.21}$$

It states that the rate of change of the internal energy of a system is equal to the sum of the heat transfer and the rate of work done by/on the system. To the term representing the rate of work (dW/dt), the following can be added: mechanical, gravitational, inertial or electromagnetic work as a result of internal forces within the system itself.

In a cyclic process of heat and work between an initial time t_i and a final time t_f, the internal energy, as a state variable, starts and ends with the same values, usually assumed to be zero in the initial and final state. Therefore, over the time of a cycle:

$$Q = -W \tag{2.22}$$

where

$$Q = \int_{t_0}^{t_F} \frac{dQ}{dt} dt \tag{2.23}$$

$$W = \int_{t_0}^{t_F} \frac{dW}{dt} dt \tag{2.24}$$

In a cyclic process that performs work by the system on the environment, the work will be considered negative (e.g., heat engine). In a system that does work on itself, the work is assumed to be positive (e.g., a heat pump). Regardless of the direction of the work, there is always a positive heat transfer from the heat source to the system and a negative heat dissipated from the system to the heat sink. The efficiency (η) of a cyclic process will be the ratio of the target work (or heat) to the heat supplied (or work supplied). Thus, for example, the efficiency of a heat engine will be given by:

$$\eta = \frac{|W|}{Q_+} \tag{2.25}$$

In this case, the less heat we need to supply from the boiler (Q_+) for a given work (W) in each cycle, the higher the efficiency. Similarly, the efficiency of a chiller will be given by:

$$\eta = \frac{Q_-}{W_+} \tag{2.26}$$

where the heat source is the system itself which dissipates heat to the surroundings by doing work on the system. The lower the mechanical work required to transfer an amount of heat out of the system, the higher the efficiency.

To understand what the efficiency in a cyclic process depends on, a known system such as ideal gas can be taken into consideration to show that the efficiency depends only on the temperature difference between the heat source and the heat sink.

2.1.5 Carnot Cycle

The *Carnot cycle* consists of the idealisation of a hypothetical heat engine in order to understand its maximum possible efficiency. The question underlying this first question is to understand if a machine can present a 100% thermal efficiency, i.e., whether all the heat supplied to a system can be transformed into work.

The first heat engines could do work by supplying heat to a system. If a gas contained in a piston was heated with a flame, the gas would expand and do work. It is necessary to heat the piston more and more for the work to continue. Obviously, a system cannot be heated indefinitely; at some point the very material that makes up the piston would eventually melt. Therefore, a cyclic process was sought where the expansion process was repeated to do continuous work. The only way to restart the expansion of the gas contained in the piston would be to reduce its temperature and, when it reaches the initial temperature, heat it up again. This is the same as stating that not all the heat supplied from a heat source to a system can be transformed into work indefinitely. Therefore, the question is how much of the heat transferred to the system undergoing a cyclic thermodynamic process will end up in the heat sink source? It follows, from the first law of thermodynamics, that the heat that is not transferred to the heat sink will be transformed into work.

The French engineer Sadi Carnot, in 1824, imagined a hypothetical machine in which the highest possible efficiency could be obtained. The hypothetical machine that Carnot idealised, referred to as the Carnot machine, consisted of a cylinder with a piston filled with an ideal gas (working substance), two tanks acting as a heat source at a relatively high temperature (T_A), and another acting as a heat sink at a relatively low temperature (T_B) with respect to the previous one. In order to obtain work, the machine performed a cyclic thermodynamic process in which there was thermal and

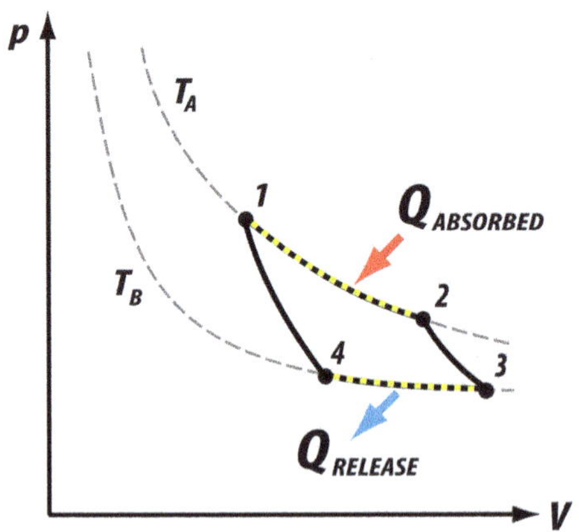

Fig. 2.1 Representation of the Carnot cycle for an ideal gas in a pV diagram. The dashed lines represent isothermal processes and the solid black lines represent adiabatic processes

mechanical equilibrium at all times to ensure its reversibility. The cylinder with the piston and the ideal gas inside it formed a closed system where there was no mass transfer with the surroundings, a system that will be referred to hereafter in general terms as the *system*. The Carnot cycle is divided into four sub-processes (Fig. 2.1):

- **Isothermal expansion** (stages 1–2): the system at a temperature (T_A) is brought into diathermic contact with a thermal reservoir at the same temperature (T_A), see Fig. 2.2. Keeping the system in thermal equilibrium with the tank at all times, the piston is released, and the ideal gas tends to expand by doing work on the piston ($W > 0$). Since there is thermal equilibrium, each dt the gas expands tends to lower its temperature and work is transferred dQ from the source to the cylinder-piston-ideal gas system, so that it remains at temperature T_A. *Joule's* experiment demonstrated the property of ideal gases, in which their internal energy is a function only of the temperature of the system and does not depend on the volume. Since there is no temperature change in the ideal gas, the change in internal energy of the ideal gas-piston-cylinder system is zero ($\Delta U = 0$). By the first law of thermodynamics all heat supplied is transformed into work during this subprocess ($Q = W$), a 100% efficient heat transfer sub-process to do work.
- **Adiabatic expansion** (stages 2–3): the system is separated from the heat source (Fig. 2.2) and the piston is left free, so that the gas expands in mechanical equilibrium. As there is no heat transfer, as the gas expands its temperature is reduced to a relatively lower temperature (T_B). In this adiabatic subprocess there is no heat transfer, but there is a change in temperature, so the work done on the outside is equivalent to the decrease in internal energy ($\Delta U = W$). There is no heat transfer, so there is no conversion of heat into work.
- **Isothermal compression** (stages 3–4): the system at one temperature is brought into contact with a thermal reservoir at constant temperature T_B which serves as a heat sink (Fig. 2.2). In thermal equilibrium, compression work is performed on the system. During the compression of the gas, in diathermic contact with the reservoir each time the gas is compressed, it tends to compress at a constant

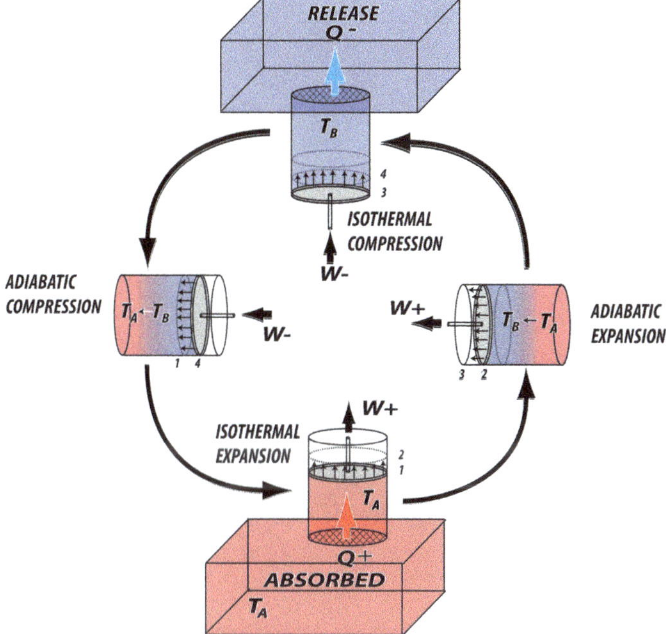

Fig. 2.2 Conceptual representation of the Carnot cycle for an ideal gas. The four sub-processes involved in the Carnot cycle are depicted: isothermal expansion, adiabatic expansion, isothermal compression and adiabatic compression. For each sub-process, reference is made to the stages highlighted in Fig. 2.1

temperature. The compressed gas tends to increase in temperature and the heat is transferred from the system to the sink source, dQ from the system to the heat sink source in such a way that the process takes place at a constant temperature T_B. Again, as an isothermal process in an ideal gas, there is no change of internal energy ($\Delta U = 0$), and the heat given up during this subprocess is equivalent to the work done on the system ($Q = W$), i.e., a 100% efficient transfer of work into heat.

- **Adiabatic compression** (stages 4–1): in the fourth sub-process, the system is isolated from the heat sink source (Fig. 2.2) and compression work is performed on the system. In mechanical equilibrium and under adiabatic conditions, each time step dt the gas is compressed, its temperature tends to increase dT until it reaches the initial temperature (T_A). There is no heat transfer, so there is no conversion of heat into work.

The efficiency of the whole cycle will be given by the ratio of the work achieved for the heat supplied Q_A from the source at temperature T_A:

$$\eta = \frac{|W|}{Q_A} \tag{2.27}$$

In a cyclic process where the starting and final stages are the same, the internal energy change as a state function is zero, therefore:

$$W = Q = Q_A + Q_B = |Q_A| - |Q_B| \tag{2.28}$$

where Q_A is the heat absorbed from the source at temperature T_A and Q_b is the heat dissipated in the heat sink source at temperature T_B. If these equations are combined:

$$\eta = \frac{|Q_A| - |Q_B|}{Q_A} = 1 - \left|\frac{Q_B}{Q_A}\right| \tag{2.29}$$

Neglecting the kinetic energy of the system and assuming the field of pressures and temperatures in the system to be homogeneous, the first law for infinitesimal changes reduces to:

$$\frac{dU}{dt} = \frac{dQ}{dt} - p\frac{dV}{dt} \tag{2.30}$$

where U and p are state functions:

$$U = U(T(t), V(t)); \quad p = p(T(t), V(t)) \tag{2.31}$$

Each type of material has a mathematical relationship between T, p, V, U and other associated state functions, so that, in general, it is necessary to obtain them experimentally. In the specific case of ideal gases:

$$U = m\gamma\frac{\varpi}{\mu}(T - T_{ref}) + U(T_{ref}) \tag{2.32}$$

$$p = \frac{m\varpi}{V\mu}T \tag{2.33}$$

where m (kg) is the total mass of the gas and μ (kg) is the mass of a gas molecule, γ (−) is the dimensionless specific heat at constant volume (values of 3/2, 5/2 and 3 for monoatomic, diatomic and gases composed of more complex molecules respectively), ϖ is the Boltzmann Constant (J K^{-1}) (1.381 × 10^{-23} J K^{-1}), T (K) is the absolute temperature and T_{ref} (K) is an arbitrary reference temperature (e.g., 298.15 K).

By knowing the equations of state of the ideal gas participating in the Carnot cycle, the contribution of Q and W in each sub-process of the complete cyclic process can be calculated (see stages of the cycle in Fig. 2.1):

$$W_{1-2} = \int_1^2 p dV = \int_1^2 \frac{m\varpi}{V\mu}T dV = m\frac{\varpi}{\mu}T_A \ln\frac{V_2}{V_1} \tag{2.34}$$

$$Q_{1-2} = W_{1-2} = m\frac{\varpi}{\mu}T_A \ln\frac{V_2}{V_1} > 0 \tag{2.35}$$

$$W_{2-3} = \int_2^3 pdV = \int_2^3 \frac{m\varpi}{V\mu}TdV = m\frac{\varpi}{\mu}(T_B - T_A) \tag{2.36}$$

$$Q_{2-3} = 0 \tag{2.37}$$

$$W_{3-4} = -\int_3^4 pdV = -\int_3^4 \frac{m\varpi}{V\mu}TdV = -m\frac{\varpi}{\mu}T_B \ln\frac{V_4}{V_3} \tag{2.38}$$

$$Q_{3-4} = W_{3-4} = m\frac{\varpi}{\mu}T_B \ln\frac{V_4}{V_3} < 0 \tag{2.39}$$

$$W_{4-1} = -\int_4^1 pdV = -\int_4^1 \frac{m\varpi}{V\mu}TdV = -m\frac{\varpi}{\mu}(T_A - T_B) \tag{2.40}$$

$$Q_{4-1} = 0 \tag{2.41}$$

The calculation of the Carnot cycle efficiency working with an ideal gas is given by:

$$\eta = 1 - \left|\frac{Q_B}{Q_A}\right| = 1 - \left|\frac{Q_{3-4}}{Q_{1-2}}\right| = 1 - \left|\frac{m\frac{\varpi}{\mu}T_B ln\frac{V_4}{V_3}}{m\frac{\varpi}{\mu}T_A ln\frac{V_2}{V_1}}\right| = 1 - \left|\frac{T_B}{T_A}\frac{ln\frac{V_4}{V_3}}{ln\frac{V_2}{V_1}}\right| \tag{2.42}$$

From the study of the adiabatic processes considering an ideal gas it can be deduced that the change in volume between stages 1 and 2 is the same as between stages 3 and 4. Therefore, the logarithms are also equal:

$$\eta = 1 - \frac{T_B}{T_A} = \frac{T_A - T_B}{T_A} \tag{2.43}$$

It is found that the total efficiency of the cycle only depends on the temperature difference between the source and the sink heat source. The efficiency of the idealised Carnot machine will be zero when the temperatures of the heat sources are equal. The efficiency will be 100% when T_B/T_A tends to zero. Although only the case of ideal gases has been considered here, *Rudolf Clausius* showed that it is independent of the working substance.

Furthermore, it follows that:

$$\frac{Q_B}{Q_A} = \frac{T_B}{T_A}; \quad Q_A = \frac{T_A}{T_B}Q_B \tag{2.44}$$

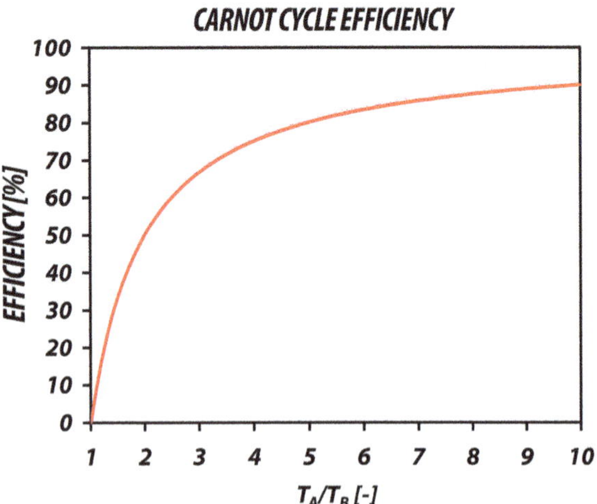

Fig. 2.3 Efficiency of the Carnot cycle according to the ratio between the temperature of the heat source (T_A) and the temperature of the heat sink (T_B)

In other words, the ratio between the temperatures of the heat source and sink will be the constant of proportionality between the heat absorbed and the heat given up. Thus, for example, if the temperature at the *hot* source is twice that of the temperature at the *cold* source ($T_A = 2T_B$) the heat transferred to the cold source will be half of the heat supplied ($Q_B = Q_A/2$) and the efficiency will be 50%, which is the same as saying that only half of the heat supplied to the system is converted into mechanical work, and the rest is transferred to the sink source. If the temperature of the heat source is five times higher than the temperature of the sink source, the efficiency would be 80% (Fig. 2.3).

During the operation of a real heat engine, the mechanical work and heating of the parts that make up the systems takes place very quickly, leading to structural deterioration of the system, turbulent flows, very heterogeneous density, pressure and temperature fields that cause an additional uncontrolled energy conversion. Friction between parts of the system, viscous fluids, heat conduction, etc. make up a total energy loss of the system that, in the end, makes the thermodynamic processes not fully reversible. However, if the heat transfer and the application of mechanical stresses are carried out slowly enough, an internal equilibrium is achieved and, therefore, the processes will be reversible. For a process to be reversible, it is generally considered to proceed slowly compared to the speed of sound, slowly enough so that the internal pressure and temperature fields of the system do not change drastically during a relaxation time of the system.

The first law of thermodynamics introduces the concept of conversion of energy from one form to another (especially between heat and work) and rules out the possibility of the existence of a machine that continuously produces work without consuming an equivalent amount of energy, which would be called a perpetual prime mover. Furthermore, the first law does not impose any restrictions on the efficiency of the transformation from one form of energy to another, nor does it indicate the spontaneity of the processes. Thus, for example, the first law does not prevent a *cold* body from transferring heat to a *hot* body, or a system that reduces its temperature and produces a decrease in its internal energy from transforming that energy into mechanical work.

2.1.6 Second Law of Thermodynamics

Experimentally, it is observed that natural, spontaneous processes are only possible in a certain direction. For example, if two bodies are placed in contact at different temperatures, the *hotter* system spontaneously gives up heat to the *cooler* one, or a gas in a container that is placed in contact with a larger container spontaneously expands. The second law of thermodynamics defines the spontaneity of a process in the sense that it involves the most disorder. The degree of disorder of a system is determined by the thermodynamic quantity *entropy* (S), which is constant (idealised theoretical models) and, if it changes, always increases with time in an isolated system, as a consequence of an irreversible process taking place within it (friction, heat loss, inelastic deformation, etc.). It is a state function and its variation (ΔS) is equal to the reversible heat (Q_{rev}) absorbed by the system during a reversible process in which the temperature remains constant divided by the absolute temperature (T_{abs}) at which the process occurs:

$$\Delta S = \frac{\Delta Q_{rev}}{T_{abs}} \tag{2.45}$$

A reversible process is one whose direction can be reversed at any time by infinitesimally modifying the conditions of the experiment, under equilibrium conditions. In practice, processes are not completely reversible because there is always a frictional force, a non-elastic deformation or a structural alteration of the system that makes it impossible to reverse the direction of the process to the initial state.

Friction, inelastic deformation, heat losses, etc. require an extra amount of energy ($Q_{irrev} > Q_{rev}$). Therefore, the entropy of a irreversible process is defined as:

$$\Delta S > \frac{\Delta Q_{rev}}{T_{abs}} \tag{2.46}$$

The *Joule experiment* consists of performing a counter-vacuum expansion of an ideal gas at low pressure. The gas is contained in an enclosure with diathermic walls in a bath at unregulated temperature, with the bath surrounded by adiabatic walls. Expansion against zero pressure produces no work. Experimentally, the bath temperature is observed to remain unchanged, indicating that the gas has neither absorbed nor given up heat. The application of the first law of thermodynamics indicates that the variation of internal energy of the gas has been zero and therefore the internal energy of that gas depends only on the temperature and not on the volume.

$$\left(\frac{\partial U}{\partial V}\right)_{Temp} = 0 \tag{2.47}$$

2.1.7 *Isoentropic Process*

An isentropic process is one in which the entropy of the system remains constant. Since in reality there are always irreversibilities that cause the entropy of thermodynamic systems to increase continuously, an isentropic process is an idealised process that fulfils two conditions: it is reversible and adiabatic. Therefore, an isentropic process is also called a reversible adiabatic process. Considering Eq. (2.45), the amount of thermal energy transferred to a system by reversible heat transfer (ΔQ_{rev}) is directly proportional to the change in entropy (ΔS); therefore, there can be no heat exchange if the entropy is to remain constant, hence the need to be an adiabatic process. Reversibility confers the ability of the macroscopic thermodynamic system to undergo physical state changes and return to the initial state, without an increase in entropy, by changing the conditions that caused those changes. If irreversibility were to occur, such as friction during compression, the heat generated by friction would have to be removed to keep the entropy constant.

2.2 Heat Transfer

Acquiring an adequate knowledge of the heat transfer phenomenon in the subsurface environment, as well as of the parameters that modulate it, is necessary for understanding shallow geothermal energy as a scientific discipline. Heat transfer is a thermodynamic process that determines the rate at which thermal energy is transferred between thermodynamic systems in their environment.

Before describing the heat transfer mechanisms in the subsurface, it is necessary to summarise some generalities of the geological systems involved. The subsurface is highly heterogeneous in terms of its geological structure and physicochemical nature. Models that reproduce its most important characteristics are needed to study it. From a physical point of view, two main components of the subsurface geological environment can be distinguished: the existence of a solid matrix, generally mineral, and a network of more or less interconnected pores. This basic differentiation determines the so-called *porous media* and constitutes the most widely used physical model of the subsurface to describe physical processes, such as mass or heat transfer. Examples of porous media are soils in the edaphological sense, sands as sediment, fractured rocks or karstified limestones. The interconnection of pores allows the flow of one or more fluids through the media (commonly water and air). In the simplest case, the pores are saturated by a liquid or gaseous fluid and, in more complex cases, a liquid and a gas coexist in the pore system. The porous media in the subsurface is characterised by a distribution of pores of irregular shapes and sizes. It is impossible to study heat transfer in the ground at the pore scale; even if the mathematical tools to do so were available, it is not possible to know the actual distribution of the interconnected pore network. Even if a model of the distribution of state variables such as fluid pressure or temperature could be found at the microscopic level, it would not be possible to verify these solutions by means of measurements at this scale. Therefore, working with porous media requires a new approach where the media can be treated as an equivalent continuum.

2.2.1 *Porous Media and Its Approximation to a Continuous Media*

A defining characteristic of porous media is that a solid matrix is distributed throughout the domain. Let l_m be the length (m) or characteristic dimension of the grains that make up the solid matrix, and let l_f be the characteristic length of the pores containing fluid. If a spherical sample of arbitrary volume of sufficiently large radius r is considered, it can be seen that the sample obtained would always contain a solid phase (Fig. 2.4a). Let V be the total volume of the sample sphere, and let V_f be the total pore volume with fluid in V, then the porosity of the sample can be calculated as ϕ:

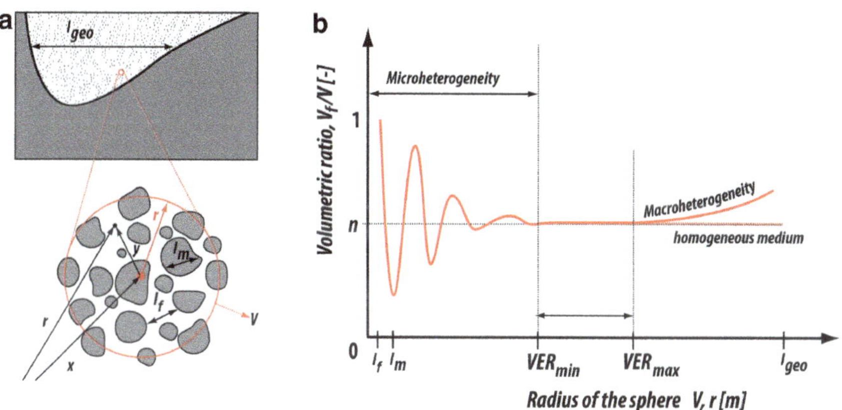

Fig. 2.4 **a** Schematic representation of the volume averaging technique on a sedimentary body with a representative macroscopic scale length. l_{geo} representative macroscopic scale length. At the microscopic scale, the characteristic length of the solid (l_m) and fluid phase (l_f). **b** Representation of the separation of scales as a function of heterogeneity at pore and geological formation scale for the definition of a representative Elementary volume (REV), modified from (Bear 1988)

$$\phi(\boldsymbol{x}) = \frac{1}{V} \int\limits_{r \in V(\boldsymbol{x})} \mathrm{a}(\boldsymbol{r}) dV = \frac{V_f(\boldsymbol{x})}{V} \tag{2.48}$$

where $\boldsymbol{x}$ is the position vector of the centre of the sample volume and $\mathrm{a}(\boldsymbol{r})$ a pore distribution function defined as:

$$\mathrm{a}(\boldsymbol{r}) = \begin{cases} 1 & if\, \boldsymbol{r}\, is\ pointing\, a\ porous\ space \\ 0 & if\, \boldsymbol{r}\, is\ pointing\, a\ solid\ phase \end{cases} \tag{2.49}$$

If $r \ll l_f$ o $r \ll l_m$ it is likely that, depending on the coordinate $(\boldsymbol{x})$ where the sample is taken, the sphere will be more likely to fall into a single phase. The graph in Fig. 2.4b shows how increasing r can vary porosity. In this graph, the small r values tend to make the porosity oscillate significantly, as porosity depends on where the sample sphere originates. From a certain r value the porosity stabilises, and independent of the sample location the porosity presents similar values. It can be said that porous media is homogeneous at a given scale. However, if r approaches the characteristic length of the geological formation l_{geo} the porosity can vary again significantly (Fig. 2.4a). From the range of r where the porosity remains approximately constant, the smaller value of that range $r_{\min}$ defines the $V(r_{\min})$ representative elementary volume (REV) of the porous media (Whitaker 1998).

2.2.2 *Heat Conduction Mechanism*

There are different physical mechanisms by which heat transfer can take place. As already mentioned in Sect. 2.1.3, three main mechanisms can be distinguished: conduction, convection and radiation. Heat in the subsurface is generally transferred as follows: (1) by the propagation of atomic or molecular movements of the rock matrix or static sediments forming sedimentary bodies (conduction), and by (2) migration of fluids through the interconnected pore network, where the internal energy intrinsic to the matter that makes up these fluids is transported as they move (convection). Fluids that do not move, such as water trapped in isolated pores without access to the network of interconnected pores, can transfer heat by conduction.

Heat conduction is one of the most important mechanisms of thermal energy flow in the subsurface. It consists of the transfer of kinetic energy between the microscopic particles that make up a system. Particles with high kinetic energy collide with those with lower kinetic energy, transferring part of their energy in the process and propagating through the system and its surroundings in physical contact. Let there be three contiguous and aligned systems (Fig. 2.5).

The systems at the extremes are considered large enough to keep their temperature constant at different temperatures ($T_A > T_B$). The amount of heat transferred between end bodies per unit time and unit cross-sectional area by conduction is called the *heat flow* (q) and can be described mathematically by *Fourier's Law* for one dimension

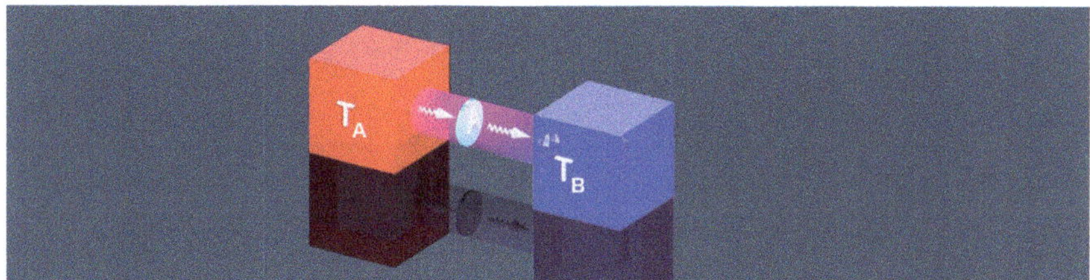

Fig. 2.5 Heat transfer by conduction

(x):

$$q''_x = -\lambda_x \frac{dT}{dx} \tag{2.50}$$

where q''_x (W m^{-2}) is the heat flux per unit cross-section in the x direction, dT/dx (K m^{-1}) is the temperature gradient in the x direction and λ_x (W K^{-1} m^{-1}) is the proportionality constant between the heat flow and the temperature gradient in the x direction, called *thermal conductivity*. Materials that are good conductors of heat, such as metals, will present high λ values, while thermal insulating materials will present low λ values. The negative sign indicates that heat is transferred in the direction of decreasing temperature. For a constant cross section A (m^{-2}), the heat flow q_x in the direction x can be expressed as:

$$q''_x = \frac{dQ_x}{dt}\frac{1}{A} = \frac{q_x}{A} = -\lambda_x \frac{dT}{dx} \tag{2.51}$$

$$q_x = -\lambda_x A \frac{dT}{dx} \tag{2.52}$$

Fourier's law in a generalised form in three dimensions is given by:

$$\begin{bmatrix} q''_x \\ q''_y \\ q''_z \end{bmatrix} = -\begin{bmatrix} \lambda_{xx} & 0 & 0 \\ 0 & \lambda_{yy} & 0 \\ 0 & 0 & \lambda_{zz} \end{bmatrix} \begin{bmatrix} \frac{\partial}{\partial x} \\ \frac{\partial}{\partial y} \\ \frac{\partial}{\partial z} \end{bmatrix} T \tag{2.53}$$

Which can be expressed more concisely as:

$$q'' = -\lambda \nabla T \tag{2.54}$$

where q'' is a vector with direction and magnitude depending on the tensor λ representing the thermal conductivity of the material in all three dimensions, ∇ is the three-dimensional vector operator nabla and $T(x, y, z)$ is the scalar field of temperatures in space. ∇T represents the temperature gradient and represents the vector

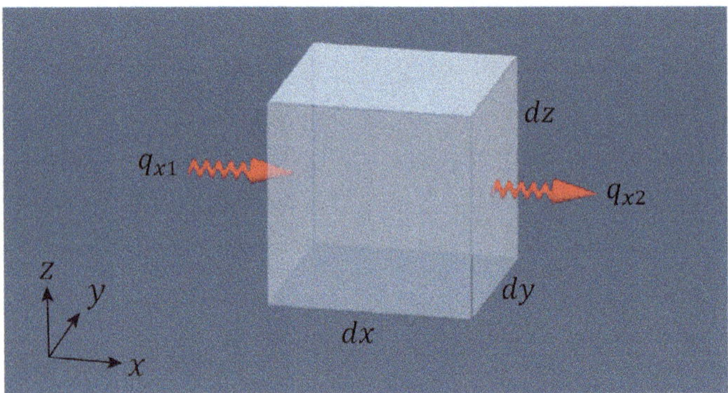

Fig. 2.6 Differential control volume for the derivation of the differential heat equation in Cartesian coordinates

whose magnitude is the maximum rate of change of temperature and points in the direction of that maximum.

To understand how heat behaves in a continuous media, it is necessary to derive the differential equation of heat conduction for a stationary, homogeneous, isotropic solid where no work is applied, but there is internal heat source/sink. Such a solid will have a control volume $dV = dx\ dy\ dz$ (Fig. 2.6). Applying the first law of thermodynamics to it, the net heat transfer will be equal to the internal energy change:

$$\frac{dU}{dt} = \frac{dQ}{dt} + \frac{dE_{fs}}{dt} \tag{2.55}$$

where dE_{fs}/dt (W) represents the heat source/sink term inside the body. The net heat transferred to the control system will be the heat flow out minus the heat flow in over a given period of time:

$$\frac{dQ}{dt} = (q_{x1} - q_{x2}) + \left(q_{y1} - q_{y2}\right) + (q_{z1} - q_{z2}) \tag{2.56}$$

Taking into account Fourier's Law, the incoming heat flow in the x direction (q_{x1}) to the control volume will be

$$q_{x1} = -\lambda_x A \frac{\partial T}{\partial x} = -\lambda_x dydz \frac{\partial T}{\partial x} \tag{2.57}$$

The output heat flow of the control volume in the x direction (q_{x2}) will depend on the heat flow at the point $x = dx$ at the outlet of the control volume. As λ_x and $A = dydz$ are constant, if the heat flux were to change along dx it would do so because the temperature gradient has changed at that point. How much has the gradient changed between $x = 0$ (e.g.) on the inlet side and $x = dx$ on the outlet side of the control volume? Since we are assessing changes of a function in an infinitesimal dx, an estimate can be made of q_{x1} at the point $x = dx$ by applying Taylor series. That is assuming that the function q_x has a smooth quadratic variation between two points separated by an infinitesimal distance. Then, by applying Taylor series we obtain a polynomial $P(x)$ very similar to q_{x2} at the point $x = dx$:

$$q_{x2} \cong P(x) = q_x(dx) + \frac{\partial q_x(dx)}{\partial x}(x - dx) + \cdots + \frac{\partial^{(n)} q_x(dx)}{\partial x}\frac{1}{n!}(x - dx)^n \tag{2.58}$$

By taking the first two terms it is assumed that the tangent line ($P(x)$) to the function q_{x2} is a sufficient approximation. Differential increments are considered here so that the value in the function and its derivative between the two points separated by dx are equal, and we obtain:

$$q_{x2} \cong q_x(dx) + \frac{\partial q_x(dx)}{\partial x}(x - dx) \cong q_x(0) + \frac{\partial q_x(0)}{\partial x}(0 - dx) \tag{2.59}$$

That is to say:

$$q_{x2} \cong q_{x1} + \frac{\partial q_{x1}}{\partial x} dx \tag{2.60}$$

$$q_{x2} \cong -\lambda_x \frac{\partial T}{\partial x} dy\, dz + \frac{\partial}{\partial x}\left(-\lambda_x \frac{\partial T}{\partial x} dy\, dz\right) dx \tag{2.61}$$

The net heat transferred to the control volume in the x direction will be:

$$q_{x1} - q_{x2} \cong -\lambda_x \frac{\partial T}{\partial x} dy\, dz + \lambda_x \frac{\partial T}{\partial x} dy\, dz - \frac{\partial}{\partial x}\left(-\lambda_x \frac{\partial T}{\partial x}\right) dx\, dy\, dz \tag{2.62}$$

$$q_{x1} - q_{x2} \cong \frac{\partial}{\partial x}\left(\lambda_x \frac{\partial T}{\partial x}\right) dx\, dy\, dz \tag{2.63}$$

This means that the net heat that is ultimately transferred between the control volume and the environment is conditioned by how much the temperature gradient along the control volume changes along dx. If the temperature gradient were constant, meaning that all heat entering is equal to the heat leaving, no heat would be stored in the control volume without a change in its internal energy and, hence, its temperature (assuming the source-sink term is zero).

Similarly for the other spatial components:

$$q_{y1} - q_{y2} \cong \frac{\partial}{\partial y}\left(\lambda_y \frac{\partial T}{\partial y}\right) dx\, dy\, dz \tag{2.64}$$

$$q_{z1} - q_{z2} \cong \frac{\partial}{\partial z}\left(\lambda_z \frac{\partial T}{\partial z}\right) dx\, dy\, dz \tag{2.65}$$

Substituting Eqs. (2.63)–(2.65) into Eq. (2.56):

$$\frac{dQ}{dt} = \left[\frac{\partial}{\partial x}\left(\lambda_x \frac{\partial T}{\partial x}\right) + \frac{\partial}{\partial y}\left(\lambda_y \frac{\partial T}{\partial y}\right) + \frac{\partial}{\partial z}\left(\lambda_z \frac{\partial T}{\partial z}\right)\right] dx\ dy\ dz \tag{2.66}$$

From Eq. (2.55), the net heat that is transferred between the control body and the surroundings is defined. The energy contributing from the interior by the source/sink term can be expressed as g (W m^{-3}):

$$\frac{dE_{fs}}{dt} = g\ dx\ dy\ dz \tag{2.67}$$

With Eqs. (2.66) and (2.67) the terms that can contribute or subtract heat to the control volume are considered. The storage capacity of the system is related to the internal energy of the system. If there is positive storage of thermal energy as sensible heat, the temperature of the system will increase and vice versa. The quantity that relates the energy required to increase by one Kelvin is the specific heat capacity c (J kg^{-1} K^{-1}). When the change of internal energy per unit mass u (J kg^{-1}) occurs at constant volume, the specific heat at constant volume is defined as c_V (J kg^{-1} K^{-1}):

$$c_V = \left.\frac{\partial u}{\partial T}\right|_V \Rightarrow u = c_V \partial T + u_{ref} \tag{2.68}$$

where u_{ref} is the arbitrary reference internal energy, which may be zero for simplicity. For an incompressible solid, as is the case here, the heat capacity at constant volume and pressure is equal to $c_V = c_p = c$. Note that u (J kg^{-1}) and U (J) therefore $U = u\ dm$. The control volume is $dV = dx\ dy\ dz$. Taking into account the density the dm (kg m^{-3}):

$$\rho = \frac{dm}{dV} \Rightarrow dm = \rho\ dx\ dy\ dz \tag{2.69}$$

$$\frac{dU}{dt} = \frac{d(udm)}{dt} = \frac{\partial T}{\partial t} c\ \rho\ dx\ dy\ dz \tag{2.70}$$

Substituting Eqs. (2.66), (2.67) and (2.70) into Eq. (2.55) gives the general heat conduction equation:

$$\frac{\partial}{\partial x}\left(\lambda_x \frac{\partial T}{\partial x}\right) + \frac{\partial}{\partial y}\left(\lambda_y \frac{\partial T}{\partial y}\right) + \frac{\partial}{\partial z}\left(\lambda_z \frac{\partial T}{\partial z}\right) + g = \frac{\partial T}{\partial t} c\rho \tag{2.71}$$

where each term has units of (W m^{-3}). This is a balance equation representing the rate of heat transfer between the system (dV) and the environment through its surfaces due to heat conduction, together with the rate of heat energy generated-injected into or subtracted from the system interior and the rate of heat energy stored in the system.

Eliminating perpendicular partial derivatives:

$$\frac{\partial}{\partial x}\left(\lambda_x \frac{\partial T}{\partial x} + \lambda_y \frac{\partial T}{\partial y} + \lambda_z \frac{\partial T}{\partial z}\right) + \frac{\partial}{\partial y}\left(\lambda_x \frac{\partial T}{\partial x} + \lambda_y \frac{\partial T}{\partial y} + \lambda_z \frac{\partial T}{\partial z}\right) + \frac{\partial}{\partial z}\left(\lambda_x \frac{\partial T}{\partial x} + \lambda_y \frac{\partial T}{\partial y} + \lambda_z \frac{\partial T}{\partial z}\right) + g = \frac{\partial T}{\partial t} c\rho \tag{2.72}$$

By means of the nabla operator ∇ and taking into account the properties of the scalar product we have:

$$\nabla = \left(\frac{\partial}{\partial x}\ \frac{\partial}{\partial y}\ \frac{\partial}{\partial z}\right)$$

$$\lambda = \left(\lambda_x\ \lambda_y\ \lambda_z\right)$$

$$\frac{\partial}{\partial x}(\lambda \nabla T) + \frac{\partial}{\partial y}(\lambda \nabla T) + \frac{\partial}{\partial z}(\lambda \nabla T) + g = \frac{\partial T}{\partial t} c\rho \tag{2.73}$$

Again using the nabla operator and considering λ as a tensor we obtain:

$$\nabla \cdot (\lambda \nabla T) + g = \rho c \frac{\partial T}{\partial t} \tag{2.74}$$

If the control volume had isotropic properties, the thermal conductivity would be constant ($\lambda_x = \lambda_y = \lambda_z$) without the sink/source term and using the Laplacian operator ∇^2

$$\nabla^2 = \nabla \cdot \nabla = \frac{\partial^2}{\partial x^2} + \frac{\partial^2}{\partial y^2} + \frac{\partial^2}{\partial z^2}$$

It would remain:

$$\nabla^2 T = \frac{\rho c}{\lambda} \frac{\partial T}{\partial t} \tag{2.75}$$

Thermal diffusivity α ($m^2\ s^{-1}$) is a thermal property of the media whose physical significance is associated with the rate at which heat propagates in a media, according to temperature changes induced by conduction. The higher the thermal diffusivity, the greater the response of the media to thermal perturbations and the faster temperature changes propagate through the media. Thermal diffusivity is defined as:

$$\alpha = \frac{\lambda}{\rho c} \tag{2.76}$$

From this definition, it is clear that the higher the thermal conductivity, the faster the media will propagate the thermal disturbances produced in its domain. However, the higher the heat capacity, the more heat the media can store, with little increase in temperature, so it will take a lot of heat to propagate the temperature disturbance.

Equation (2.75) can be found expressed in terms of thermal diffusivity:

$$\nabla^2 T = \frac{1}{\alpha}\frac{\partial T}{\partial t} \tag{2.77}$$

In an isotropic media without generation of internal energy and under stationary conditions ($\partial T/\partial t = 0$), the general heat conduction equation will be:

$$\nabla^2 T = 0 \tag{2.78}$$

Equation (2.78) is known as the *Laplace* equation. What this equation tells us is that when the heat transfer reaches a steady state, where the temperature at no point in the domain varies with time, the term in the equation that symbolises storage disappears. That is, the heat going in is equal to the heat coming out. Furthermore, the second derivative of the temperature field being zero means that the thermal gradient is constant over the whole space (no source/sink term), and the temperature distribution would be represented by a plane of constant slope.

2.2.3 Heat Convection Mechanism

Heat convection is the process by which a moving fluid transports stored heat as internal energy of the fluid itself. The advection heat flux q_x^{adv} (W m^{-2}) in the x direction can be expressed as:

$$q_x^{adv} = \frac{dU}{dt} = v_x \rho_f c_f \left(T - T_{ref}\right) \tag{2.79}$$

where v_x (m^3 s^{-1} m^{-2}) is the fluid velocity in the x direction, ρ_f (kg m^{-3}) is the density of the fluid and c_f (J kg^{-1} K^{-1}) is the heat capacity of the fluid. T_{ref} is the reference temperature which can be zero Kelvin for simplicity. It follows that:

$$\frac{dU}{dt} = \frac{d(u\ dm)}{dt} = \frac{\partial T}{dt} c\ \rho\ dx\ dy\ dz \tag{2.80}$$

Heat convection occurs as a result of the simultaneous combination of heat conduction acting on a fluid capable of transporting heat from one region to another in the direction of fluid motion. The convective heat flux per unit cross-section for one dimension is:

$$q''_x = -\lambda_x \frac{dT}{dx} + v_x \rho_f c_f T \tag{2.81}$$

For three-dimensional problems where, in addition to conduction there is a fluid causing heat convection, the energy balance in a control volume (Fig. 2.6), expressed in Eq. (2.55) must take into account this new component in the rate of thermal energy transferred with the environment across its surface. The heat flux entering in the direction x in the control volume will be:

$$q_{x1} = -\lambda_x \frac{dT}{dx}\, dy\, dz + v_x \rho_f c_f T\, dy\, dz \tag{2.82}$$

Using the Taylor series, we obtain an approximation to the outflow in the x direction:

$$q_{x2} \cong -\lambda_x \frac{\partial T}{\partial x} dy\, dz + \frac{\partial}{\partial x}\left(-\lambda_x \frac{\partial T}{\partial x} dy\, dz\right) dx + v_x \rho_f c_f T\, dy\, dz \\ - \frac{\partial}{\partial x}(v_x \rho_f c_f T\, dy\, dz) dx \tag{2.83}$$

The net flow in the x direction within the control volume shall be

$$q_{x1} - q_{x2} \cong \frac{\partial}{\partial x}\left(\lambda_x \frac{\partial T}{\partial x}\right) dx\, dy\, dz + \frac{\partial}{\partial x}(v_x \rho_f c_f T) dx\, dy\, dz \tag{2.84}$$

Similarly for the rest of the space addresses:

$$q_{y1} - q_{y2} \cong \frac{\partial}{\partial y}\left(\lambda_y \frac{\partial T}{\partial y}\right) dx\, dy\, dz + \frac{\partial}{\partial y}(v_y \rho_f c_f T) dx\, dy\, dz \tag{2.85}$$

$$q_{z1} - q_{z2} \cong \frac{\partial}{\partial z}\left(\lambda_z \frac{\partial T}{\partial z}\right) dx\, dy\, dz + \frac{\partial}{\partial z}(v_z \rho_f c_f T) dx\, dy\, dz \tag{2.86}$$

With Eqs. (2.84), (2.85) and (2.86) we can obtain the heat transferred by conduction–convection through the surfaces of the control volume:

$$\frac{dQ}{dt} = \left[\frac{\partial}{\partial x}\left(\lambda_x \frac{\partial T}{\partial x}\right) + \frac{\partial}{\partial x}(v_x \rho_f c_f T) + \frac{\partial}{\partial y}\left(\lambda_y \frac{\partial T}{\partial y}\right) + \frac{\partial}{\partial y}(v_y \rho_f c_f T) \right. \\ \left. + \frac{\partial}{\partial z}\left(\lambda_z \frac{\partial T}{\partial z}\right) + \frac{\partial}{\partial z}(v_z \rho_f c_f T)\right] dx\, dy\, dz \tag{2.87}$$

Applying again the first law of thermodynamics and keeping the storage and source/sink terms used for the conduction case, the three-dimensional heat conduction–convection equation is obtained:

$$\frac{\partial}{\partial x}\left(\lambda_x \frac{\partial T}{\partial x}\right) + \frac{\partial}{\partial y}\left(\lambda_y \frac{\partial T}{\partial y}\right) + \frac{\partial}{\partial z}\left(\lambda_z \frac{\partial T}{\partial z}\right) + \frac{\partial}{\partial x}(v_x \rho_f c_f T) + \frac{\partial}{\partial y}(v_y \rho_f c_f T) \\ + \frac{\partial}{\partial z}(v_z \rho_f c_f T) + g = \frac{\partial T}{\partial t} c\rho \tag{2.88}$$

Using the nabla operator ∇ and considering λ as a tensor, Eq. (2.88) can be expressed as

$$\nabla \cdot (\lambda \nabla T) + \nabla (\boldsymbol{v} \rho_f c_f T) + g = \rho c \frac{\partial T}{\partial t} \tag{2.89}$$

where $\boldsymbol{v}$ is the fluid velocity vector.

2.2.4 *Hydrodynamic Heat Dispersion*

Around the 1950s, the first studies appeared related to the quality of aquifers, shown from field and laboratory experiments (Bear 1988). The studies showed that chemical species dissolved in groundwater were transported through the porous media not only in the direction of groundwater flow but also perpendicularly. This could not be explained by groundwater advection and molecular diffusion alone. The spreading of the solutes during transport in porous media was called *dispersion.*

The convection of heat at the pore scale due to fluids moving through a network of interconnected pores represents a complex three-dimensional problem. The fluid velocity in the pores between particles of porous media is not uniform, and there is also a certain discontinuity between the velocity and temperature fields due to the solid particles. If all the mineral grains of the solid matrix of the porous media were equal and uniformly distributed in the three axes of space (Fig. 2.7a), the fluid velocity would be constant throughout the media and all flow lines would be parallel to each other. However, the nature of porous media in the subsurface presents a degree of heterogeneity (Fig. 2.7a) that makes the fluid flow lines irregular and tend to potentially cross each other. Since the velocity will be changing and the trajectories will tend to be different, there is the potential for fluid molecules to interact with other molecules of different internal energy, leading to energy transfer during mixing.

A given volume of fluid under such conditions of thermal equilibrium with a solid matrix could move through the porous media while retaining its internal energy and initial temperature. The heat transfer due to hydrodynamic mixing of an interstitial fluid at the pore scale is called *thermal dispersion.* The phenomenon of hydrodynamic dispersion also affects the transport of solutes in porous media (Dagan 1989; Gelhar 1993). There are different sub-processes inherent in the nature of this media that contribute to the hydrodynamic mixing listed below (Slattery 1984). Part of the mixing is due to the following: (1) blockages in the pore network that cause tortuosity of the flow system. Tortuosity causes two fluid molecules flowing through a pore network at identical velocity to have a different distance between them at different times. (2) The dead-end porosity zones where it does not intervene in the flow system since there are no flow lines through that zone. (3) Hydrodynamic mixing may occur due to recirculation caused by low pressure zones generated by flow restrictions. (4) Loss of fluid velocity when approaching the solid matrix wall causes mixing of fluid

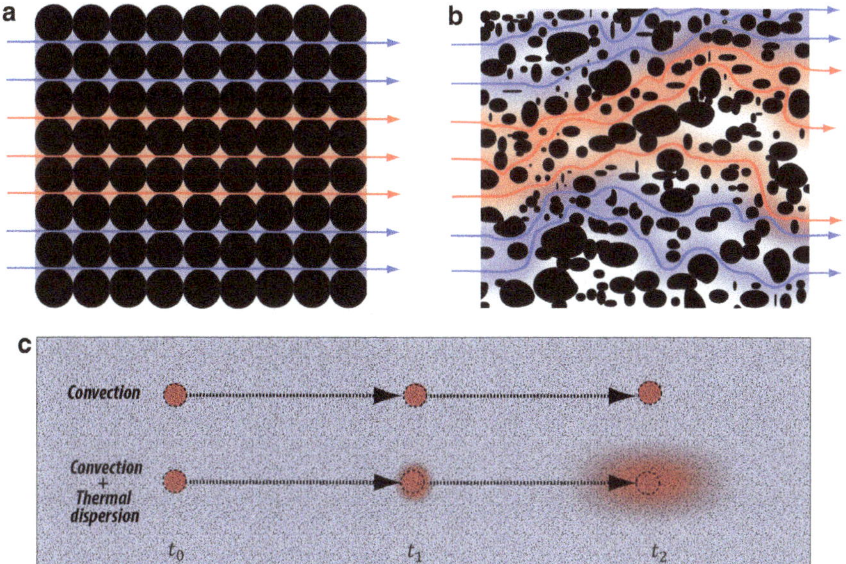

Fig. 2.7 Flow paths of a fluid through a **a** porous media idealised by equal and uniformly distributed solid particles and a **b** heterogeneous porous media at pore scale. **c** Effect of macroscopic-scale thermal dispersion of a heat pulse at a time t_0 and its evolution in time as it travels by convection through the porous media up to a time t_2

particles moving at different velocities. (5) The existence of eddies/turbulence when the flow becomes turbulent.

Although the thermal dispersion effect takes place at the pore scale, it is at larger scales that the heterogeneity of the porous media is manifest. At the geological formation scale, it is observed that a heat pulse transported in a porous media by convection will present a *dilution* effect of the advancing front from a net front to a diffuse gradual front (Fig. 2.7c). Therefore, to obtain a model of heat transport in the porous media that reproduces the effect of hydrodynamic heat dispersion, it is necessary to create a bridge between the real (microscopic) dispersive media and the fictitious (macroscopic) equivalent continuous media. To do this, a new component can be included in the energy balance of the control volume defined before, so that in the one-dimensional problem, the heat transferred per unit area by hydrodynamic scattering q_x^{dis} (W m^{-2}) is:

$$q_x^{dis} = D_x \rho_f c_f v_x \frac{\partial T}{\partial x} \tag{2.90}$$

where D_x (m) is the dispersion coefficient. Again carrying out the heat balance in a control volume, as in the previous cases, and also taking into account the thermal dispersion, we see that heat transfer at the inlet front to the control volume in the x direction will be:

$$q_{x1} = -\lambda_x \frac{dT}{dx} \, dy \, dz - D_x \rho_f c_f v_x \frac{\partial T}{\partial x} dy \, dz + v_x \rho_f c_f T \, dy \, dz \tag{2.91}$$

The heat transfer out of the control volume in the x direction, by applying Taylor's series is obtained:

$$\begin{aligned} q_{x2} \cong & -\lambda_x \frac{\partial T}{\partial x} dy \, dz + \frac{\partial}{\partial x}\left(-\lambda_x \frac{\partial T}{\partial x} dy \, dz\right) dx - D_x \rho_f c_f v_x \frac{\partial T}{\partial x} dy \, dz \\ & - \frac{\partial}{\partial x}\left(D_x \rho_f c_f v_x \frac{\partial T}{\partial x} dy \, dz\right) dx \\ & + v_x \rho_f c_f T \, dy \, dz - \frac{\partial}{\partial x}(v_x \rho_f c_f T \, dy \, dz) dx \end{aligned} \tag{2.92}$$

The net heat flow in the x direction in the control volume would be

$$q_{x1} - q_{x2} \cong \frac{\partial}{\partial x}\left(\lambda_x \frac{\partial T}{\partial x}\right) dV + \frac{\partial}{\partial x}\left(D_x \frac{\partial T}{\partial x} \rho_f c_f v_x\right) dV + \frac{\partial}{\partial x}(v_x \rho_f c_f T) dV \tag{2.93}$$

Analogously for the rest of the space directions and considering that $dx \, dy \, dz = dV$:

$$q_{y1} - q_{y2} \cong \frac{\partial}{\partial y}\left(\lambda_y \frac{\partial T}{\partial y}\right) dV + \frac{\partial}{\partial y}\left(D_y \frac{\partial T}{\partial y} \rho_f c_f v_y\right) dV + \frac{\partial}{\partial y}(v_y \rho_f c_f T) dV \tag{2.94}$$

$$q_{z1} - q_{z2} \cong \frac{\partial}{\partial z}\left(\lambda_z \frac{\partial T}{\partial z}\right) dV + \frac{\partial}{\partial z}\left(D_z \frac{\partial T}{\partial z} \rho_f c_f v_z\right) dV + \frac{\partial}{\partial z}(v_z \rho_f c_f T) dV \tag{2.95}$$

From Eqs. (2.93), (2.94) and (2.95), the heat transferred by conduction–convection-dispersion through the surface of the unit volume can be obtained:

$$\begin{aligned} \frac{dQ}{dt} = & \left[\frac{\partial}{\partial x}\left(\lambda_x \frac{\partial T}{\partial x}\right) + \frac{\partial}{\partial x}\left(D_x \frac{\partial T}{\partial x} \rho_f c_f v_x\right) + \frac{\partial}{\partial x}(v_x \rho_f c_f T) + \frac{\partial}{\partial y}\left(\lambda_y \frac{\partial T}{\partial y}\right)\right. \\ & + \frac{\partial}{\partial y}\left(D_y \frac{\partial T}{\partial y} \rho_f c_f v_y\right) + \frac{\partial}{\partial y}(v_y \rho_f c_f T) + \frac{\partial}{\partial z}\left(\lambda_z \frac{\partial T}{\partial z}\right) \\ & \left. + \frac{\partial}{\partial z}\left(D_z \frac{\partial T}{\partial z} \rho_f c_f v_z\right) + \frac{\partial}{\partial z}(v_z \rho_f c_f T)\right] dV \end{aligned} \tag{2.96}$$

$$\begin{aligned} \frac{dQ}{dt} = & \left[\frac{\partial}{\partial x}\left(\lambda_x \frac{\partial T}{\partial x}\right) + \frac{\partial}{\partial y}\left(\lambda_y \frac{\partial T}{\partial y}\right) + \frac{\partial}{\partial z}\left(\lambda_z \frac{\partial T}{\partial z}\right) + \frac{\partial}{\partial x}\left(D_x \frac{\partial T}{\partial x} \rho_f c_f v_x\right)\right. \\ & + \frac{\partial}{\partial y}\left(D_y \frac{\partial T}{\partial y} \rho_f c_f v_y\right) + \frac{\partial}{\partial z}\left(D_z \frac{\partial T}{\partial z} \rho_f c_f v_z\right) + \frac{\partial}{\partial x}(v_x \rho_f c_f T) \end{aligned}$$

$$+\frac{\partial}{\partial y}\left(v_y\rho_f c_f T\right)+\frac{\partial}{\partial z}\left(v_z\rho_f c_f T\right)\Bigg]dx\,dy\,dz \tag{2.97}$$

$$\begin{aligned}\frac{dQ}{dt}=&\Bigg[\frac{\partial}{\partial x}\left(\lambda_x\frac{\partial T}{\partial x}+\lambda_y\frac{\partial T}{\partial y}+\lambda_z\frac{\partial T}{\partial z}\right)+\frac{\partial}{\partial y}\left(\lambda_x\frac{\partial T}{\partial x}+\lambda_y\frac{\partial T}{\partial y}+\lambda_z\frac{\partial T}{\partial z}\right)\\&+\frac{\partial}{\partial z}\left(\lambda_x\frac{\partial T}{\partial x}+\lambda_y\frac{\partial T}{\partial y}+\lambda_z\frac{\partial T}{\partial z}\right)\\&+\frac{\partial}{\partial x}\left(D_x\frac{\partial T}{\partial x}\rho_f c_f v_x+D_y\frac{\partial T}{\partial y}\rho_f c_f v_y+D_z\frac{\partial T}{\partial z}\rho_f c_f v_z\right)\\&+\frac{\partial}{\partial y}\left(D_x\frac{\partial T}{\partial x}\rho_f c_f v_x+D_y\frac{\partial T}{\partial y}\rho_f c_f v_y+D_z\frac{\partial T}{\partial z}\rho_f c_f v_z\right)\\&+\frac{\partial}{\partial z}\left(D_x\frac{\partial T}{\partial x}\rho_f c_f v_x+D_y\frac{\partial T}{\partial y}\rho_f c_f v_y+D_z\frac{\partial T}{\partial z}\rho_f c_f v_z\right)\\&+\frac{\partial}{\partial x}\left(v_x\rho_f c_f T\right)+\frac{\partial}{\partial y}\left(v_y\rho_f c_f T\right)+\frac{\partial}{\partial z}\left(v_z\rho_f c_f T\right)\Bigg]dx\,dy\,dz\end{aligned} \tag{2.98}$$

$$\begin{aligned}\frac{dQ}{dt}=&\Bigg[\frac{\partial}{\partial x}(\lambda\nabla T)+\frac{\partial}{\partial y}(\lambda\nabla T)+\frac{\partial}{\partial z}(\lambda\nabla T)+\frac{\partial}{\partial x}\left(\boldsymbol{D}\nabla T\rho_f c_f\boldsymbol{v}\right)\\&+\frac{\partial}{\partial y}\left(\boldsymbol{D}\nabla T\rho_f c_f\boldsymbol{v}\right)+\frac{\partial}{\partial z}\left(\boldsymbol{D}\nabla T\rho_f c_f\boldsymbol{v}\right)\\&+\frac{\partial}{\partial x}\left(\boldsymbol{v}\rho_f c_f T\right)+\frac{\partial}{\partial y}\left(\boldsymbol{v}\rho_f c_f T\right)+\frac{\partial}{\partial z}\left(\boldsymbol{v}\rho_f c_f T\right)\Bigg]dx\,dy\,dz\end{aligned} \tag{2.99}$$

$$\frac{dQ}{dt}=\left[\nabla(\lambda\nabla T)+\nabla\left(\boldsymbol{D}\nabla T\rho_f c_f\boldsymbol{v}\right)+\nabla\left(\boldsymbol{v}\rho_f c_f T\right)\right]dx\,dy\,dz \tag{2.100}$$

$$\frac{dQ}{dt}=\left[\nabla\left(\lambda\nabla T+\boldsymbol{D}\nabla T\rho_f c_f\boldsymbol{v}\right)+\nabla\left(\boldsymbol{v}\rho_f c_f T\right)\right]dx\,dy\,dz \tag{2.101}$$

$$\frac{dQ}{dt}=\left[\nabla\left(\left(\lambda+D\rho_f c_f\boldsymbol{v}\right)\nabla T\right)+\rho_f c_f\boldsymbol{v}\nabla T\right]dx\,dy\,dz \tag{2.102}$$

Again applying the first law of thermodynamics, and keeping the storage and source/sink terms used for the conduction case, the three-dimensional equation of conduction–convection-heat dispersion is obtained:

$$\underbrace{\nabla(\lambda\nabla T)}_{\textbf{Condition}}+\underbrace{\nabla\left(\boldsymbol{D}\nabla T\rho_f c_f\boldsymbol{v}\right)}_{\textbf{Dispersion}}+\underbrace{\nabla\left(\boldsymbol{v}\rho_f c_f T\right)}_{\textbf{Advection}}+g=\underbrace{\rho c\frac{\partial T}{\partial t}}_{\textbf{Storage}} \tag{2.103}$$

In a more compact form:

$$\nabla\left(\left(\lambda+D\rho_f c_f\boldsymbol{v}\right)\nabla T\right)+\rho_f c_f\boldsymbol{v}\nabla T+g=\rho c\frac{\partial T}{\partial t} \tag{2.104}$$

2.2.5 Conduction–Convection-Heat Dispersion in a Porous Media

To model thermal processes in a water-saturated porous media, it will be necessary to take into account heat conduction, both through the solid matrix and the fluid phase, and heat advection by groundwater flow, including the effects of hydrodynamic thermal dispersion. Although phase change processes could be considered (Williams and Smith 1989) they are generally not significant in the exploitation of shallow geothermal energy resources and are therefore neglected.

If a differential control volume is composed of a porous media (Fig. 2.8), it is clear in Eq. (2.104) that each of the terms will need to reflect the specificities of the porous media.

The equivalent volumetric heat capacity (C_e) (J m^{-3} K^{-1}) of porous material saturated with water is usually calculated as the weighted arithmetic mean of the values for the matrix (c_m) and water (c_w), by their partial volumes as a function of porosity ϕ according to:

$$C_e = \rho_e c_e = \phi \rho_w c_w + \rho_m c_m (1 - \phi) \tag{2.105}$$

where ρ_e (kg m^{-3}) and c_e (J kg^{-3} K^{-1}) represent the equivalent density and equivalent specific heat capacity of the porous media. ρ_w (kg m^{-3}) and ρ_m (kg m^{-3}) represent the density of the water and the solid matrix; c_w (J kg^{-3} K^{-1}) and c_m (J kg^{-3} K^{-1}) represent the specific heat capacity of the water and the solid matrix, respectively.

Proceeding in the same way as in the previous case, the equivalent thermal conductivity (λ_e) (W m^{-1} K^{-1}) of the porous media as a function of porosity will be given by:

$$\lambda_e = \phi \lambda_w + \lambda_m (1 - \phi) \tag{2.106}$$

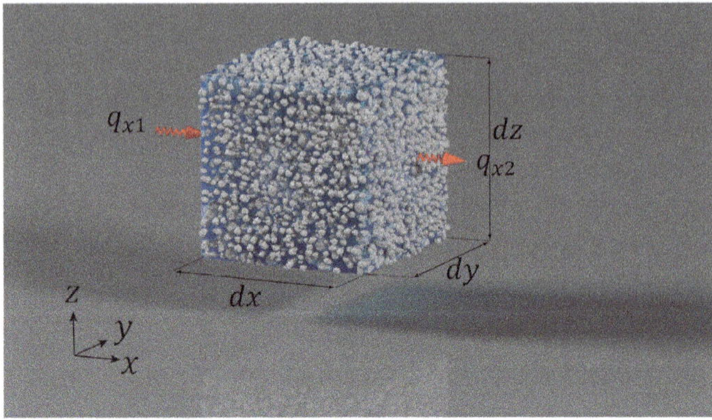

Fig. 2.8 Differential control volume of a porous media

where λ_w (W m^{-1} K^{-1}) and λ_m (W m^{-1} K^{-1}) represent the thermal conductivity of the water and the solid matrix respectively.

Substituting the equivalent properties of the porous media described above for those of the continuous media in Eq. (2.104), we obtain the general equation of conduction–convection-heat dispersion in the porous media saturated in water as the interstitial fluid:

$$\rho_e c_e \frac{\partial T}{\partial t} = \nabla((\lambda_e + D\rho_w c_w v)\nabla T) + \rho_w c_w v_D \nabla T + g \quad (2.107)$$

where $\rho_e c_e = \phi\rho_w c_w + \rho_m c_m(1-\phi)$ represents the equivalent volumetric heat capacity of the porous media (J m^{-3} K^{-1}); ϕ (−) is the porosity c_w (J kg^{-3} K^{-1}) and c_m (J kg^{-3} K^{-1}) represent the specific heat capacity of the water and the solid matrix respectively, ρ_w (kg m^{-3}) and ρ_m (kg m^{-3}) are the density of water and solid matrix,$T(x, y, z, t)$ is the scalar field of temperatures in space and time, ∇ is the three-dimensional vector operator nabla, $\lambda_e = \phi\lambda_w + \lambda_m(1-\phi)$ is the equivalent or effective thermal conductivity (W m^{-1} K^{-1}) of the porous media, D (m) is the coefficient of hydrodynamic thermal dispersion, v_D (m^3 m^{-2} s^{-1}) is the *Darcy* velocity vector of the groundwater in the porous media, and g is the source/sink term (W m^{-3}).

2.3 Parameters of Interest in Shallow Geothermal Energy

The main parameters involved in quantitative determinations taken in shallow geothermal energy, as well as some indicative values for them, are listed below:

2.3.1 Thermal Conductivity λ (W m^{-1} K^{-1})

Thermal conductivity is an intrinsic property of materials that defines their ability to diffuse heat through themselves. It represents the constant of proportionality between the heat flux per unit area passing through a material and the existing thermal gradient under stationary conditions. According to Fourier's law (Eq. 2.50) in one dimension, λ can be expressed as:

$$\lambda = \rho c \frac{ql}{A\Delta T} \quad (2.108)$$

where q (W) is the heat flux through a cross-section of area A (m^2) under a temperature gradient defined by an increase in temperatures ΔT (K) over a distance l (m) (see Fig. 2.4). Tables 2.1 and 2.2 show values of thermal conductivities of some minerals and materials of interest in shallow geothermal energy. Of all the minerals present in the rocks, quartz has one of the highest thermal conductivities, only surpassed

Table 2.1 Table of thermo-physical properties of different minerals of interest in shallow geothermal energy

Mineral	Thermal conductivity λ (W m^{-1} K^{-1})		Density ρ (kg m^{-3})		Heat capacity c (kJ kg^{-1} K^{-1})	
	Min	Max	Min	Max	Min	Max
Quartz ($-\alpha$)	7.69	7.70	2648		0.70	0.74
Quartz	6.5		2648		0.75	
Olivine	3.16	5.21	3213	4393	0.55	0.84
Garnet	3.31	5.48	4318		0.74	
Pyroxene	4.1	5.10	3209	3277	0.67	0.69
Amphibole	2.81	3.00	3080		0.75	
Mica	0.7	2.32	2831	2900	0.76	0.78
Feldspar	1.68	2.31	2560	2760	0.63	0.75
Kaolinite	2.60		2610	2680	0.93	
Ilita	1.90		2600	2900	0.81	0.82
Chlorite	4.20	5.92	2600	3300	0.60	
Smectite	1.90		2000	2600	0.86	
Clay minerals	1.70	5.95	1920	2450	0.54	0.95
Calcite	3.25	3.90	2710		0.79	0.80
Dolomite	5.30	5.51	2866		0.86	0.88
Halite	5.30	7.20	2163		0.79	0.84
Anhydrite	0.90		2963		0.75	
Plaster	1.00	1.30	2305		1.07	

Source Rohsenow et al. (1998) and Schön (2011). In cases where a maximum and minimum value of the property does not exist or is not available, the only available value centred between the two columns of the property is shown

by accessory minerals such as rutile or spinel, which present extreme values (8–13 W m^{-1} K^{-1}). The lowest λ values are found in the group of phyllosilicates, such as biotite. The size of the crystals that make up a rock sample influences its thermal conductivity. This happens in the case of quartz or calcite, decreasing its value as the size decreases. Thermal conductivity is a very important property in shallow geothermal energy. For example, the efficiency of a geothermal heat exchanger in a closed circuit is proportional to the thermal conductivity of the ground, where a granite with a thermal conductivity of 3.30 W m^{-1} K^{-1} is three times better than a clay soil with a thermal conductivity of 1.10 W m^{-1} K^{-1} (Eskilson 1987).

The thermal conductivity of solids and liquids is temperature dependent (Pitts and Sissom 1998). In the case of solids, it is experimentally observed that their thermal conductivity as a function of temperature $\lambda(T)$ can be expressed as:

$$\lambda(T) = \lambda(T_0)(1 + a(T - T_0)) \tag{2.109}$$

Table 2.2 Table of thermo-physical properties of different rocks, sediments and materials of interest in shallow geothermal energy

	Thermal conductivity		Density		Heat capacity		Porosity	
	λ (W m^{-1} K^{-1})		ρ (kg m^{-3})		c (kJ kg^{-1} K^{-1})		ϕ (−)	
	Min	Max	Min	Max	Min	Max	Min	Max
Rock								
Granite	1.25	4.45	2630	2750	0.67	1.55	5E−04	9E−03
Basalt	1.50	2.50	2800	3000	0.84	1.28	6E−03	1E−02
Granodiorite	1.35	3.40	2530	2940	0.84	1.26	6E−04	8E−03
Limestone	0.62	6.26	2300	2900	0.82	1.72	6E−03	0.35
Dolomite	1.60	6.30	2800	2900	0.84	1.55	1E−03	0.30
Anhydrite	1.00	6.05	2800	2900	0.81	0.94	1E−04	0.15
Plaster	1.29	1.29	2300	2800	0.85	1.09	1E−04	0.15
Sandy	0.90	6.50	2160	2800	0.75	1.60	0.05	0.30
Siltstone	0.61	2.10	2200	2880	0.91	1.52	0.06	0.35
Shale	0.55	4.25	2400	2800	0.88	1.44	0.10	0.43
Marls	0.50	4.00	2243	2830	0.78	1.50	0.02	0.35
Sediment								
Sand	0.10	2.75	1280	2150	1.97	1.20	0.20	0.60
Clay	0.60	2.60	1070	1600	0.84	1.00	0.33	0.60
Soil	0.40	0.86	1600	2050	1.80	1.90	0.30	0.50
Geothermal heat exchangers								
Polyethylene	0.33		960		2.10			
Thermoactive cement	0.80	1.50	1100	1400	2.00	2.20		
Water (20 °C)	0.60		1000		4.166			
Water +25% Ethylene glycol	0.50		1050		3.79			
Ice (0 °C)	2.20		917		2.04			
Air (dry 20 °C)	0.03		1.275E−03		1.005			

Source Rohsenow et al. (1998) and Schön (2011). In cases where a maximum and minimum value of the property is not available or does not exist, the only available value centred between the two columns of the property is shown

where T_0 is the reference temperature, $\lambda(T_0)$ is the value of the thermal conductivity at a given reference temperature and a (K^{-1}) is a constant characteristic of each material obtained experimentally. The constant a is positive, so that as the temperature increases, the thermal conductivity tends to increase. The opposite is true for liquids (except in the case of water); increasing the temperature of a liquid tends to decrease its thermal conductivity. As the temperature of water increases, its thermal conductivity increases up to 150 ºC, after which it decreases, as in other liquids. Water has one of the highest thermal conductivities of all liquids.

In magmatic and metamorphic rocks, the thermal conductivity depends on their mineral composition, the existence of fractures and the type of fracture filling. The alignment of crystal axes and/or fractures can generate anisotropy in the thermal properties of the porous media of up to 40% (Schön 2011). This is especially true for gneisses and schists, where thermal conductivity depends on whether the measurement is made perpendicular or parallel to the schistosity. It has been observed that the positive correlation of the thermal conductivity of these types of rocks and their quartz content is unequivocal (Touloukian et al. 1981). The thermal conductivity of rocks in general tends to decrease with increasing temperature, and there are different empirical equations that relate these variables for magmatic rocks (Seipold 1998). Volcanic rocks have vacuoles and high porosity (e.g., pumice) causing thermal conductivity to decrease.

The effect of porosity has a particular impact on the thermal properties of porous media in sedimentary rocks. In general, sedimentary rocks and sedimentary materials show considerable variation in thermal properties within the same classificatory typology, reflecting the complexity of this type of material in terms of mineral composition, structure and texture.

When the material through which heat conduction occurs is a porous media (soils, sediment and rocks), i.e., a multiphase media, the thermal conductivity used in the mathematical models requires using an effective or equivalent thermal conductivity (λ_e) that takes into account the solid matrix and pore fluids for heat transfer. There are several models in which a media is made up of n components that can be idealised, in the simplest case, as a stack of layers. In this layer model, each layer represents the relative volume of each component (Fig. 2.9). If the layers are parallel to the heat flow, parallel models are considered, and if the heat flow is perpendicular to the layers, series models are considered.

In a generalized model for n parallel components (Fig. 2.9a), the equivalent thermal conductivity will be determined by the volume-weighted arithmetic mean for the relative volume of each phase (Dagan 1989):

$$\lambda_e = \frac{V_i\lambda_1 + V_i\lambda_2 + V_i\lambda_3 + \cdots + V_n\lambda_n}{V_T} = \sum_{i=1}^{i=n}\left(\frac{V_i}{V_T}\lambda_i\right) \tag{2.110}$$

In a generalized model for n components in series (Fig. 2.9b), the equivalent thermal conductivity will be determined by the volume-weighted harmonic mean for the relative volume of each phase:

$$\lambda_e = \left[\sum_{i=1}^{i=n} V_i\lambda_i^{-1}\right]^{-1} \tag{2.111}$$

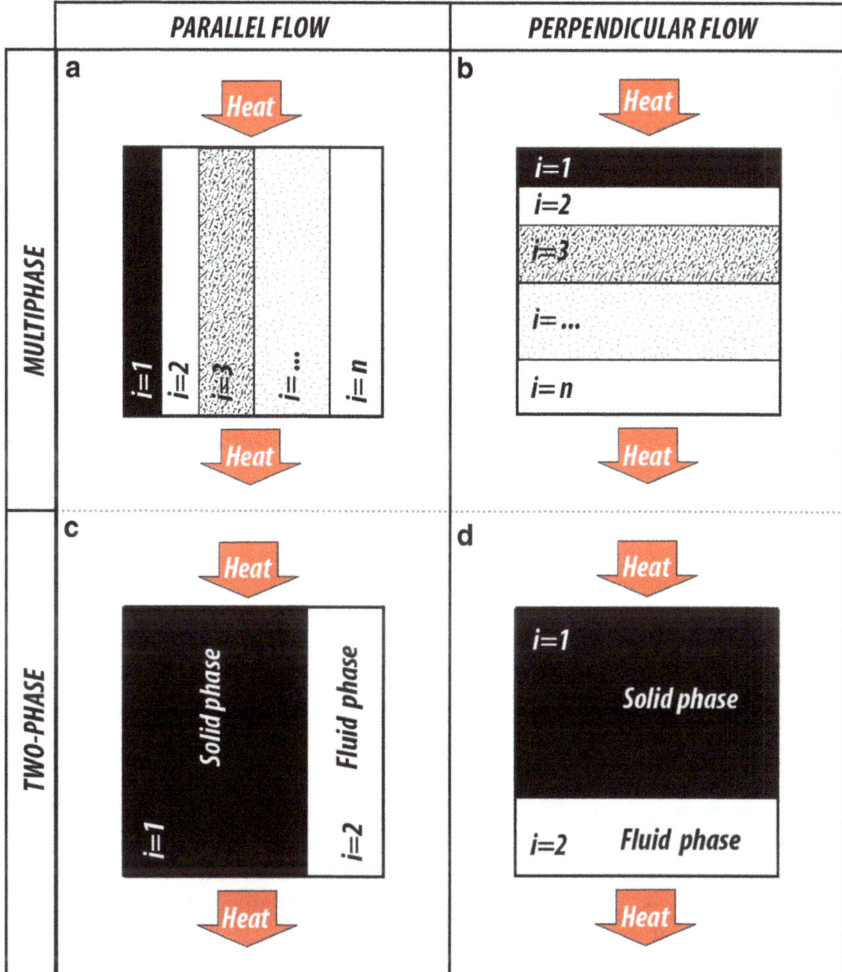

Fig. 2.9 Layer models for the calculation of the effective thermal conductivity of a porous media. **a** Multiphase parallel model. **b** Multiphase series model. **c** Two-phase parallel model. **d** Two-phase series model

The thermal conductivity calculated with Eq. (2.110) will always be higher than that calculated with Eq. (2.111), since in the case of parallel flow the heat will preferentially flow through the more conductive layer and residually through the less conductive layer. In the case of the series model, all heat must pass through the less conductive layer, which will limit heat transfer. In general, it is considered that the parallel model (weighted arithmetic mean) would give the most representative equivalent conductivity value and vice versa; the conductivity value calculated with the series model (weighted harmonic mean) would be less representative. Both could be considered as the limiting values for the range of variations that could occur.

In shallow geothermal energy it is usually sufficient to work with the porous media using a two-phase rock-water model. Therefore, it will be useful to consider $n = 2$ that by introducing the porosity, the two-phase model in parallel will be obtained:

$$\lambda_e = \lambda_w \phi + \lambda_m (1 - \phi) \tag{2.112}$$

where λ_w (W m^{-1} K^{-1}) is the thermal conductivity of water, λ_m (W m^{-1} K^{-1}) is the thermal conductivity of the solid matrix, and ϕ (−) is the porosity. Similarly, for two-phase series models the equivalent thermal conductivity is:

$$\lambda_e = \frac{1}{\frac{1-\phi}{\lambda_m} + \frac{\phi}{\lambda_f}} \tag{2.113}$$

Other experimentally derived models include the weighted geometric mean (Balling et al. 1981; Sass et al. 1971):

$$\lambda_e = \prod_{i=1}^{i=n} \left(\lambda_i{}^{V_i}\right) = (\lambda_m)^{1-\phi} \left(\lambda_f\right)^{\phi} \tag{2.114}$$

This model provides an intermediate value between the extreme values provided by the arithmetic and harmonic weighted averages. Figure 2.10 shows the dependence of the effective thermal conductivity on the porosity depending on the interstitial fluid and the calculation method.

Figure 2.10 shows the effect of changing the porosity of a porous media, with water or air as the interstitial fluid. As porosity appears and increases, the equivalent thermal conductivity decreases, as a phase with lower thermal conductivity than the solid matrix. In the case of Fig. 2.10, a porous media with a solid matrix composed of quartz (6.5 W m^{-1} K^{-1}), equivalent to a sandstone, and a mineral of average thermal conductivity (2.5 W m^{-1} K^{-1}) has been considered.

The range of variation of thermal conductivity is relatively small compared to other material properties, such as hydraulic conductivity. While most mineral materials are in the range of 1–4 W m^{-1} K^{-1}, hydraulic conductivity can vary by more than eight orders of magnitude. In porous rocks and sediments, porosity and moisture content affect thermal conductivity and can be a determining variable (Kappelmeyer and Haenel 1974). This is because the fluids in the porous media have generally lower thermal properties than the matrix-forming solids. The greater this difference (contrast), the more important the effect of porosity and moisture will be (Fig. 2.10).

The thermal conductivity of liquid water is 20 times that of air and, therefore, water saturation has an important impact on effective thermal conductivity of porous media (deVries 1966). Figure 2.11 presents some experimental results, showing how sediment of high porosity, initially with pores totally occupied by air (0.03 W m^{-1} K^{-1}), are saturated with water (0.6 W m^{-1} K^{-1}) and consequently increase their equivalent thermal conductivity. Some specific studies on the thermal properties of soils (Abu-Hamdeh and Reeder 2000; Clark 1966) show the influence of density, moisture

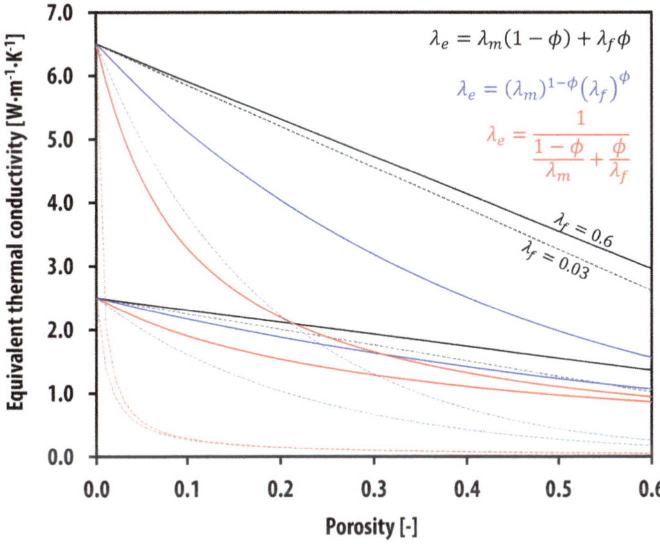

Fig. 2.10 Effective thermal conductivity as a function of the material porositycalculated according to the weighted arithmetic mean (black), weighted geometric mean (blue) and harmonic mean (red). Two different solid matrix phases are considered: quartz (6.5 W m^{-1} K^{-1}) and an average composition (2.5 W m^{-1} K^{-1}). Two thermal conductivities of the fluid phase are considered: 0.6 W m^{-1} K^{-1} for water (solid line) and 0.003 W m^{-1} K^{-1} for air (dashed line)

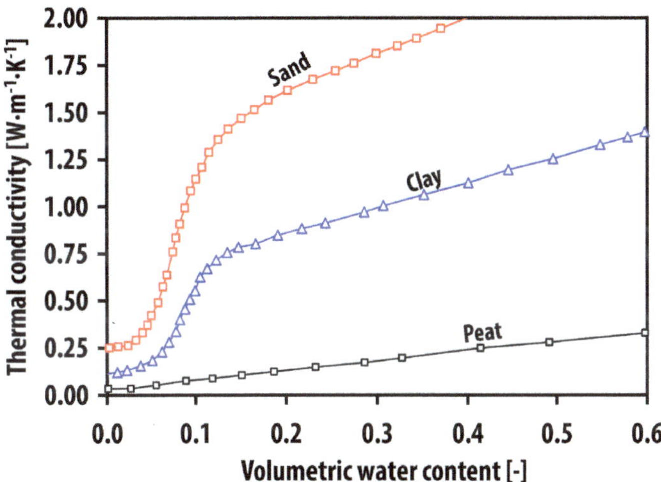

Fig. 2.11 Equivalent thermal conductivity of sand, clay and peat in relation to volumetric moisture content (van Duin 1963)

content and soil type. In this type of material, the organic matter content is decisive, given its low thermal conductivity (0.29 W m^{-1} K^{-1}).

2.3.2 *Thermal Resistivity* R *(K W^{-1})*

Equation (2.52) (one-dimensional Fourier's Law) can be rewritten as:

$$q = \frac{T_2 - T_1}{\Delta x/(\lambda A)} = \frac{T_2 - T_1}{R} \tag{2.115}$$

where R (K W^{-1}) is the thermal resistivity. A heat flow from one body at temperature T_2 to another body T_1 will find a thermal resistance in the media with magnitude R.

2.3.3 *Thermal Expansion ϵ (K^{-1})*

It is well known that most materials expand or contract with temperature. Thermal expansion describes the volumetric change of a material as a result of a change in its temperature. It is generally accepted that the thermal expansion of minerals and rocks is relatively small in absolute terms (Chekhonin et al. 2012). However, the differential behaviour of different materials in expanding/contracting with temperature changes are responsible for structural changes of the sediment and even for causing structural damage (Somerton 1992). This is of great importance for certain types of shallow geothermal installations like geostructures.

When calculating thermal expansion, it is necessary to consider whether the body to be expanded can expand freely or is constrained. The latter is most common in the subsurface and is, therefore, of interest in shallow geothermal energy. Constrained bodies that are unable to expand experience an internal stress as their temperature changes. The internal stress is calculated assuming that the body is free to expand, and the stress that would have to be applied to the body to return to initial volume is equivalent to the internal stress. The axial internal thermal expansion (ε) (−) is given by (Fjær et al. 2008):

$$\varepsilon = \frac{\Delta l}{l_0} = -\epsilon \Delta T \tag{2.116}$$

where ϵ (K^{-1}) is the coefficient of linear thermal expansion and represents the proportionality constant between the linear expansion/contraction increment (Δl) (m) normalised by the initial situation (l_0) (m) and the temperature increase experienced ΔT (K). The negative sign ensures that the coefficient has a positive value for most cases where a temperature increase causes an expansion. There is not much data on linear expansion coefficients in sedimentary rocks, but it is noted that a common value is about 10^{-5} K^{-1} (Fjær et al. 2008). This is approximately an expansion/contraction of about ten thousandths of a millimetre in a rock of initial length of ten centimetres when changing its initial temperature by ten Kelvin. When considering the three-dimensional space, the coefficient of volumetric thermal expansion for isotropic materials is three times the coefficient of linear thermal expansion, since there are three dimensions in space.

2.3.4 Density ρ (kg m^{-3})

The mass density ρ (kg m^{-3}) in a homogeneous material is defined as the amount of mass m (kg) in a volume V (m^3) when the volume under consideration tends to zero, expressed mathematically as:

$$\rho = \lim_{\Delta V \to 0} \frac{\Delta m}{\Delta V} \tag{2.117}$$

In a heterogeneous media the density depends on the domain where this quantity is observed and on the sample volume chosen. The density in this case can be described as:

$$\rho = \lim_{\Delta V \to REV} \frac{\Delta m}{\Delta V} \tag{2.118}$$

where REV is the representative elemental volume of a porous media (see Fig. 2.4). A pragmatic way of defining the porosity of a porous media is to consider the density as an intensive variable with the additive property where a porous media of n components will have the bulk or volumetric density:

$$\rho = \sum_{i=1}^{i=n} \left(\rho_i \frac{V_i}{V_T} \right) \tag{2.119}$$

where ρ_i and V_i are the density and volume of the i component respectively. V_T is the total volume. In the case of a two-phase porous system where the volume fraction is expressed as porosity ϕ:

$$\rho = (1 - \phi)\rho_s + \phi\rho_f \tag{2.120}$$

When considering other fluids occupying the porosity, having saturation χ the density is given by:

$$\rho = (1 - \phi)\rho_s + \phi(\chi_w \rho_w + \chi_h \rho_h + \chi_{ga} \rho_{ga}) \tag{2.121}$$

where the subscripts w, h and ga refer to water, oil and gas, respectively.

The density of rocks and sedimentary materials depends on the mineral composition of the solid matrix (see Table 2.1), the porosity and the density of the included fluids. The mineralogical composition largely determines the density of igneous rocks. It is generally accepted that the density of felsic (acidic) rocks increases as they become more mafic (basic). Increasing porosity of rocks decreases their density and increases with water saturation for a given porosity. Table 2.2 shows some density values of rocks and materials of interest in shallow geothermal energy.

2.3.5 Specific Heat Capacity c (J kg^{-1} K^{-1})

Specific heat capacity c (J kg^{-1} K^{-1}), also referred to as specific heat, represents the amount of heat Q (J) required to change the temperature of a body of one kilogram by one Kelvin. It is generally expressed as:

$$c = \frac{Q}{m\Delta T} \tag{2.122}$$

If it is found more appropriate to work in terms of volume, the same concept can be expressed as volumetric heat capacity C (J m^{-3} K^{-1}) taking into account its density (ρ) (kg m^{-3}):

$$C = \rho c \tag{2.123}$$

When working with shallow geothermal energy problems, it is usually sufficient to work with the pore media asuming a two-phase rock-water model. It is common to use the two-phase model to calculate the equivalent volumetric heat capacity of the porous media (C_e) (J m^{-3} K^{-1}) as an arithmetic mean weighted by the volume fraction:

$$C_e = \rho_e c_e = \phi \rho_w c_w + \rho_s c_s (1 - \phi) \tag{2.124}$$

where ρ_e (kg m^{-3}) and c_e (J kg^{-3} K^{-1}) are the density and the effective or equivalent specific heat capacity of the two-phase porous media, c_w and c_s represent the specific heat capacity (J kg^{-3} K^{-1}) of the water and solid matrix respectively, ρ_w and ρ_s represent the density (kg m^{-3}) of the water and the solid matrix respectively, and ϕ (−) is the porosity. In Tables 2.1 and 2.2, specific heat capacity values are given for different minerals, rocks and materials of interest in shallow geothermal energy. It is worth highlighting the high heat capacity of water, about four times higher than the rest of the solid materials that make up the subsoil. As seen for thermal conductivity, the range of variation of the specific heat capacity is limited to a range in minerals of 500–1000 J kg^{-1} K^{-1}. Depending on the porosity and the fluid associated with it, this value may increase or decrease within the same order of magnitude, especially if the fluid in the porous system is water (4180 J kg^{-1} K^{-1}).

According to the first law of thermodynamics (Eq. 2.9) heat (energy in transit) transferred to a system can increase its internal energy (ΔU) and/or perform work (W):

$$Q = \Delta U + W \tag{2.125}$$

It follows from this statement that measurement of specific heat capacity depends on the conditions under which the heat transfer takes place. In the case of gases, work will be done if a change in pressure induced by the heat transfer succeeds in changing the volume of the system (Eq. 2.16). If the volume of the system is changed, it will do work on its surroundings, and additional energy transfer should be considered to do that work. Therefore, in order to measure the specific heat capacity of gases, isochoric conditions ($\Delta V = 0$) during heat transfer are required. By introducing a given gas into a container of fixed volume, heat can be transferred and its temperature change measured to estimate its specific heat capacity at constant volume (c_V) (J kg^{-1} K^{-1}). Under these conditions no work is done and the heat (dQ) (J) transferred to a system will become entirely part of its internal energy (du) (J) (Eq. 2.9):

$$dQ = du \tag{2.126}$$

Combining the first law of thermodynamics (Eq. 2.9) with the definition of heat capacity (Eq. 2.122), in an isochoric process the specific heat capacity at constant volume will be obtained, which will be given by:

$$du = m\, c_V\, dT \tag{2.127}$$

Considering $du = dUm$ with dU (J kg^{-1}) as the internal energy per unit mass, it follows that:

$$c_V = \left(\frac{\partial U}{\partial T}\right)_V \tag{2.128}$$

The derivative of the internal energy as a function of state depends on other variables (Eq. 2.19) and is therefore calculated as a partial derivative.

In the case of solids and liquids, it is usual to measure their heat capacity under conditions of constant atmospheric pressure, i.e., by an isobaric process. In a given heat transfer to a system where its pressure remains constant, the volume must increase, so the system expands performing work, according to the first law of thermodynamics:

$$dQ = dU + dW \tag{2.129}$$

where Q (J) is the heat transferred to a system, U (J) is the internal energy of the system and W (J) is the work done by the system. Taking into account the definition of specific heat per unit mass at constant pressure expressed as:

$$dQ = m\, c_P\, dT \tag{2.130}$$

where Q (J) is the heat transferred to a system. Assuming that the pressure is constant, the work will be given by $dW = p\, dV$ and substituting dQ from Eqs. (2.129) and (2.130):

$$dU + p\, dV = m\, c_P\, dT \tag{2.131}$$

$$m\, c_P = \frac{dU + p\, dV}{dT} \tag{2.132}$$

$$m\, c_P = \left(\frac{\partial U}{\partial T}\right)_P + p\left(\frac{\partial V}{\partial T}\right)_P \tag{2.133}$$

Since the enthalpy (H) (J) is $H = U + pV$, by substitution and per unit of mass:

$$c_P = \left(\frac{\partial(H - pV)}{\partial T}\right)_P + p\left(\frac{\partial V}{\partial T}\right)_P \tag{2.134}$$

what exactly it is:

$$c_P = \left(\frac{\partial H}{\partial T}\right)_P - p\left(\frac{\partial V}{\partial T}\right)_P + p\left(\frac{\partial V}{\partial T}\right)_P \tag{2.135}$$

$$c_P = \left(\frac{\partial H}{\partial T}\right)_P \tag{2.136}$$

which is the same as

$$c_P = \left(\frac{\partial U}{\partial T}\right)_P + p\left(\frac{\partial V}{\partial T}\right)_P \tag{2.137}$$

Equation (2.136) is generally presented in the literature. However, Eq. (2.137) shows how the first summand term represents the fraction of heat contributed to the system and used to change its internal energy (and temperature change) and the second summand represents the fraction of heat used to carry out the volume change at constant pressure.

In the case of solids and liquids, the coefficient of volumetric thermal expansion ($\partial V/\partial T$) is very low (Sect. 2.3.3), so the work done will also be very low. In general problems in shallow geothermal energy, where mineral solids and water as a fluid saturate the pore space, the system pressure exerted during thermal expansion into the atmosphere with a displacement of thousandths of a millimetre can be considered negligible. Consequently:

$$c_P = \left(\frac{\partial H}{\partial T}\right)_P \cong \left(\frac{\partial U}{\partial T}\right)_V = c_V \tag{2.138}$$

Between liquids and solids, the volume variation that occurs is negligible, so that $\Delta U \cong \Delta H$. In shallow geothermal energy problems, it shall be considered for practical purposes that $c_V = c_P = c$. If there is a phase change during heat transfer, the associated latent heat should be considered in the enthalpy change.

The specific heat capacity is temperaturedependent but, for the temperature range in which shallow geothermal energy systems operates, its value may change by less than 1%. This can be seen in the case of water; at constant pressure its specific heat capacity varies by 1% as its temperature rises from zero degrees Celsius to 100 °C.

2.3.6 Thermal Diffusivity α (m^2 s^{-1})

When heat transfer occurs mainly through the conduction mechanism, it is convenient to introduce another property related to thermal conductivity: it is thermal diffusivity α (m^2 s^{-1}) expressed as:

$$\alpha = \frac{\lambda}{\rho c} \tag{2.139}$$

where λ (W m^{-1} K^{-1}) is the thermal conductivity, ρ (kg m^{-3}) is the density and c (J kg^{-3} K^{-1}) is the specific heat capacity of the media. The physical meaning of thermal diffusivity is associated with the rate of heat propagation in a media in which temperature changes occur. It represents heat flow by conduction with respect to energy storage. The higher the thermal diffusivity of a media, the greater the response of the media to temperature perturbations and the faster those temperature perturbations will propagate through that media. Considering a semi-infinite media where, at its boundary condition, the initial temperature is reduced from room temperature to zero degrees Celsius, maintaining it at that temperature, the temperature will be reduced through the media as heat dissipates over the edge of the cooled media. Depending on the thermal diffusivity of the media, it will take more or less time to reduce the initial temperature. For example, reducing the initial temperature by half at 30 cm from the cooled edge in a silver media (170 × 10^6 m^2 s^{-1}) will take 9.5 min; if it is granite (0.22 × 170 × 10^6 m^2 s^{-1}) it would take 5.5 days. Therefore, the propagation of a heat front by conduction will be proportional to the thermal conductivity and inversely proportional to its heat capacity and density. In the case of shallow geothermal energy problems, the thermal diffusivity is important in the design of heat exchangers, allowing to assess their thermal conductivities and resistivities.

2.3.7 *Viscosity μ (Pa s)*

Viscosity is a measure of the internal friction in a fluid and represents its resistance to deformation. Viscous forces are forces that oppose the movement of one portion of a fluid relative to another. A fluid with a high viscosity will require a greater force to deform it than one with a low viscosity. *Newton's Law of Viscosity* states that when a shear stress is applied to a fluid, the rate of deformation of the fluid is directly proportional to the applied stress, is linear and passes through the origin. When a fluid follows this law, the fluid is considered to have Newtonian behaviour. Consequently, the friction in a one-dimensional flow of a Newtonian fluid can be expressed as the ratio of a shear stress to a fluid according to:

$$\tau_{xy} = \mu \frac{dv_x}{dy} \tag{2.140}$$

where τ_{xy} (Pa) is the shear stress exerted at a point in the fluid, dv_x/dy (s^{-1}) represents the velocity gradient perpendicular to the shear plane where the shear stress is calculated, and μ (Pa s, Poise) is the constant of proportionality between the two quantities, called dynamic viscosity. Water is a Newtonian fluid and its viscosity does not depend on the forces acting on it but only on its temperature and pressure. As the temperature increases, the viscosity of gaseous fluids increases, while for liquid fluids its viscosity decreases. The viscosity of water goes from 0.28 to 1.79 mPa s as its temperature decreases from 100 to 0 °C. This represents a relatively small and generally negligible change in viscosity when modelling groundwater flow and heat transport in shallow geothermal energy problems. However, in the design of closed-loop geothermal heat exchangers, viscosity has relevant effects on the flow of liquids through the pipes that constitute the geothermal heat exchangers, affecting their efficiency.

In heat transfer problems, the dynamic viscosity (μ) by the density (ρ) of the fluid ratio is calculated to obtain the so-called kinematic viscosity υ ($m^2\ s^{-1}$, Stokes):

$$\upsilon = \frac{\mu}{\rho} \tag{2.141}$$

Note that the kinematic viscosity has the same units as the thermal diffusivity (Eq. 2.139), which is why it has also been called the *momentum diffusivity.*

2.3.8 *Reynolds Number Re (−)*

In fluid dynamics, the *Reynolds number* (−) is used to indicate the degree of turbulence in a fluid flow. *Re* is used in fluid dynamics to indicate the degree of turbulence in a fluid flow. The dimensionless number is defined as (Reynolds 1883):

$$Re = \frac{Ql}{\upsilon A} \tag{2.142}$$

where Q (m^3 s^{-1}) is the volumetric flow rate, l (m) is the characteristic distance, υ (m^2 s^{-1}) is the kinematic viscosity, and A (m^2) is the cross section. In granular porous media l is usually related to the grain size distribution or to the intrinsic permeability as $\sqrt{k}$ (Ward 1964), where k (m^2) is the intrinsic permeability. The Reynolds number expresses the ratio between the interstitial forces of the fluid and the viscous forces. For a pipe it can be expressed as:

$$Re = \frac{\rho v D_{tb}}{\mu} \tag{2.143}$$

where ρ (kg m^{-3}) is the density of the fluid, v (m s^{-1}) is the average fluid velocity, D_{tb} (m) is the pipe diameter and μ (Pa s) is the dynamic viscosity. The flow shall be considered to be laminar ($Re < 2.300$), transient ($2.300 \le Re \le 4.000$) or turbulent ($Re > 4.000$).

2.3.9 Fourier Number Fo (−)

The *Fourier Number* is defined as the ratio of the heat conduction rate (Q_{con}) and the rate of storage (Q_{alm}/t) of internal energy of a system. According to Eq. (2.1) (with $l = \Delta x$) and 2.50 (with $m = \rho V$) this ratio can be expressed as:

$$\frac{Q_{con}}{\frac{Q_{alm}}{t}} = \frac{\lambda \frac{\Delta T}{l} A}{\rho\ V\ c \frac{\Delta T}{t}} = \frac{\lambda\ \Delta T\ A\ t}{\rho\ A\ l\ c \Delta T} = \frac{\lambda}{\rho c}\frac{t}{l^2} = \frac{\alpha t}{l^2} \tag{2.144}$$

This ratio defines the dimensionless Fourier number Fo (−) as:

$$Fo = \frac{\alpha t}{l^2} \tag{2.145}$$

where α (m^2 s^{-1}) is the thermal diffusivity, t (s) is the characteristic time in which the heat front propagates by conduction over a distance of (m), (s) is the characteristic time in which the heat front propagates over the characteristic distance l (m).

2.3.10 Peclet Number Pe (−)

The *dimensionless Peclet number* is defined as the ratio of the advection rate to the conduction (diffusion) rate. Expressed mathematically as:

$$Pe = \frac{vl}{\alpha} \tag{2.146}$$

where v (m s^{-1}) is the average fluid velocity, l (m) is the characteristic distance and $\alpha = \lambda/\rho c$ is the thermal diffusivity (m^2 s^{-1}). Where λ (W m^{-1} K^{-1}) is the thermal conductivity, ρ (kg m^{-3}) is the density and c (J kg^{-3} K^{-1}) is the specific heat capacity. The Peclet number characterises the heat transfer mechanism in a media. Values of Pe less than unity indicate that the heat flow will be dominated by heat conduction. Values Pe greater than unity indicate heat flow dominated by heat convection.

2.3.11 *Porosity* ϕ *(−)*

Porosity ϕ is defined as the volume fraction of pores in a porous media. That is, the ratio of the volume of pores (m^3) to the total volume of porous media. For V_{po} (m^3) existing in a total volume of porous media V_T (m^3), the total volume is defined by the concept of elemental reference volume (ERV). The mathematical expression for porosity is:

$$\phi = \frac{V_{po}}{V_T} \tag{2.147}$$

Porosity plays an important role in shallow geothermal energy problems. It influences the thermal and hydraulic properties of the ground and consequently modulates the heat transfer mechanisms (conduction and convention),increasing the porosity with water as pore fluid decreases the thermal conductivity of the porous media and increases its heat capacity. An increase in porosity tends to increase groundwater storage and permeability, with very important consequences for heat convection.

2.4 Fluid Mechanics in Porous Media

2.4.1 *Darcy's Law*

In 1856 the engineer *Henry Darcy* published the results of the optimisation of the design of sand filters to supply drinking water to the population of Dijon, France (Simmons 2008). To carry out this optimisation, he used an experimental device, shown in Fig. 2.12. The experimental results obtained can be summarised as follows: the flow rate of water Q (m^3 s^{-1}) passing through a sand filter (a porous media), is proportional to the hydraulic gradient ∇h (−) multiplied by the cross sectional area A (m^2) according to:

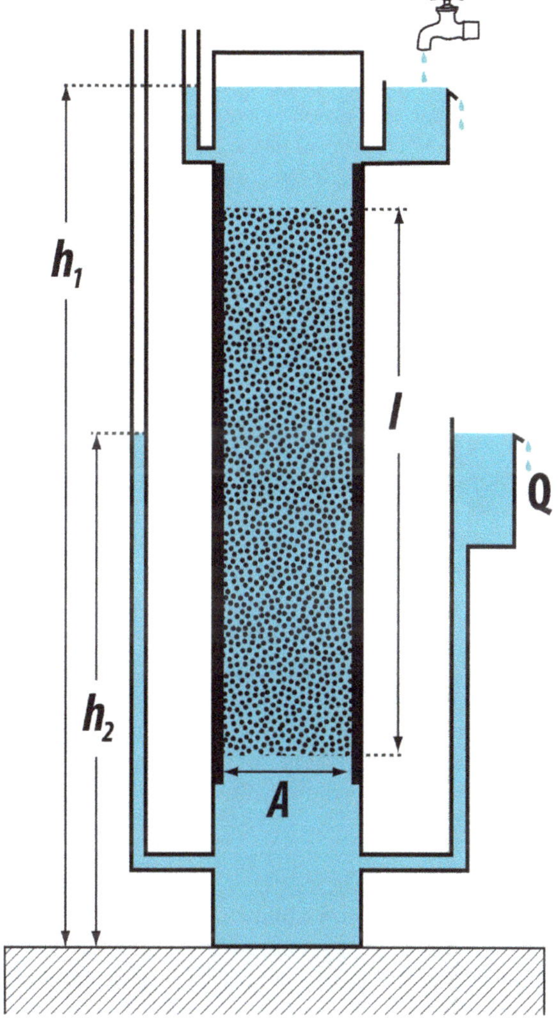

Fig. 2.12 Schematic diagram of the sand filter used by Henry Darcy in his experiments

$$Q = -KA\frac{h_2 - h_1}{l} = -KA\nabla h \tag{2.148}$$

where K (m s^{-1}) is the constant of proportionality between the flow rate and the hydraulic gradient called hydraulic permeability. The negative sign ensures that Q has a positive value since the water flows in the direction of decreasing hydraulic level, $h_1 > h_2$.

Expressing the water flow rate per unit cross-sectional area $v_D = \mathrm{Q}/A$ (m s^{-1}) gives Darcy's Law, for one dimension (x):

$$v_{Dx} = -K_x \frac{dh}{dx} \tag{2.149}$$

The flow rate per unit area is known as the Darcy velocity (v_D). Note that the mathematical expression is equivalent to Fourier's Law (Eq. 2.50). Darcy's Law expressed in vector form in the three dimensions $[x, y, z]$ is given by:

$$\begin{bmatrix} v_{Dx} \\ v_{Dy} \\ v_{Dz} \end{bmatrix} = -\begin{bmatrix} K_{xx} & 0 & 0 \\ 0 & K_{yy} & 0 \\ 0 & 0 & K_{zz} \end{bmatrix} \begin{bmatrix} \frac{\partial h}{\partial x} \\ \frac{\partial h}{\partial y} \\ \frac{\partial h}{\partial z} \end{bmatrix} \tag{2.150}$$

expressed in a compact form:

$$v_D = -K\nabla h \tag{2.151}$$

where K is a tensor with zero non-principal components. Since it is directional, it does not affect the frontwards-backwards flow. It follows that for anisotropic media $K_{xx} \neq K_{yy} \neq K_{zz}$.

The hydraulic permeability (K) is a specific parameter of a freshwater flow through a porous media. However, if other fluids are used, it is a property of the porous media which, in turn, has a combination of properties of its constituent solids and fluids (Hubbert 1940, 1957). In particular, the permeability of the porous media K (m s^{-1}) depends on the specific gravity ρg of the fluid [where ρ (kg m^{-3}) is the density of the fluid and g the acceleration of gravity (m s^{-2})], μ (Pa s) is the dynamic viscosity of the fluid and k (m^2) is the intrinsic permeability:

$$K = \frac{k\rho g}{\mu} \tag{2.152}$$

The intrinsic permeability (k), in contrast to the hydraulic conductivity (K) is independent of the fluid properties and depends only on the structure of the solid matrix. Different laboratory experiments using quartz or glass microspheres of uniform diameter, simulating a porous granular media, have found that k is proportional to the mean particle diameter squared.

Note that density and dynamic viscosity depend on temperature (especially dynamic viscosity), so that this dependence can affect hydrogeological behaviour in certain situations, for example in aquifers with thermal waters (Sánchez Navarro et al. 2004). In shallow geothermal energy, the change in hydraulic conductivity values due to changes in temperature is generally neglected. The study of the effect of temperature on dynamic viscosity in the calculation of thermal impacts by shallow geothermal systems has shown that the effect is negligible in most cases, being considered only in environments of high hydraulic conductivity with relatively *hot* water injection ((Lo Russo et al. 2018).

The Darcy velocity v_D (m s^{-1}), flow rate per unit cross-sectional area, is not actual fluid velocity because fluid flows only through the network of interconnected pores and not through the entire cross-section A. The *effective porosity* ϕ_{ef} is the volume fraction of the interconnected pores (those through which a fluid flows) in the porous media expressed as:

$$\phi_{ef} = \frac{V_{co}}{V_T} \tag{2.153}$$

where V_{co} (m^3) represents the volume of interconnected pores in a total volume of porous media. V_T (m^3) is the total volume defined by the concept of the elemental reference volume (ERV). The ratio representing effective porosity is retained in the cross-sectional area and, consequently, the real average velocity v_r (m s^{-1}) in the porous media would be:

$$v_r = \frac{v_D}{\phi_{ef}} \tag{2.154}$$

where v_D (m s^{-1}) is the Darcy velocity and ϕ_{ef} (−) is the effective porosity. Since $0 < \phi_{ef} < 1$ the v_r will always be greater than the Darcy velocity.

Darcy's law, although initially obtained experimentally, can be derived from the *Navier–Stokes equations,* assuming a laminar flow regime and other conditions characteristic of the porous media generally fulfilled. As the fluid velocities increase, the linear relationship between the flow rate and the hydraulic gradient becomes non-linear. This is due to the fact that lot of energy is lost due to friction when switching to a turbulent regime. The turbulent regime in the subsurface environment is only reached in very specific situations, such as in the vicinity of a groundwater well or sporadically in pipelines in karstified areas. The transition to non-Darcy flow in porous media is considered to be around a dimensionless Reynolds number of five (Bear 1979).

2.4.2 General Groundwater Flow Equation

The general groundwater flow equation is derived from the concept of conservation of mass; the procedure is equivalent to that followed in Sect. 2.2.2 for the case of heat conduction. Considering a representative elemental volume (REV) of the porous media $dV = dx\ dy\ dz$ (Fig. 2.13) where the mass conservation principle is applied, the change in stored mass (dm_{int}/dt) (kg s^{-1}) inside it over time will be equal to the net mass exchange with the environment (dm_{ext}/dt) (kg s^{-1}) across the surface area of the control volume, plus the source/sink term $\left(dm_{fs}/dt\right)$ (kg s^{-1}) of mass creation/destruction:

$$\frac{dm_{int}}{dt} = \frac{dm_{ext}}{dt} + \frac{dm_{fs}}{dt} \tag{2.155}$$

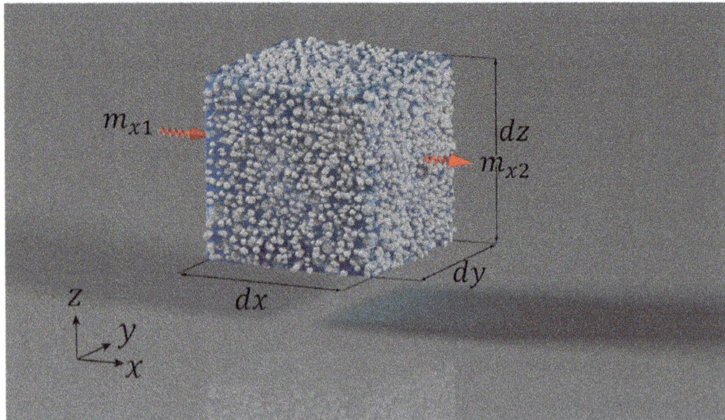

Fig. 2.13 Differential control volume for the derivation of the differential equation of groundwater flow in Cartesian coordinates

The net mass exchange with the environment of the control volume would consist of the mass flow, as groundwater outflow minus inflow in a given period of time:

$$\frac{dm_{ext}}{dt} = (m_{x1} - m_{x2}) + (m_{Y1} - m_{Y2}) + (m_{Z1} - m_{Z2}) \tag{2.156}$$

The mass entering the control volume would be the volumetric flow rate of groundwater multiplied by the fluid density. Darcy's Law can be applied to obtain it in the x direction as m_{x1}:

$$m_{x1} = -K_x A \frac{\partial h}{\partial x} \rho_w = -K dy\, dz \rho_w \frac{\partial h}{\partial x} \tag{2.157}$$

where ρ_w is the density of the groundwater. The outflow mass flux of the control volume in the x direction (m_{x2}) will depend on the mass flux at the outflow point of the volume, when the value of coordinate $x = dx$. By applying a Taylor series, we obtain a polynomial $P(x)$ very similar to m_{x2}:

$$m_{x2} \cong P(x) = m_x(dx) + \frac{\partial m_x(dx)}{\partial x}(x - dx) + \cdots + \frac{\partial^{(n)} m_x(dx)}{\partial x} \frac{1}{n!}(x - dx)^n \tag{2.158}$$

Taking the first two terms:

$$m_{x2} \cong m_{x1} + \frac{\partial m_{x1}}{\partial x} dx \tag{2.159}$$

Applying Darcy's Law:

$$m_{x2} \cong -K_x \frac{\partial h}{\partial x} dy\, dz\, \rho_w + \frac{\partial}{\partial x}\left(-K_x \frac{\partial h}{\partial x} dy\, dz\, \rho_w\right) dx \tag{2.160}$$

The quantity of mass as groundwater transferred to the control volume in the x direction would be:

$$m_{x1} - m_{x2} \cong -K_x \frac{\partial h}{\partial x} dy\, dz\, \rho_w + K_x \frac{\partial h}{\partial x} dy\, dz\, \rho_w - \frac{\partial}{\partial x}\left(-K_x \frac{\partial h}{\partial x}\right) dx\, dy\, dz\, \rho_w \tag{2.161}$$

$$m_{x1} - m_{x2} \cong \frac{\partial}{\partial x}\left(K_x \frac{\partial h}{\partial x}\right) dx\, dy\, dz\, \rho_w \tag{2.162}$$

Similarly for the other spatial components:

$$m_{y1} - m_{y2} \cong \frac{\partial}{\partial y}\left(K_y \frac{\partial h}{\partial y}\right) dx\, dy\, dz\, \rho_w \tag{2.163}$$

$$m_{z1} - m_{z2} \cong \frac{\partial}{\partial z}\left(K_z \frac{\partial h}{\partial z}\right) dx\, dy\, dz\, \rho_w \tag{2.164}$$

Substituting Eqs. (2.162)–(2.164) into Eq. (2.156) and assuming equality:

$$\frac{dm_{ext}}{dt} = \left[\frac{\partial}{\partial x}\left(K_x \frac{\partial h}{\partial x}\right) + \frac{\partial}{\partial y}\left(K_y \frac{\partial h}{\partial y}\right) + \frac{\partial}{\partial z}\left(K_z \frac{\partial h}{\partial z}\right)\right] dx\, dy\, dz\, \rho_w \tag{2.165}$$

The mass created or destroyed within the control volume by the source/sink term $\left(dm_{fs}/dt\right)$ (kg s^{-1}) can be expressed per unit volume as g (kg m^{-3}):

$$\frac{dm_{fs}}{dt} = \text{g}\, dx\, dy\, dz\, \rho_w \tag{2.166}$$

The storage capacity of the system is related to the effective porosity $\left(\phi_{ef}\right)$ of the control volume. The effective porosity is the volume fraction that can be filled or emptied with groundwater. The mass change in the control volume over time will be:

$$\frac{dm_{int}}{dt} = \frac{\partial\left(\phi_{ef}\rho_w\right)}{\partial t} dx\, dy\, dz \tag{2.167}$$

The change of effective porosity with time is related to the compressibility of the solid matrix of the porous media. If the fluid pressure in the pore increases, the pore can deform elastically and change the effective porosity volume. It follows that the variation of effective porosity is a function of hydrostatic pressure. The *piezometric level* h (m) is defined as:

$$h = \frac{p}{\rho_w g} + z \tag{2.168}$$

Assuming that ϕ_{ef} is a function of pressure or hydraulic level, $\phi_{ef}(h)$, then the hydraulic level is time-varying, $\phi_{ef}(h(t))$. The effective porosity variation over time can now be derived.

The density of water also depends on whether it compresses with increasing pressure. The pressure is equivalent to working with the state variable that is the hydraulic level which, in turn, changes over time, $\rho_w(h(t))$. Applying the chain rule:

$$\frac{\partial(\phi_{ef}\rho_w)}{\partial t} = \rho_w \frac{\partial \phi_{ef}}{\partial h}\frac{\partial h}{\partial t} + \phi_{ef}\frac{\partial \rho_w}{\partial h}\frac{\partial h}{\partial t} \tag{2.169}$$

Meaning that the product of the density of the water by how much the porosity changes as the hydraulic level does (pore hydrostatic pressure) multiplied by how much the hydraulic level varies over time, plus the effective porosity by how much the density of the water changes as the hydraulic level does multiplied by how much the hydraulic level varies over time.

$$\frac{\partial(\phi_{ef}\rho_w)}{\partial t} = \frac{\partial h}{\partial t}\left(\rho_w \frac{\partial \phi_{ef}}{\partial h} + \phi_{ef}\frac{\partial \rho_w}{\partial h}\right) \tag{2.170}$$

If the control volume is considered to be at coordinate $z = 0$ and $h = p/(\rho_w g) + z$, where h (m) is the hydraulic level, p (Pa) is the hydrostatic pressure, g (m s^{-2}) is the gravitational acceleration and z is the reference height coordinate, it is possible to calculate the derivative of h:

$$\partial h = \frac{\partial p}{\rho_w g} \tag{2.171}$$

Substituting:

$$\frac{\partial(\phi_{ef}\rho_w)}{\partial t} = \frac{\partial h}{\partial t}\left(\rho_w \frac{\partial \phi_{ef}}{\partial p}\rho_w g + \phi_{ef}\frac{\partial \rho_w}{\partial p}\rho_w g\right) \tag{2.172}$$

The compressibility of the porous media can be defined as ω_m (m kg^{-1} s^{-2}) as

$$\frac{\partial \phi_{ef}}{\partial p} = \omega_m \tag{2.173}$$

The compressibility of water can be defined as ω_w (m kg^{-1} s^{-2}):

$$\frac{1}{\rho_w}\frac{\partial \rho_w}{\partial p} = \omega_w \tag{2.174}$$

Then:

$$\frac{\partial(\phi_{ef}\rho_w)}{\partial t} = \frac{\partial h}{\partial t}(\rho_w\omega_m\rho_w g + \phi_{ef}\rho_w\omega_w\rho_w g) \tag{2.175}$$

$$\frac{\partial(\phi_{ef}\rho_w)}{\partial t} = \frac{\partial h}{\partial t}\rho_w\rho_w g(\omega_m + \phi_{ef}\omega_w) \tag{2.176}$$

The specific storage coefficient S_s (m^{-1}) can be defined as:

$$S_s = \rho_w g(\omega_m + \phi_{ef}\omega_w) \tag{2.177}$$

Then:

$$\frac{\partial(\phi_{ef}\rho_w)}{\partial t} = \frac{\partial h}{\partial t}\rho_w S_s \tag{2.178}$$

Therefore, taking into account Eqs. (2.167) and (2.178):

$$\frac{dm_{int}}{dt} = \frac{\partial h}{\partial t}S_s\, dx\, dy\, dz\, \rho_w \tag{2.179}$$

Substituting Eqs. (2.179), (2.165) and (2.166) into Eq. (2.155) gives Eq. (2.155):

$$\begin{aligned}\frac{\partial h}{\partial t}S_s\, dx\, dy\, dz\, \rho_w = &\left[\frac{\partial}{\partial x}\left(K_x\frac{\partial h}{\partial x}\right) + \frac{\partial}{\partial y}\left(K_y\frac{\partial h}{\partial y}\right)\right.\\ &\left.+\frac{\partial}{\partial z}\left(K_z\frac{\partial h}{\partial z}\right)\right]dx\, dy\, dz\, \rho_w + g\, dx\, dy\, dz\, \rho_w\end{aligned} \tag{2.180}$$

Then the general groundwater flow equation is obtained as:

$$\frac{\partial}{\partial x}\left(K_x\frac{\partial h}{\partial x}\right) + \frac{\partial}{\partial y}\left(K_y\frac{\partial h}{\partial y}\right) + \frac{\partial}{\partial z}\left(K_z\frac{\partial h}{\partial z}\right) + g = \frac{\partial h}{\partial t}S_s \tag{2.181}$$

Taking into account that the partial derivatives in perpendicular directions are zero, the above equation can be expressed as:

$$\begin{aligned}&\frac{\partial}{\partial x}\left(K_x\frac{\partial h}{\partial x} + K_y\frac{\partial h}{\partial y} + K_z\frac{\partial T}{\partial z}\right) + \frac{\partial}{\partial y}\left(K_x\frac{\partial h}{\partial x} + K_y\frac{\partial h}{\partial y} + K_z\frac{\partial h}{\partial z}\right)\\ &+\frac{\partial}{\partial z}\left(K_x\frac{\partial h}{\partial x} + K_y\frac{\partial h}{\partial y} + K_z\frac{\partial h}{\partial z}\right) + g = \frac{\partial h}{\partial t}S_s\end{aligned} \tag{2.182}$$

By means of the nabla operator ∇ and taking into account the properties of the scalar product:

$$\nabla = \left(\frac{\partial}{\partial x}\ \frac{\partial}{\partial y}\ \frac{\partial}{\partial z}\right)$$

$$K = \begin{pmatrix} K_x & K_y & K_z \end{pmatrix}$$

$$\frac{\partial}{\partial x}(K\nabla h) + \frac{\partial}{\partial y}(K\nabla h) + \frac{\partial}{\partial z}(K\nabla h) + g = \frac{\partial h}{\partial t}S_s \tag{2.183}$$

Again using the nabla operator and considering K as a tensor:

$$\nabla \cdot (K\nabla h) + g = \frac{\partial h}{\partial t}S_s \tag{2.184}$$

If the control volume presents isotropic properties the thermal conductivity would be independent of the direction ($K_x = K_y = K_z$), without the sink/source term and using the Laplacian operator ∇^2:

$$\nabla^2 = \nabla \cdot \nabla = \frac{\partial^2}{\partial x^2} + \frac{\partial^2}{\partial y^2} + \frac{\partial^2}{\partial z^2} \tag{2.185}$$

$$\nabla^2 h = \frac{S_s}{K}\frac{\partial h}{\partial t} \tag{2.186}$$

The expressions obtained from the general groundwater flow equation govern the hydraulic regime of the subsurface. Understanding the behaviour of groundwater is necessary to understand the thermal regime of the subsurface, since the mechanism of forced convection or advection may be dominant in relation to heat flow in the subsurface. Inevitably the hydraulic regime of the physical media must be resolved if the behaviour of a shallow geothermal system, including its performance, is to be predicted. It is common to work with numerical models in which the codes commonly used solve Eq. (2.181) in the whole domain, to obtain the groundwater velocity in each cell of the considered domain and, subsequently, calculate the heat transport taking into account this velocity. This allows the heat transfer corresponding to the advection and hydrodynamic thermal dispersion component to be obtained.

References

Abu-Hamdeh NH, Reeder RC (2000) Soil thermal conductivity effects of density, moisture, salt concentration, and organic matter. Soil Sci Soc Am J 64(4):1285–1290. https://doi.org/10.2136/sssaj2000.6441285x

Balling N, Kristiansen J, Breiner N, Poulsen KD, Rasmussen R (1981) Geothermal measurements and subsurface temperature modelling in Denmark. GeoSkrifter 16

Bear J (1979) Hydraulics of groundwater. McGraw-Hill International Book Co

Bear J (1988) Dynamics of fluids in porous media. Dover

Chekhonin E et al (2012) When rocks get hot: thermal properties of reservoir rocks. Oilfield Rev 24(3):20–37

Clark SP Jr (1966) Handbook of physical constants. The Geological Society of America, New York

Dagan G (1989) Solute transport at the local (formation) scale. In: Dagan G (ed) Flow and transport in porous formations. Springer, Berlin, pp 263–350. https://doi.org/10.1007/978-3-642-75015-1_4

deVries DA (1966) Thermal properties of soils. In: van Wijk WR (ed) Physics of plant environment. North-Holland Publishing, Amsterdam

Eskilson P (1987) Thermal analysis of heat extraction boreholes. University of Lund, Lund, Sweden

Fjær E, Holt RM, Horsrud P, Raaen AM, Risnes R (2008) Elasticity. In: Fjær E, Holt RM, Horsrud P, Raaen AM, Risnes R (Eds) Developments in petroleum science. Elsevier, pp 1–54. https://doi.org/10.1016/S0376-7361(07)53001-3 (Chapter 1)

Gelhar LW (1993) Stochastic subsurface hydrology. Prentice-Hall

Hubbert MK (1940) The theory of ground-water motion. J Geol 48(8):785–944

Hubbert MK (1957) Darcy's law and the field equations of the flow of underground fluids. Int Assoc Sci Hydrol Bull 2(1):23–59. https://doi.org/10.1080/02626665709493062

Kappelmeyer O, Haenel R (1974) Geothermics with special reference to application. Gebrüder Borntraeger

Lo Russo S, Taddia G, Cerino Abdin E (2018) Modeling the effects of the variability of temperature-related dynamic viscosity on the thermal-affected zone of groundwater heat-pump systems. Hydrogeol J 26(4):1239–1247. https://doi.org/10.1007/s10040-017-1714-x

Pitts DR, Sissom LE (1998) Heat transfer. In: Schaum's outline. McGraw-Hill, NY

Reynolds O (1883) XXIX. An experimental investigation of the circumstances which determine whether the motion of water shall be direct or sinuous, and of the law of resistance in parallel channels. Philos Trans R Soc Lond 174:935–982. https://doi.org/10.1098/rstl.1883.0029

Rohsenow WM, Hartnett JP, Cho YI (1998) Handbook of heat transfer. McGraw-Hill

Sánchez Navarro JÁ, Coloma López P, Perez-Garcia A (2004) Evaluation of geothermal flow at the springs in Aragón (Spain), and its relation to geologic structure. Hydrogeol J 12(5):601–609. https://doi.org/10.1007/s10040-004-0330-8

Sass JH, Lachenbruch AH, Munroe RJ (1971) Thermal conductivity of rocks from measurements on fragments and its application to heat-flow determinations. J Geophys Res (1896–1977) 76(14):3391–3401. https://doi.org/10.1029/JB076i014p03391

Schön SJ (2011) Thermal properties. In: Schön JH (ed) Handbook of petroleum exploration and production. Elsevier, pp 337–372. https://doi.org/10.1016/S1567-8032(11)08009-8 (Chapter 9)

Seipold U (1998) Temperature dependence of thermal transport properties of crystalline rocks—a general law. Tectonophysics 291(1):161–171. https://doi.org/10.1016/S0040-1951(98)00037-7

Simmons CT (2008) Henry Darcy (1803–1858): immortalised by his scientific legacy. Hydrogeol J 16(6):1023. https://doi.org/10.1007/s10040-008-0304-3

Slattery JC (1984) Flow phenomena in porous media, by Robert A. Greenkorn, published by Marcell Dekker (1983), 560 pages, $75.00. AIChE J 30(1):172–173. https://doi.org/10.1002/aic.690300126

Somerton WH (1992) Thermal expansion of rocks, developments in petroleum science. Elsevier, pp 29–38. https://doi.org/10.1016/S0376-7361(09)70024-X (Chapter IV)

Touloukian YS, Judd WR, Roy RF (1981) Physical properties of rocks and minerals. McGraw-Hill

van Duin RHA (1963) The influence of soil management on the temperature wave near the soil surface. Technical bulletin. Institute for land and water management research. [s.n.] Wageningen

Ward JC (1964) Turbulent flow in porous media. J Hydraul Div 90(5):1–12

Whitaker S (1998) The method of volume averaging. Springer, Netherlands

Williams PJ, Smith MW (1989) The Frozen earth: fundamentals of geocryology. In: Studies in polar research. Cambridge University Press, Cambridge. https://doi.org/10.1017/cbo9780511564437

Chapter 3
Underground Thermal Regime

In order to understand the possibilities and advantages of using shallow geothermal energy, it is necessary to understand the origin and magnitude of heat transfer processes in the subsurface. This chapter provides an overview of the spatial and temporal temperature distribution in the underground, i.e., its thermal regime. The most important sources and sinks of heat in nature for the shallow subsurface domain are the sun, the atmosphere and the earth's core. In the following subsections of this chapter, the energy balance in the atmosphere-earth system and the concept of the terrestrial geothermal gradient will be described, as well as other boundary conditions, in order to subsequently study the temperature profile in the subsurface.

3.1 Energy Balance of the Earth-Atmosphere System

On the earth's surface, thermal energy flows continuously between the ground and the atmosphere by means of the different heat transfer mechanisms: radiation, conduction and convection.

The radiation heat transfer mechanism refers to the ability of the electromagnetic spectrum to transfer energy through photons in space, even in a vacuum. The propagation of this energy follows a wave model and is transmitted with a wavelength that defines the electromagnetic spectrum. The range of the spectrum of interest in the energy balance of the atmosphere is from 0.1 to 100 μm, which includes part of the ultraviolet range (0.1–0.4 μm), the visible range (0.4–0.7 μm) and the infrared range (0.7–100 μm). When studying the atmosphere, it is common to distinguish between shortwave radiation (0.1–3.0 μm) and longwave radiation (3.0–100 μm), thus differentiating longwave radiation, the part of the spectrum that interacts most with water vapour (H_2O), carbon dioxide (CO_2) and ozone (O_3). From the sun, the earth's atmosphere receives annually a continuous radiation flux of 1367 W m^{-2} (Wehrli 1985) measured outside the atmosphere in a surface plane perpendicular to the solar flux. If the decrease in radiation due to the angle of incidence is taken into

A. García Gil et al., *Shallow Geothermal Energy*, Springer Hydrogeology,
https://doi.org/10.1007/978-3-030-92258-0_3

account, the average total radiation $\left(\overline{J}^{T}_{ext}\right)$ received over the year is about a quarter of the indicated value, about 338 W m^{-2} (29.2 MJ m^{-2} day^{-1}). As a function of the average radiation received, Fig. 3.1 shows the total net radiation absorbed by the terrain, differentiating between shortwave and longwave radiation. Of the total average radiation received by the atmosphere, 50% is reflected $\left(J^{r}_{n}, J^{r}_{a}\right)$ or absorbed $\left(J^{*}_{n}, J^{*}_{a}\right)$ by clouds and atmospheric gases. Of the remaining 50% that is able to impact the earth's surface, 3% is reflected $\left(J^{r}_{t}\right)$ and 47% is absorbed by the terrain $\left(J^{*}_{t}\right)$. The earth, like any body with internal energy at a temperature other than zero Kelvin, emits long-wave radiation $\left(\Upsilon^{emi}_{t}\right)$ which, because of its surface temperature, is within the long-wave (infrared) radiation range and is equivalent to 114% of $\overline{J}^{T}_{ext}$. Of the radiation emitted by the earth's surface, 96% of the radiation emitted by the Earth's surface is absorbed by clouds and gases in the atmosphere $\left(\Upsilon^{*}_{a+n}\right)$. Part of the thermal energy absorbed in the clouds and gases of the atmosphere returns in the form of longwave radiation to the ground $\left(\Upsilon^{abs}_{t}\right)$ and some is emitted into outer space (Υ_{Ex}). In total, the surface of the ground emits 18% of $\overline{J}^{T}_{ext}$. The net radiation energy $\left(J^{net}\right)$ is the sum of J^{*}_{t}, Υ^{emi}_{t} and Υ^{abs}_{t}, giving a value of 29% of the total extra-terrestrial radiation received.

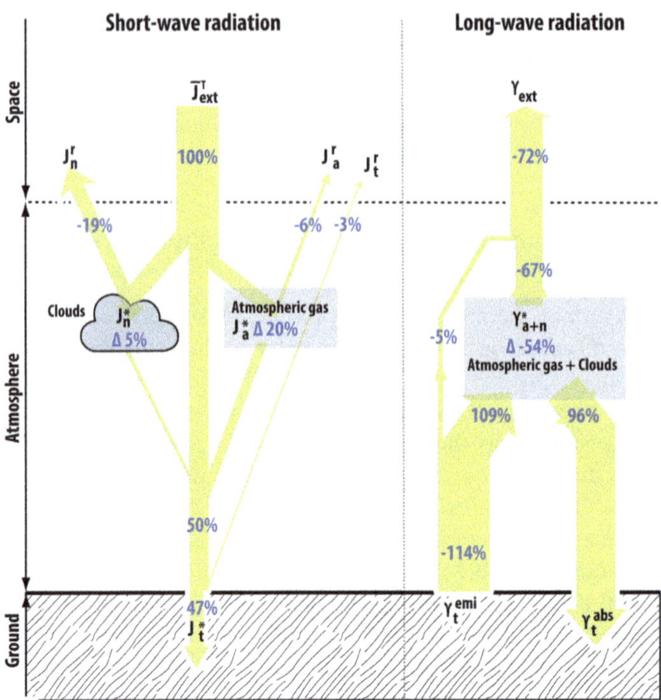

Fig. 3.1 Schematic diagram of the radiation energy flow between outer space and the earth's surface. Values expressed as percentages of the average annual solar radiation (338 W m^{-2}). Adapted from Oke (1987)

In addition to the mechanism of heat transfer by radiation, thermal conduction and convection must also be taken into consideration. Heat conduction between the ground and the air in the atmosphere is practically negligible (Oke 1987). Heat transfer by the convection mechanism, however, is very important. Convection transports heat to and from the atmosphere in the form of latent heat (Q_E) and/or sensible heat (Q_H). Latent heat refers to the thermal energy that is transferred from the ground to the atmosphere through the endothermic process of direct water evaporation or transpiration through existing plants. The sum of both processes is called evapotranspiration. Latent heat transport is closely linked to water vapour. When a drop of water evaporates at the surface, the heat of evaporation passes from the ground to the atmosphere in the form of a water vapour molecule and by convection (free or forced). This latent heat is transported until it is released when the vapour condenses in clouds, heating the environment where condensation occurs. The flow of latent heat through evapotranspiration is the most important mechanism for the transfer of energy (also mass) between the geosphere, hydrosphere, biosphere and atmosphere, determining the overall energy balance (Sellers et al. 1996; Song et al. 2012). It is estimated that more than half of the net radiative energy $\left(J^{net}\right)$ received by the earth's surface on a continental scale returns to the atmosphere as latent heat (Brutsaert 1982).

Sensible heat refers to the thermal energy transferred between the surface and the air of the atmosphere due to the thermal gradient between the two. The sensible heat is moved from the *hot* surface to the air in contact with the surface by turbulent flow until it is transferred to the surrounding air. The opposite happens when the *hot* air cools the surface. With thermal gradient there can also be a heat transfer (Q_G) by conduction from the ground surface to the underlying ground layers. This involves an outflow and inflow of thermal energy from the ground to the geological or geothermal reservoir. In winter the subsoil gives up heat to the ground and in summer it receives heat from the ground.

If all heat transport mechanisms and the principle of energy conservation are taken into account, the energy balance in the soil (Fig. 3.2) can be obtained by:

$$J^{net} = Q_H + Q_E + Q_G \tag{3.1}$$

After one year, the total net radiation of any wavelength $\left(J^{net}\right)$ brings to the ground 29% of the total extraterrestrial radiation received (98 W m^{-2}). This is equal to the sum of the heat dissipated to the atmosphere through convection as sensible heat (Q_H) due to the atmospheric wind, approximately 81 W m^{-2}, or as latent heat (Q_E), approximately 17 W m^{-2}, due to evapotranspiration and condensation of water, plus the heat transferred from the subsurface to the ground (Q_G) through conduction.

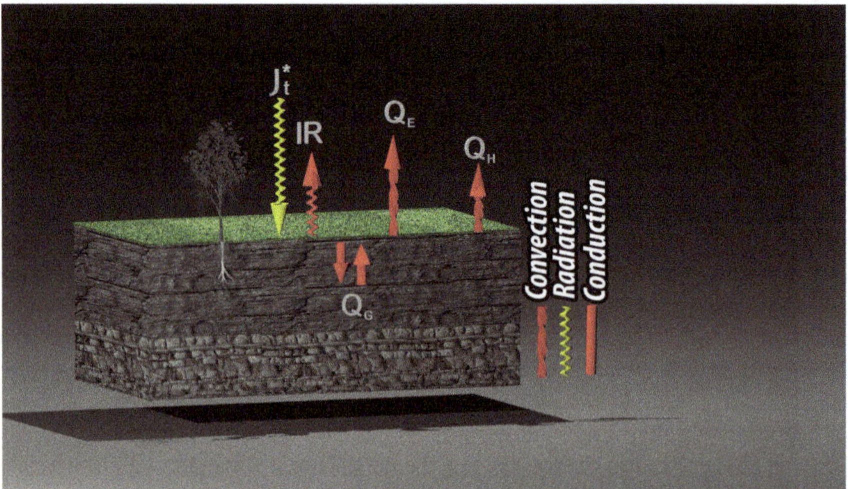

Fig. 3.2 Thermal energy balance in the soil. The energy absorbed by the soil from shortwave radiation (J_t^*), the heat emitted in the form of long-wave infrared radiation (IR) long-wave radiation, latent heat (Q_E) and sensible heat (Q_H) transferred by turbulent convection to the atmosphere and the heat (Q_G) transferred by conduction from the ground to and from the subsurface

3.2 Deep Geothermal Upward Heat Flow

The earth's core is estimated to be at a temperature of approximately 4000–7000 °C (Anzellini et al. 2013), more than a hundred times higher than the temperature that can exist in the atmosphere. The temperature at the earth's core is attributed to the remaining heat from accretion during its formation, internal friction due to mass flows, and radioactive decay of heavy unstable elements existent in the core. On a planetary scale, heat flux from the core to the earth's surface averages 0.04–0.11 W m^{-2} in the continental lithosphere (Chapman 1986), and there can be significant local variations, with much larger heat fluxes at plate boundaries, volcanoes and *hot-spot* zones. Heat flow through the liquid outer core and much of the mantle occurs via free convection. Convection cells make the upwards movement magmatic fluid, generated by density difference due to thermal expansion. In the lithosphere, presenting rigid behaviour, heat is transferred by conduction, passing about 100 km through the upper mantle and the earth's crust up to the planet's surface. Crustal materials have relatively constant thermal conductivities, so that the upward flow of heat in the crust by conduction generates a relatively constant temperature gradient called the geothermal gradient, with an average value in the continental domain of 18–30 °C km^{-1}.

The upward heat flux from the core is eventually transferred to the atmosphere. This means that, as a rough estimation, the net heat transfer between the subsurface and the atmosphere (Q_G) at the end of the year is practically zero, although positive, and equal to the heat flux transferred from the earth's core to the atmosphere that will finally will be transferred to outer space.

3.2.1 Underground Temperature Profile

The temperature of the atmosphere depends mainly on the solar radiation received and on convection processes, forming a complex atmospheric dynamic which, together with the annual seasonality, means that air temperature in a given location varies considerably throughout the year. However, it can be experimentally verified, by measuring the ground temperature at different depths, that this seasonal variability of temperatures is damped until it disappears at a certain depth (Fig. 3.3).

The interaction of the geothermal gradient with atmospheric dynamics generates a characteristic subsurface temperature profile (CSTP) (Fig. 3.4). The depth at which the annual seasonal temperature variation disappears is between 10 and 20 m depth. This depth marks the boundary where the shallow or transitional terrain domain ends and the purely geothermal thermal domain begins (Parsons 1970). The shallow domain has a transient hydraulic and thermal regime. Temperature or water table or unsaturated zone oscillations in this domain may have a periodic component of hourly magnitude at the water table to several days at greater depths. As for the temperature distribution in this domain, if profiles were measured over a period of several years to obtain the average annual vertical profile, the magnitude and direction of the thermal gradient would depend on the climatic trend over the decades or centuries preceding the monitoring period and on the thermal and hydraulic properties of the terrain.

The depth where the geothermal domain begins, below the thermal influence of the atmosphere, tends to have a stationary thermal and hydraulic regime. The temperature gradient is influenced by groundwater movement and is not equivalent to the geothermal gradient that would exist under static hydrodynamic conditions. It is highly dependent on the climatology of the site, especially on extreme climatic events such as glaciation, solar radiation, air temperature, air velocity, annual period, shading, the thermal properties of the terrain and the existing geothermal gradient. Therefore, the geothermal domain is free from temperature and seasonal fluctuations of the atmosphere.

The temperature at the earth's surface tends to be higher than the atmospheric air temperature, generally below two degrees Celsius on cloudy days and up to seven degrees Celsius higher on clear days (Gallo et al. 2011). Another widespread characteristic of subsurface temperatures is that the temperature in the geothermal domain is approximately one to two degrees Celsius above the annual mean atmospheric temperature (Carson 1963; Mihalakakou et al. 1992; Parsons 1970; Penrod et al. 1960). This is due to the upward geothermal heat flow and the contrast in thermal conductivity between the ground and the atmosphere. The thermal conductivity of gases is about 100 times lower than that of solids.

In general, unless otherwise specified, shallow geothermal installations are assumed to operate in the subsurface outside the range of atmospheric temperature

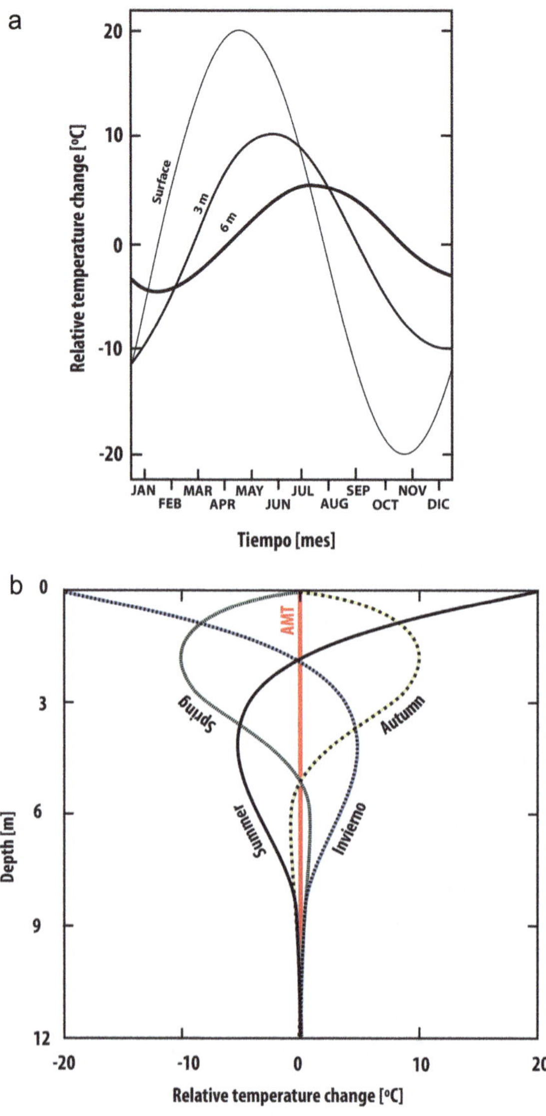

Fig. 3.3 Spatio-temporal distribution of relative temperature changes in the subsurface. Temperature change as a function of time **a** and depth as a function of relative temperature change **b** are shown

variations, i.e., in geothermal domain conditions only influenced by the geothermal gradient. However, the geothermal gradient is only representative above 100 m depth, as at shallower depths the measured thermal gradients are more irregular due to different thermal processes affecting the near-surface environment. As detailed above, atmospheric weathering produces seasonal changes in temperature's surface

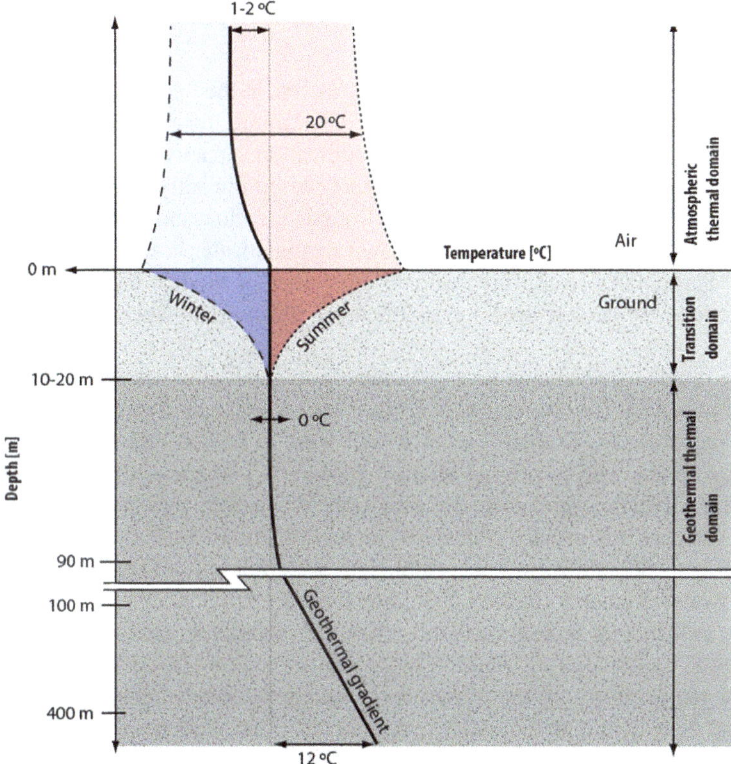

Fig. 3.4 Characteristic subsurface temperature profile (CSTP) with the spatio-temporal distribution of temperatures in the atmosphere and in the ground over the course of a year. The thick black line indicates the annual mean temperature, the dashed lines with long and short dashes represent the annual minimum and maximum temperature respectively

that can propagate up to the first 10–20 m in the subsurface. Groundwater flow can transport advection heat and modify these gradients. The deep propagation of climatic variability in atmospheric temperature over decades to a few centuries (Gupta and Roy 2007) generates palaeoclimatic thermal signals. For example, a thermal wave with a period of 1000 years can be detected as an anomaly of a few tenths of a degree Celsius at a depth of 400 m (Pollack and Huang 2000). Palaeoclimatic thermal signals tend to be blurred by groundwater flow, although they tend not to disappear completely. In general, the deeper sections of the vertical temperature profiles are a reflection of past climates. This thermal *footprint* of climate is a necessary correction in geophysics to determine the geothermal gradients that allow deep geothermal heat fluxes to be properly studied (Kohl 1998).

3.3 Regional Groundwater Flow and Heat Advection

There are different natural and anthropic processes in the underground environment that are responsible for modifying the CSTP. Among them, regional groundwater flow is one of the processes of greatest modification capacity.

Regional flow on a sedimentary basin scale can significantly modify CSTP profiles down to great depths (Beck et al. 1991). Groundwater flow and heat transport models are commonly used to establish conceptual regional water flow models (Cartwright 1970). Particularly useful are regional cross-section models to explain observed temperatures in certain areas of sedimentary basins (Domenico and Palciauskas 1973).

In groundwater recharge areas, recharge water will tend to have a mean annual temperature 1–2 °C lower than CSTP, i.e., cooler water than in areas without groundwater recharge. In discharge areas it will tend to be the opposite; groundwater discharge areas tend to present warmer, especially if there is a fractured environment that facilitates rapid groundwater ascent. When deep groundwater rises rapidly to the surface, it is usually considered as hydrothermal water (>30 ºC). There are studies using aerial measurement campaigns with thermal sensors to detect areas of groundwater discharge (Becker 2006; Mejías et al. 2012).

The influence of groundwater movement on ground temperature depends on the magnitude and spatial distribution of groundwater flow. Local flows will tend to modify temperatures in the atmospheric and transitional thermal domain, while regional flows will tend to modify temperatures in the geothermal domain.

Using numerical models, Smith et al. (1989) studied heat redistribution due to groundwater flow and demonstrated its ability to cause significant disturbances through advection in the standard conductive heat regime of the earth's crust. This is especially significant in high relief areas. Regional groundwater flow is more likely to modify the heat conductive regime, due to the larger ranges of variation and shapes of the piezometric level and the fact that steep topography favours the vertical components of groundwater flow, causing its deep circulation and, consequently, its significant heating.

Figure 3.5a illustrates a conductive thermal regime scenario in a hydrogeological cross-section of a mountainous terrain. This example corresponds to a numerical modelling where a mountain massif of low hydraulic conductivity (10^{-18} m^2) and a very low hydraulic conductivity (10^{-22} m^2) basement with an upward geothermal heat flux of 0.06 W m^{-2} were considered. The figure shows the geothermal heat flow isolines clearly dominated by the conductive heat component, where the groundwater flow isolines do not interact with the heat isolines. It can be seen from this figure that, in this type of situation, groundwater flow is so slow that the upward heat is transferred to higher materials before it is significantly displaced by groundwater advection into the discharge zone. Four heat flow isolines are depicted in the figure, which determine a yellow heat flow tube.

Figure 3.5b shows the same system, but in this case a mountain massif of intermediate hydraulic conductivity (10^{-16} m^2) was considered, thus increasing groundwater

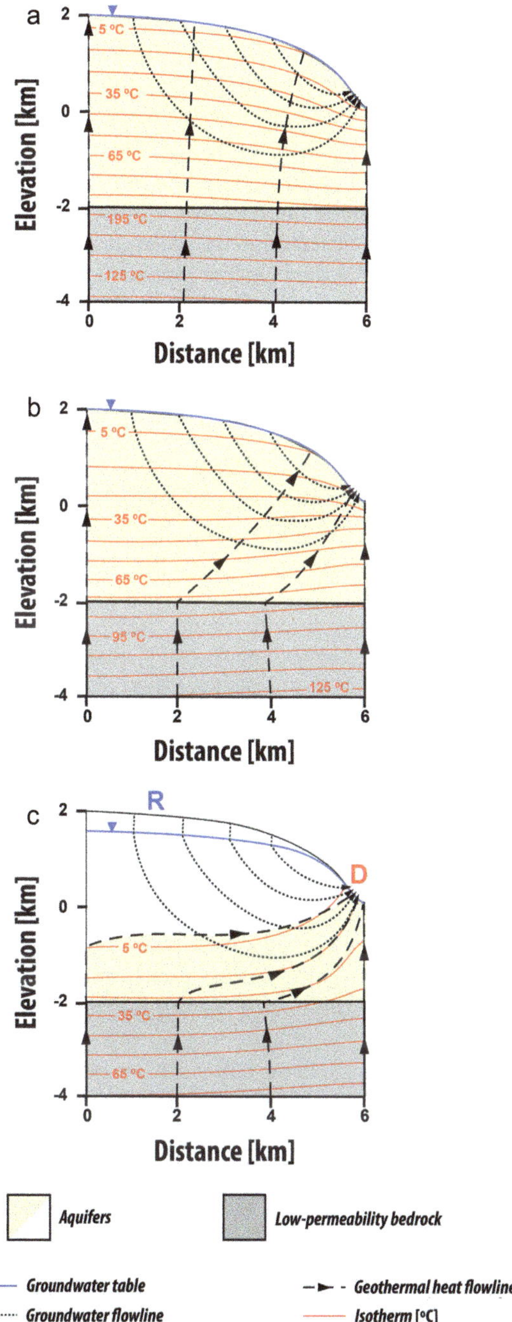

Fig. 3.5 Effects of groundwater flow on the regional thermal regime for low permeability **a**, medium hydraulic conductivity **b** and high hydraulic conductivity **c** systems. A recharge environment R and discharge environment D are highlighted. Modified from Smith et al. (1989)

flux by two orders of magnitude. The isotherms in this case show a slightly different pattern, indicating a disturbance in the conductive flux as the advective term increases. In this case, the leftmost heat flux tube (the first two kilometres of distance) moves from the conductive basement zone to a conductive-advective zone and expands considerably. This means that, with the same heat flux, much more ground must be heated, resulting in a decrease in temperatures with respect to the conductive scenario and is represented by the isotherms in that zone at lower elevation. In other words, a reduction in temperatures is observed in the recharge zone. The opposite happens in the heat flux pipe on the right; the heat flux pipe reduces its section when passing to the conductive-advective domain, so that the ascendent heat affects less land and, therefore, its temperature is higher, as shown by the rising isotherms in that area, corresponding to the groundwater discharge zone.

Finally, Fig. 3.5c shows a third scenario, where a mountain massif of high hydraulic conductivity (10^{-15} m^2) was considered, thus increasing the groundwater flow by three orders of magnitude with respect to the first scenario. The first consequence is that the system is able to discharge all the recharge (constant in all models) and the water table is several hundred metres below the topographic surface. The advective heat flux is so important that all heat pipes (yellow colour) representing the basal geothermal heat flux end in the discharge zone (zone D). A new heat flux tube (white) is created, whose temperatures will be determined by the energy balance of the ground in the recharge zones (zone R) and not by the upward conductive geothermal flow. As in scenario B, the isotherms decrease with elevation in the recharge zone and increase in the discharge zone. In addition, mixing with the lower temperature downward groundwater causes temperatures to be much lower, at 50 °C at two kilometres depth in the discharge zone, comparing scenario A with scenario C. This example shows a pronounced effect of advection on the temperature field when there are zones of vertical groundwater flow, resulting in a significant distortion of the isotherms to horizontal positions perpendicular to the flow (Parsons 1970).

Scenario C in Fig. 3.5 helps to explain variations in CSTP profiles as a consequence of advection heat transfer in groundwater circulation in recharge and discharge zones (Anderson 2005). Such hydrogeological environments generate impacts on CSTP profiles, such as those shown in Fig. 3.6. Recharge water (e.g., zone R in Fig. 3.5c) percolates through the ground, transferring heat by advection from the surface, causing seasonal atmospheric variability to propagate to greater depths, modifying the CSTP (Fig. 3.6a) and presenting lower average annual ground temperatures closer to the average atmospheric air temperature. In recharge zones, the influence of climate probably determines the temperature at the phreatic surface, whereas groundwater flow influences are of less relevance. In discharge zones (e.g., zone D in Fig. 3.5c), the upward heat flux is higher, so that the influence of seasonal variability of the atmospheric temperatures decreases in depth and the CSTP profiles are substantially modified (Fig. 3.6b), as well as having higher average temperatures than expected. Advective upward flow in discharge zones is likely to determine the near-surface groundwater surface temperature, where *cold* or *warm* springs will appear, depending on the regional flow pattern itself. Groundwater temperatures in discharge zones are likely to be influenced by climatic factors but may also reflect the

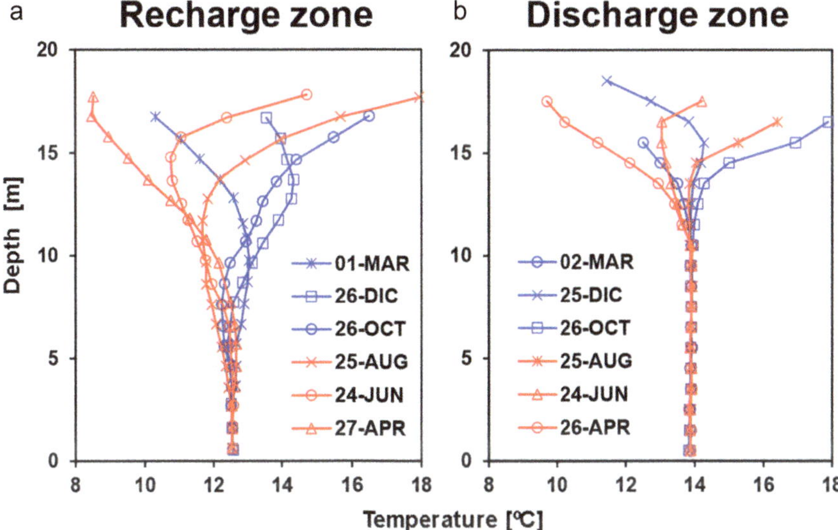

Fig. 3.6 Effect of groundwater recharge (**a**) and discharge (**b**) on the characteristic subsurface temperature profile (CSTP). Modified from Taniguchi (1993)

thermal environment through which the groundwater has circulated during its path from the recharge zone to the discharge zone (Parsons 1970). In extreme cases, when the geothermal domain meets the surface, the springs that may exist in that area will tend to be a few degrees *warmer* than those springs fed by shallower groundwater flow paths.

Consequently, the regional geothermal gradient will remain uniform at depths where the conductive regime dominates. At a given depth (depth *a* in Fig. 3.7), where a regional groundwater flow exists, the normal geothermal gradient will be affected. As can be seen in Fig. 3.6, the depth at which the influence of surface temperature fluctuation disappears is affected in the discharge zones. In the recharge zones, downward heat advection from the surface causes thermal disturbances to propagate to greater depths (depth *c* in Fig. 3.7) than would be the case without heat advection by groundwater flow (depth *a* in Fig. 3.7). In groundwater discharge zones, the geothermal domain gains space by displacing the depth of surface thermal oscillation. The upward groundwater flow is able to reduce the surface domain by a few metres (depth *b* in Fig. 3.7).

The existence of preferential flow zones, such as permeable fracture zones, has a major impact on regional groundwater flow and thermal regime of the terrain. Figure 3.8 shows the thermal effects of a fracture zone, with hydraulic conductivity four orders of magnitude higher than the rock massif that conditions the topography, resulting in an asymmetric valley. In a scenario (Fig. 3.8a) in which the rock massif has a low hydraulic conductivity (10^{-16} m^2), a scenario with a dominantly conductive regime is generated, with greater importance of advection around the fracture zones,

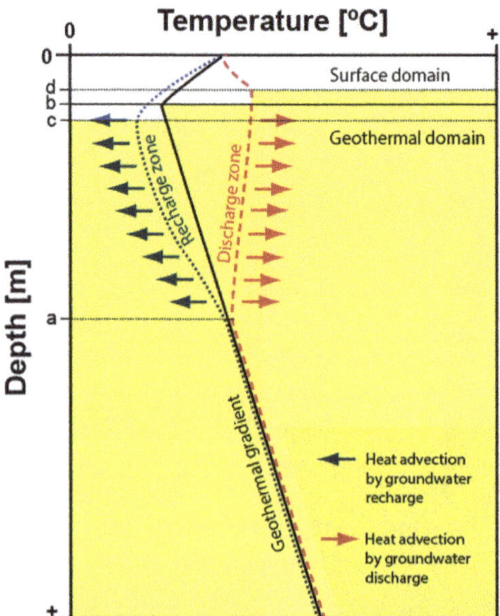

Fig. 3.7 Effect of groundwater recharge and discharge on characteristic subsurface temperature profile (CSTP) and depth to the surface domain

causing the isotherms to increase in height. Although the central heat tubes are found towards the preferential flow zone, the heat tubes cover the entire recharge zone. If the rock massif is considered to have higher hydraulic conductivity (10^{-15} m^2) (Fig. 3.8b), there is a clear advective disruption of the existent conductive flow, where all heat tubes end up being collected by the fracture zone (Bodner et al. 1985; Painter et al. 2003). The massive inflow of groundwater cools the entire rock massif and mixes with the *hot* rising water.

The authors of this study (Smith et al. 1989) highlighted the fact that the spring temperature in the fracture zone was strongly dependent on the magnitude of the basal geothermal heat flux and the hydraulic conductivity of the rock massif. With a basal heat flux of 0.06 W m^{-2} and a very low hydraulic conductivity (10^{-18} m^2) of the rock massif considered, the spring had a temperature of 12 °C. As the hydraulic conductivity of the fracture zone was increased, the spring temperature increased up to an average value of 10^{-16} m^2 where a maximum upwelling temperature of 23 °C was reached. If the hydraulic conductivity of the rock massif was increased, the spring temperature decreased. This is due to the fact that, as the hydraulic conductivity increases, more heat tubes are intercepted by the fracture and, as the hydraulic conductivity increases further, heat pipes from the surface are intercepted, which means that cold water enters the fracture and mixes with the surrounding water, thus lowering the temperature. Therefore, a *window* (range) of hydraulic conductivities make hydrothermal springs/surgences possible. The same pattern is observed with a higher basal geothermal flow; with 0.12 W m^{-2} an upwelling temperature of 42 °C is

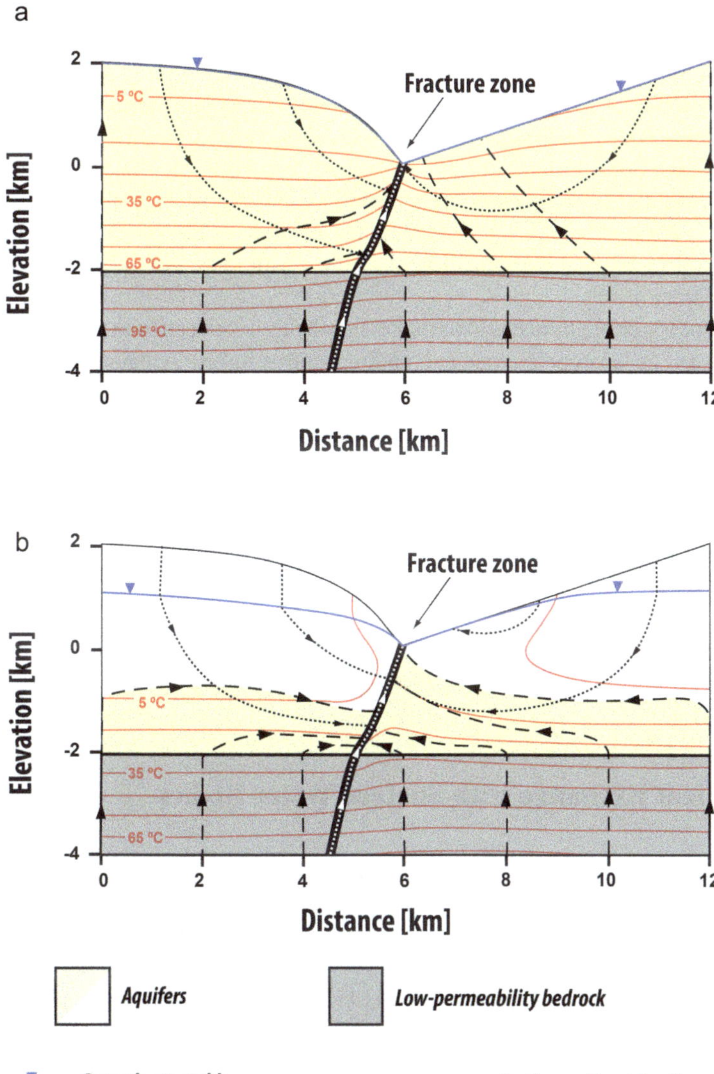

Fig. 3.8 Effects of groundwater flow on the regional thermal regime with preferential flow zones (e.g., fracture zone) for low permeable (**a**) and permeable (**b**) systems. Data from Smith et al. (1989)

reached at an average hydraulic conductivity of 10^{-16} m^2 of the mountain massif. The hydraulic conductivity of the fracture zone plays an important role in determining the upwelling temperature.

Several studies have shown that, in boreholes intercepting fracture planes, where preferential flow zones exist, a thermal anomaly caused by relatively *cold* or *hot* groundwater flows into or out of the fracture can be found. These anomalies are clearly reflected in the temperature profiles and it is even possible to distinguish isolated fracture planes that give specific temperature peaks (Silliman and Robinson 1989). Additionally, continuous fracture zones with more diffuse anomalies can be found (Drury 1989).

Groundwater as a fluid could transport heat by free convection in areas of large heat fluxes, such as in oceanic crustal spreading centres linked to magmatism. However, free convection is rarely found in continental sedimentary basins where very large basal heat fluxes and geothermal gradients together with very strong permeable layers would be required (Sharp et al. 1988). There is no empirical evidence for convection cells in the subsurface shallow geothermal domain environment. Perhaps some exceptional evidence is presented by Aziz et al. (1973) in the form of spatial patterns in groundwater temperatures in sandstone units in the Gulf of Mexico basin, under normal geothermal gradients. From a theoretical point of view, there are several studies (Domenico and Schwartz 1998) in which the critical hydraulic conductivity necessary for initiation of the free convection process for a given thickness of permeable layers and given geothermal gradients is presented.

3.4 Heat Exchange with Surface Water Bodies

In areas where there is a surface water body (rivers, lakes, coastal waters, transitional waters or artificial water bodies), heat transfer between the atmosphere and the ground takes on special character. The fact that (1) the water temperature in surface water bodies is closely related to the weekly or monthly mean atmospheric temperature (Van Vliet et al. 2011), together with the fact that (2) the daily surface temperature oscillations disappear in the first 1.5 m depth, makes the ground temperature profiles in these specific areas very similar to the CSTP temperature profiles where the surface is not flooded. Therefore, it could be said that the CSTP profiles can be modified by the presence of surface water bodies. When there is a good hydraulic connection between the surface water body and the groundwater body, usually rivers (Lapham 1989) or lakes (Lee 1985), the bidirectional water flow (inflow to and outflow from the ground) generally involves a transfer of thermal energy by heat advection.

The river-aquifer relationship is complex and up to five sub-environments can be distinguished, depending on the type of river section (Conant 2004): hydraulic short-circuit spring/discharge, high discharge, low-moderate discharge, no discharge and recharge. In influent rivers, surface water recharge occurs in the aquifer, and the temperature profile may change, both in humid regions (Conant 2004; Silliman and Booth 1993) and in arid regions (Constantz et al. 2003). In some cases, the change in the river-aquifer relationship may shift seasonally from effluent to influent and affect the CSTP (Allander 2003; Constantz et al. 2001, 2002). CSTP over the hyporheic zone of the rivers themselves would be sensitive to these water exchanges (Conant 2004; Evans et al. 1995; Storey et al. 2003; White et al. 1987). Research works related to the efficiency of recharge ponds (Jaynes 1990) have shown that even temperature affects recharge rates, being higher during the day when water viscosity is reduced and water flows easily.

The surface water-aquifer relationship can vary over time, e.g., effluent rivers receiving groundwater for almost the entire year (Fig. 3.9a). During flood events the river level rises by several metres so that the flow is reversed and the surface water recharges the aquifer, downwards and laterally (Fig. 3.9b), a process known as *bank storage*. When the flood event ceases, the accumulated water is slowly discharged back into the river.

Influent rivers produce a constant recharge in the aquifer, modifying the thermal regime in the first metres of depth under the hyporheic zone (Fig. 3.9, zone ZH). Temperature profiles similar to the CSTP are seen and they distinguish themselves by presenting vertical temperature profiles with greater penetration into the ground

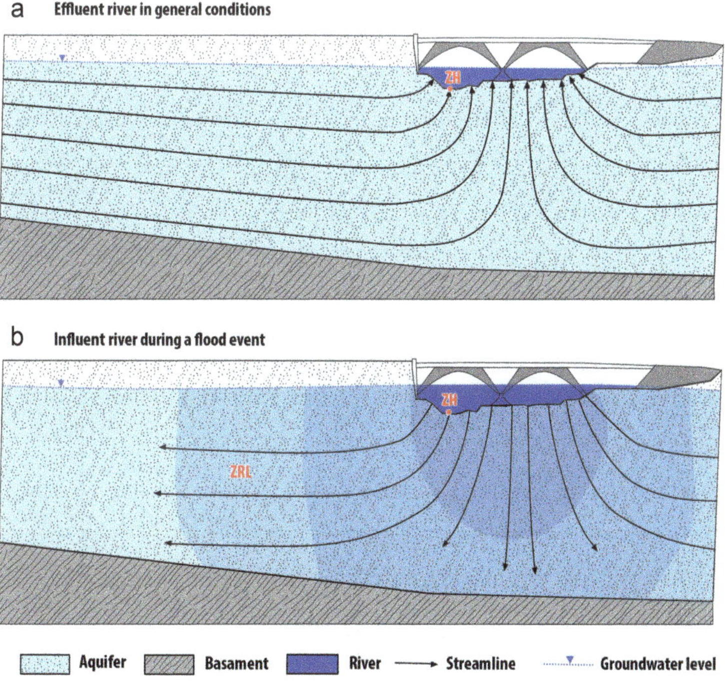

Fig. 3.9 Conceptual scheme of groundwater recharge and discharge in the vicinity of a river hydraulically connected to a surficial aquifer. It shows the general situation where the river is effluent (**a**) and the situation during a river flood event where the river is influent (**b**). The hyporheic zone (ZH) and the lateral recharge zone (ZRL) are highlighted

(Fig. 3.10a), even greater than the groundwater temperature profiles in surface recharge zones (Fig. 3.6a). Seasonality is well reflected in these temperature profiles, becoming clearly inverted at depth, more markedly so than in temperature profiles in other situations.

The inflow of surface water, either during flood events or continuously in effluent rivers, into contiguous hydraulically connected aquifers is referred to as *lateral recharge,* and it produces thermal anomalies within the aquifer (Fig. 3.9b, ZLH zone) that can propagate up to several hundred metres in highly transmissive aquifers, subject to extraordinary flood events. Figure 3.10b shows the relative thermal impacts as a function of distance from the river bed in the horizontal along a hydraulically

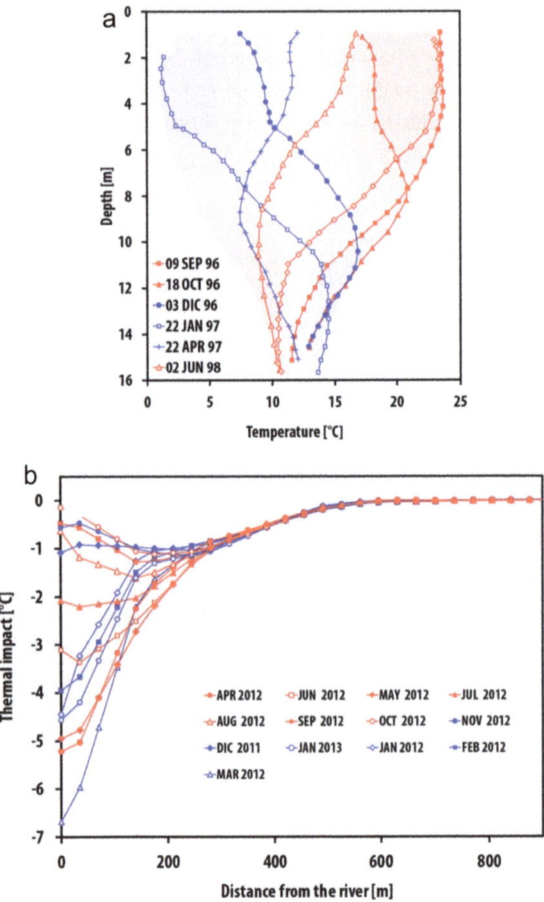

Fig. 3.10 **a** Groundwater temperature profiles measured below the hyporheic zone (see Fig. 3.9, zone ZH) of a river (Bartolino and Niswonger 1999). **b** Thermal impact of the river-aquifer relationship as lateral recharge (see Fig. 3.9, zone ZRL) calculated for a highly transmissive detrital aquifer in the vicinity of a river (García-Gil et al. 2014a)

well-connected, highly transmissive aquifer. In this particular case, floods occur in February–March during the snowmelt season. This means that the recharge water has a very low temperature, cooling the areas of the aquifer relatively close to the river.

Other areas where heat transfer between groundwater and surface water occurs are submarine groundwater discharge zones in coastal aquifers (Burnett et al. 2003; Mejías et al. 2012), lakes (Krabbenhoft and Babiarz 1992) and wetlands (Hunt et al. 1996). In coastal aquifers, vertical groundwater temperature profiles have specific characteristics related to the saline interface (Taniguchi 2000). In arctic areas where permafrost is present (Wankiewicz 1984), this river-aquifer relationship has additional specificities.

3.5 Heat Exchange with Urban Structures

Cities often generate *hot* thermal anomalies with respect to their natural environment (Oke 1973). This so-called *urban heat island* (UHI) effect can be identified at atmospheric, surface and subsurface levels (Menberg et al. 2013; Taniguchi et al. 2009; Zhu et al. 2010). While atmospheric and surface thermal anomalies could be referred to as surface urban heat islands, the *subsurface urban heat island (*SUHI) is a special case (Fig. 3.11).

Anthropogenic activity on the surface and underground affects ground temperature in the surface, transition domain and in the geothermal domain. Urbanisation removes vegetation cover from the surface where evapotranspiration occurs and, together with climate change, is responsible for ground warming. Taniguchi et al.

Fig. 3.11 Conceptual scheme of the surface and subsurface urban heat island (SUHI) effect

(1999) assumed a warming of 0.025 °C per year for the last 100 years to explain the temperature minima in vertical profiles measured in cities and determine a rate of heat transfer produced from the city to the ground. Different studies identify urban ground heating up to 60 m depth (Salem et al. 2004). Studies on the impact of urbanisation and climate change on groundwater temperature are limited, but they are of importance for the study of possible consequences of global climate change on water resources (Sophocleous 2004).

The study of surface urban heat islands from ground surface temperatures, calculated from satellite data, has been extensively investigated in cities around the world (Chakraborty et al. 2015; Hu and Brunsell 2015; Peng et al. 2012; Pongrácz et al. 2010). SUHI requires the interpolation of point groundwater temperature measurements in piezometers and boreholes (Ferguson and Woodbury 2007; Lubis et al. 2013; Taniguchi et al. 2009). This task is generally not straightforward, as boreholes are expensive to drill and pre-existing boreholes are often of insufficient depth to study SUHIs, which are generally less than 20 m. Nevertheless, the interrelationship between these SUHI performance levels is clear and has been positively correlated with both atmospheric temperatures in the city (Cheon et al. 2014) and groundwater temperatures (Menberg et al. 2014), as well as correlations between ground surface and/or atmospheric temperatures with groundwater temperatures (Benz et al. 2016). Although it is possible to reproduce the variance up to 98% of borehole temperatures from long temperature time series in rural areas, in urban areas the estimation of ground temperatures is more complicated and ground temperatures can only be predicted for the first few metres of depth (<4 m) (Zhan et al. 2014). This difficulty is due to the existence of anthropogenic heat source/sink sources of various types that add complexity to the thermal regime in the subsurface. Some of these most important anthropogenic elements are (Fig. 3.12) above-ground buildings and their foundations in the subsoil, paving and waterproofing of the urban surface, sewage systems, water supply or sewage network leaks, underground or road transport tunnels, district air conditioning networks, re-injection of industrial thermal discharges or shallow geothermal systems (Huang et al. 2000, 2009; Köhler et al. 2015; Kooi 2008; Smerdon

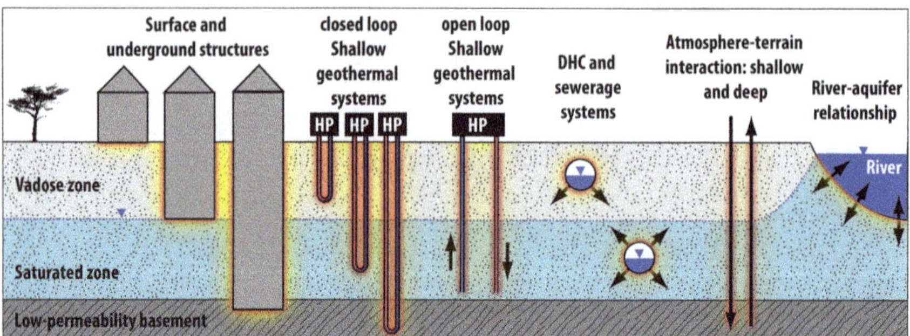

Fig. 3.12 Conceptual scheme of the surface and ground heat island effect (SUHI). HP: Geothermal heat pump. Modified from Köhler et al. (2015)

et al. 2004). These heat sources/sinks consist of elements that can be classified as points, lines or areas. Isolated buildings and shallow geothermal systems could be considered as points. Metro lines, sewage-supply networks and district heating and cooling networks would be line elements and area elements would be related to densely built-up areas (industrial or residential).

The depth of underground structures and diffuse heat transfer from buildings observed in many urban areas is the main component of the increase in groundwater temperatures in urban areas, i.e., SUHI (Epting and Huggenberger 2013; Ferguson and Woodbury 2007; Taniguchi et al. 1999). It is generally accepted that heat losses from a building through the ground are significant, up to 50% of the annual heat loads, especially in low-rise buildings (Deru 2003). When basements or foundations do not reach the water table, heat loss from buildings occurs mainly through heat conduction. If the building comes into contact with the groundwater table, heat losses are up to ten times higher and occur mainly by advection due to groundwater flow (Heinrich and Dahlem 1999). Currently, half the world's population lives in cities and this is expected to reach 70% by 2050 (Un-Habitat 2012). Urban development is increasing and there is a tendency to build increasingly deeper (Bobylev 2009) due to the physical limitations of above-ground land use. Therefore, additional heat transfers to the subsoil and to urban shallow aquifers is expected.

The quantification of thermal impacts on the ground derived from underground structures is crucial to understand the thermal regime in urban environments. Heat advection and knowledge of the hydraulic regime, as seen above, are of great importance for understanding the thermal regime. Therefore, any subsurface structure that also conditions the water flow will be affecting the thermal regime. The hydraulic impacts generated by underground structures (Attard et al. 2016a) must be taken into account in order to understand the general thermal regime in the subsoil of a city.

The study of the thermal impact of building basements in the spatio-temporal distribution of temperatures in the subsurface depends on three main factors (Fig. 3.12): (1) the past history of air conditioning loads of the building, (2) the distance to the building, and (3) the density and depth of existing buildings in an area (Epting et al. 2017b). This same study showed that in a detrital aquifer, an isolated building fully penetrating the aquifer can impact groundwater by up to just over two degrees Celsius in water below the structure. The importance of heating degree days at the beginning of a winter heating season is highlighted. If there are large heating loads at that time, the impacts are greater. Heat fluxes through the walls of buildings below ground surface can vary significantly, depending on the depth of penetration of the building into the water table, the groundwater velocity that produces heat advection and the construction characteristics. Values ranging from -0.1 to $10\ \mathrm{W\ m^{-2}}$ have been reported in the literature (Menberg et al. 2013): in the city of Basel from 0.2 to $0.9\ \mathrm{W\ m^{-2}}$ (Epting et al. 2017b), in the city of Karlsruhe from $3.61 \pm 3.37\ \mathrm{W\ m^{-2}}$ and in the city of Cologne from $0.57 \pm 0.47\ \mathrm{W\ m^{-2}}$ (Benz et al. 2015). Other authors estimate the heat transfer range from 0 to $20\ \mathrm{W\ m^{-2}}$ (Rees et al. 2000; Thomas and Rees 1998). However, the calculated values are closer to the $\approx 2\ \mathrm{W\ m^{-2}}$ of (Ferguson and Woodbury 2004) which considered a temperature inside the buildings of 20 °C and outside the ground of 5.0–7.2 °C (Fig. 3.13).

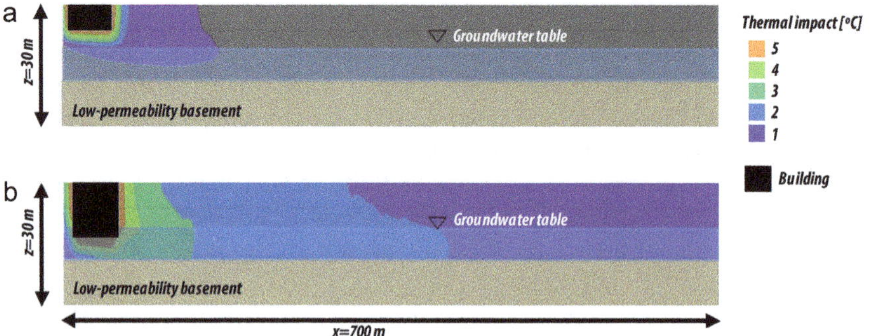

Fig. 3.13 Relative thermal impact of buildings in unsaturated zone (**a**) and saturated zone (**b**). Data from Epting et al. (2017b)

Increased groundwater velocity enhances heat dissipation of buildings. Heat is transported more efficiently by advection than by conduction and, consequently, the temperature outside the immediate vicinity of the building wall is lower. This causes the thermal gradient between the inside of the building and the ground to be higher, thus favouring higher heat fluxes. Advection by groundwater movement means that the derived thermal impacts, the temperature plumes, are lower and also decrease logarithmically with distance from the building. In urban environments with permeable aquifers and groundwater flows greater than 1 m day^{-1} the thermal impacts produced by buildings tend to be punctual with small and narrow thermal plumes. In contrast, in dense urban environments with lower groundwater velocities, thermal impacts tend to be diffuse. Regarding the magnitude of the impact, for the hydrogeological and urban conditions of the city of Basel (Switzerland), it was shown that no significant thermal impacts are expected more than 10 m from buildings (Epting et al. 2017b).

When studying the thermal impact of underground structures on the ground on a city scale, it can be observed how the thermal impacts of different structures interact. When underground structures are linear, the thermal impact generated is highly dependent on the orientation of the linear underground structure with the direction of groundwater flow. Figure 3.14 shows the plan and cross-sectional views of the thermal plumes generated by 11 deep buildings and two metro lines in the city of Lyon (France), after 20 years of heat loss with the associated urban aquifer. The Lyon urban aquifer has background temperatures of about 15 °C and is composed of two distinct aquifer levels: a package of more permeable recent alluvial materials in the first 15 m and a less permeable aquifer level in the lower part, composed of molasse (Attard et al. 2016b). Structures such as an underground metro line can transfer up to 30% of the heat generated by the line to the ground (Ampofo et al. 2006). In the case of the city of Lyon, the thermal impact generated by these structures covers 14% of the area of interest assessed. This is a significant impact when compared to the thermal impacts of punctual shallow geothermal installations, where it can represent 2.9–9.9% (Herbert et al. 2013). In addition, Fig. 3.14 shows an urban setting where

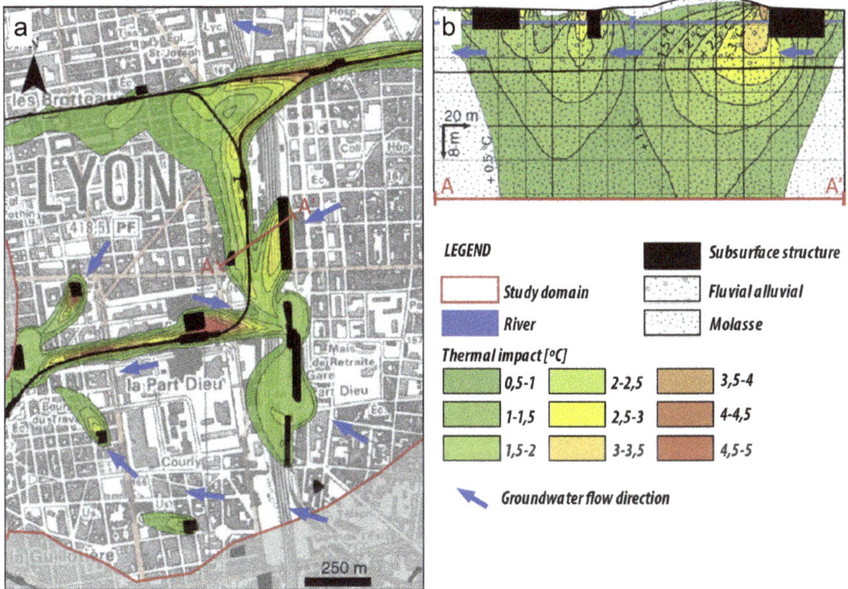

Fig. 3.14 Plan view (**a**) and cross-sectional view (**b**) of the thermal impact generated by point and linear underground structures. Modified from Attard et al. (2016b)

the regional groundwater flow is significantly modified by existing groundwater structures (Attard et al. 2016a). In this study, similar to that of the city of Basel, it was calculated by numerical modelling that, after one year, urban groundwater structures transfer a total of 0.37 GWh km^{-2} year^{-1}. In Basel, a very similar value of 0.47 GWh km^{-2} year^{-1} was calculated (Epting et al. 2013). Other calculations based on analytical formulas of heat transport in GIS platforms for the German cities of Karlsruhe and Cologne yielded higher values of 9.42 and 3.41 GWh km^{-2} year^{-1} respectively (Benz et al. 2015). Such studies can demonstrate how underground structures can create significant impacts on aquifers, in the case of the city of Lyon from 14 to 20 times the volume of the underground structure under consideration.

The impact of climate change-induced ground surface warming has been quantified and the effect of heat transfer by buildings has been found to be much more important, leaving this first aspect as a secondary factor (Epting and Huggenberger 2013). The large number of factors and processes that can modify the temperature of the subsoil and groundwater beneath cities makes it difficult to identify the source of a thermal anomaly when measuring the temperature in an individual borehole.

References

Allander KK (2003) Trout creek—evaluating ground-water and surface water exchange along an alpine stream, Lake Tahoe, California. In: Stonestrom DA, Constantz J (eds) Heat as a tool for studying the movement of ground water near streams USGS, Reston, pp 35–45

Ampofo F, Maidment GG, Missenden JF (2006) Review of groundwater cooling systems in London. Appl Therm Eng 26(17):2055–2062. https://doi.org/10.1016/j.applthermaleng.2006.02.013

Anderson MP (2005) Heat as a ground water tracer. Ground Water 43(6):951–968. https://doi.org/10.1111/j.1745-6584.2005.00052.x

Anzellini S, Dewaele A, Mezouar M, Loubeyre P, Morard G (2013) Melting of iron at earth's inner core boundary based on fast X-ray diffraction. Science 340(6131):464. https://doi.org/10.1126/science.1233514

Attard G, Rossier Y, Winiarski T, Cuvillier L, Eisenlohr L (2016a) Deterministic modelling of the cumulative impacts of underground structures on urban groundwater flow and the definition of a potential state of urban groundwater flow: example of Lyon, France. Hydrogeol J 24(5):1213–1229. https://doi.org/10.1007/s10040-016-1385-z

Attard G, Rossier Y, Winiarski T, Eisenlohr L (2016b) Deterministic modeling of the impact of underground structures on urban groundwater temperature. Sci Total Environ 572:986–994. https://doi.org/10.1016/j.scitotenv.2016.07.229

Aziz K, Bories SA, Combarnous MA (1973) The influence of natural convection in gas, oil and water reservoirs. PETSOC-73-02-05 12(02):8. https://doi.org/10.2118/73-02-05

Bartolino JR, Niswonger RG (1999) Numerical simulation of vertical ground-water flux of the Rio Grande from ground-water temperature profiles, central New Mexico. 99-4212. Reston, VA. https://doi.org/10.3133/wri994212

Beck AE, Garven G, Stegena L (1991) Hydrogeological regimes and their subsurface thermal effects. Wiley

Becker MW (2006) Potential for satellite remote sensing of ground water. Groundwater 44(2):306–318. https://doi.org/10.1111/j.1745-6584.2005.00123.x

Benz SA, Bayer P, Menberg K, Jung S, Blum P (2015) Spatial resolution of anthropogenic heat fluxes into urban aquifers. Sci Total Environ 524–525:427–439. https://doi.org/10.1016/j.scitotenv.2015.04.003

Benz SA, Bayer P, Goettsche FM, Olesen FS, Blum P (2016) Linking surface Urban heat Islands with groundwater temperatures. Environ Sci Technol 50(1):70–78. https://doi.org/10.1021/acs.est.5b03672

Bobylev N (2009) Mainstreaming sustainable development into a city's Master plan: a case of Urban underground space use. Land Use Policy 26(4):1128–1137. https://doi.org/10.1016/j.landusepol.2009.02.003

Bodner DP, Blanchard PE, Sharp JM (1985) Variations in gulf coast heat flow created by groundwater flow

Brutsaert W (1982) Evaporation into the atmosphere: theory, history and applications. Springer, Netherlands

Burnett WC, Bokuniewicz H, Huettel M, Moore WS, Taniguchi M (2003) Groundwater and pore water inputs to the coastal zone. Biogeochemistry 66(1):3–33. https://doi.org/10.1023/b:biog.0000006066.21240.53

Carson JE (1963) Analysis of soil and air temperatures by Fourier techniques. J Geophys Res (1896–1977) 68(8):2217–2232. https://doi.org/10.1029/JZ068i008p02217

Cartwright K (1970) Groundwater discharge in the illinois basin as suggested by temperature anomalies. Water Resour Res 6(3):912–918. https://doi.org/10.1029/WR006i003p00912

Chakraborty SD, Kant Y, Mitra D (2015) Assessment of land surface temperature and heat fluxes over Delhi using remote sensing data. J Environ Manage 148:143–152. https://doi.org/10.1016/j.jenvman.2013.11.034

Chapman DS (1986) Thermal gradients in the continental crust. Geol Soc Lond Spec Publ 24(1):63. https://doi.org/10.1144/gsl.sp.1986.024.01.07

Cheon J-Y, Ham B-S, Lee J-Y, Park Y, Lee K-K (2014) Soil temperatures in four metropolitan cities of Korea from 1960 to 2010: implications for climate change and urban heat. Environ Earth Sci 71(12):5215–5230. https://doi.org/10.1007/s12665-013-2924-8

Conant B Jr (2004) Delineating and quantifying ground water discharge zones using streambed temperatures. Groundwater 42(2):243–257. https://doi.org/10.1111/j.1745-6584.2004.tb02671.x

Constantz J, Stonestorm D, Stewart AE, Niswonger R, Smith TR (2001) Analysis of streambed temperatures in ephemeral channels to determine streamflow frequency and duration. Water Resour Res 37(2):317–328. https://doi.org/10.1029/2000wr900271

Constantz J, Stewart AE, Niswonger RG, Sarma L (2002) Analysis of temperature profiles for investigating stream losses beneath ephemeral channels. Water Resour Res 38(12):52-1–52-13. https://doi.org/10.1029/2001wr001221

Constantz J, Cox MH, Su GW (2003) Comparison of heat and bromide as ground water tracers near streams. Groundwater 41(5):647–656. https://doi.org/10.1111/j.1745-6584.2003.tb02403.x

Deru M (2003) Model for ground-coupled heat and moisture transfer from buildings. United States. https://doi.org/10.2172/15004062

Domenico PA, Palciauskas VV (1973) Theoretical analysis of forced convective heat transfer in regional ground-water flow. GSA Bull 84(12):3803–3814. https://doi.org/10.1130/0016-7606(1973)84%3c3803:taofch%3e2.0.co;2

Domenico PA, Schwartz FW (1998) Physical and chemical hydrogeology. Wiley

Drury M (1989) Fluid flow in crystalline crust: detecting fractures by temperature logs. In: Beck AE, Garven G, Stegena L (eds) Hydrogeological regimes and their subsurface thermal effects, pp 129–135. https://doi.org/10.1029/GM047p0129

Epting J, Huggenberger P (2013) Unraveling the heat island effect observed in urban groundwater bodies—definition of a potential natural state. J Hydrol 501:193–204. https://doi.org/10.1016/j.jhydrol.2013.08.002

Epting J, Händel F, Huggenberger P (2013) Thermal management of an unconsolidated shallow urban groundwater body. Hydrol Earth Syst Sci 17(5):1851–1869. https://doi.org/10.5194/hess-17-1851-2013

Epting J et al (2017) The thermal impact of subsurface building structures on urban groundwater resources—a paradigmatic example. Sci Total Environ 596–597:87–96. https://doi.org/10.1016/j.scitotenv.2017.03.296

Evans EC, Greenwood MT, Petts GE (1995) Thermal profiles within river beds. Hydrol Process 9(1):19–25. https://doi.org/10.1002/hyp.3360090103

Ferguson G, Woodbury AD (2004) Subsurface heat flow in an urban environment. J Geophys Res Solid Earth **109**(B2). https://doi.org/10.1029/2003jb002715

Ferguson G, Woodbury AD (2007) Urban heat island in the subsurface. Geophys Res Lett 34(23):L23713. https://doi.org/10.1029/2007gl032324

Gallo K, Hale R, Tarpley D, Yu Y (2011) Evaluation of the relationship between air and land surface temperature under clear- and cloudy-sky conditions. J Appl Meteorol Climatol 50(3):767–775. https://doi.org/10.1175/2010jamc2460.1

García-Gil A, Vázquez-Suñe E, Garrido E, Sánchez-Navarro JA, Mateo-Lázaro J (2014) The thermal consequences of river-level variations in an urban groundwater body highly affected by groundwater heat pumps. Sci Total Environ. https://doi.org/10.1016/j.scitotenv.2014.03.123

Gupta H, Roy S (2007) Basic concepts. In: Gupta H, Roy S (eds) Geothermal energy. Elsevier, Amsterdam, pp 15–30. https://doi.org/10.1016/B978-044452875-9/50002-2 (Chapter 2)

Heinrich H, Dahlem K-H (1999) The effect of groundwater flow on the heat loss from buildings to the ground. In: 11th thermo conference. Budapest

Herbert A, Arthur S, Chillingworth G (2013) Thermal modelling of large scale exploitation of ground source energy in urban aquifers as a resource management tool. Appl Energy 109:94–103. https://doi.org/10.1016/j.apenergy.2013.03.005

Hu L, Brunsell NA (2015) A new perspective to assess the urban heat Island through remotely sensed atmospheric profiles. Remote Sens Environ 158:393–406. https://doi.org/10.1016/j.rse.2014.10.022

Huang S, Pollack HN, Shen P-Y (2000) Temperature trends over the past five centuries reconstructed from borehole temperatures. Nature 403(6771):756–758. https://doi.org/10.1038/35001556
Huang S, Taniguchi M, Yamano M, Wang C-H (2009) Detecting urbanization effects on surface and subsurface thermal environment—a case study of Osaka. Sci Total Environ 407(9):3142–3152. https://doi.org/10.1016/j.scitotenv.2008.04.019
Hunt RJ, Krabbenhoft DP, Anderson MP (1996) Groundwater inflow measurements in Wetland systems. Water Resour Res 32(3):495–507. https://doi.org/10.1029/95wr03724
Jaynes DB (1990) Temperature variations effect on field-measured infiltration. Soil Sci Soc Am J 54(2):305–312. https://doi.org/10.2136/sssaj1990.03615995005400020002x
Kohl T (1998) Palaeoclimatic temperature signals—can they be washed out? Tectonophysics 291(1):225–234. https://doi.org/10.1016/S0040-1951(98)00042-0
Köhler M et al (2015) Numerical evaluation and optimization of depth-oriented temperature measurement for the investigation of thermal influences on groundwater resources. Energy Proc 76:371–380. https://doi.org/10.1016/j.egypro.2015.07.844
Kooi H (2008) Spatial variability in subsurface warming over the last three decades; insight from repeated borehole temperature measurements in The Netherlands. Earth Planet Sci Lett 270(1):86–94. https://doi.org/10.1016/j.epsl.2008.03.015
Krabbenhoft DP, Babiarz CL (1992) The role of groundwater transport in aquatic mercury cycling. Water Resour Res 28(12):3119–3128. https://doi.org/10.1029/92wr01766
Lapham WW (1989) Use of temperature profiles beneath streams to determine rates of vertical ground-water flow and vertical hydraulic conductivity, vol 2337. https://doi.org/10.3133/wsp2337
Lee DR (1985) Method for locating sediment anomalies in lakebeds that can be caused by groundwater flow. J Hydrol 79(1):187–193. https://doi.org/10.1016/0022-1694(85)90192-1
Lubis RF et al (2013) Assessment of urban groundwater heat contaminant in Jakarta, Indonesia. Environ Earth Sci 70(5):2033–2038. https://doi.org/10.1007/s12665-013-2712-5
Mejías M et al (2012) Methodological study of submarine groundwater discharge from a karstic aquifer in the Western Mediterranean Sea. J Hydrol 464–465:27–40. https://doi.org/10.1016/j.jhydrol.2012.06.020
Menberg K, Bayer P, Zosseder K, Rumohr S, Blum P (2013) Subsurface urban heat islands in German cities. Sci Total Environ 442:123–133. https://doi.org/10.1016/j.scitotenv.2012.10.043
Menberg K, Blum P, Kurylyk BL, Bayer P (2014) Observed groundwater temperature response to recent climate change. Hydrol Earth Syst Sci 18(11):4453–4466. https://doi.org/10.5194/hess-18-4453-2014
Mihalakakou G, Santamouris M, Asimakopoulos D (1992) Modelling the earth temperature using multiyear measurements. Energy Build 19(1):1–9. https://doi.org/10.1016/0378-7788(92)90031-B
Oke TR (1973) City size and the urban heat Island. Atmosp Environ 7(8):769–779 (1967). https://doi.org/10.1016/0004-6981(73)90140-6
Oke TR (1987) Boundary layer climates. Routledge
Painter S, Winterle J, Armstrong A (2003) Using temperature to test models of flow near Yucca Mountain, Nevada. Groundwater 41(5):657–666. https://doi.org/10.1111/j.1745-6584.2003.tb02404.x
Parsons ML (1970) Groundwater thermal regime in a glacial complex. Water Resour Res 6(6):1701–1720. https://doi.org/10.1029/WR006i006p01701
Peng S et al (2012) Surface urban heat Island across 419 global big cities. Environ Sci Technol 46(2):696–703. https://doi.org/10.1021/es2030438
Penrod EB, Elliott JM, Brown WK (1960) Soil temperature variation (1952–1956) at Lexington. Kentucky 90(5):275–283
Pollack HN, Huang S (2000) Climate reconstruction from subsurface temperatures. Annu Rev Earth Planet Sci 28(1):339–365. https://doi.org/10.1146/annurev.earth.28.1.339

Pongrácz R, Bartholy J, Dezső Z (2010) Application of remotely sensed thermal information to urban climatology of Central European cities. Phys Chem Earth Parts A/B/C 35(1):95–99. https://doi.org/10.1016/j.pce.2010.03.004

Rees SW, Adjali MH, Zhou Z, Davies M, Thomas HR (2000) Ground heat transfer effects on the thermal performance of earth-contact structures. Renew Sustain Energy Rev 4(3):213–265. https://doi.org/10.1016/S1364-0321(99)00018-0

Salem ZE-S, Taniguchi M, Sakura Y (2004) Use of temperature profiles and stable isotopes to trace flow lines: Nagaoka Area, Japan. Groundwater 42(1):83–91. https://doi.org/10.1111/j.1745-6584.2004.tb02453.x

Sellers PJ et al (1996) A revised land surface parameterization (SiB2) for atmospheric GCMS. Part I: model formulation. J Clim 9(4):676–705. https://doi.org/10.1175/1520-0442(1996)009<0676:arlspf>2.0.co;2

Sharp JM et al (1988). Diagenetic processes in Northwestern Gulf of Mexico sediments. In: Chilingarian GV, Wolf KH (eds) Developments in sedimentology. Elsevier, pp 43–113. https://doi.org/10.1016/S0070-4571(08)70006-2 (Chapter 1)

Silliman SE, Booth DF (1993) Analysis of time-series measurements of sediment temperature for identification of gaining vs. losing portions of Juday Creek, Indiana. J Hydrol 146:131–148. https://doi.org/10.1016/0022-1694(93)90273-C

Silliman S, Robinson R (1989) Identifying fracture interconnections between boreholes using natural temperature profiling: I. Conceptual basis. Groundwater 27(3):393–402. https://doi.org/10.1111/j.1745-6584.1989.tb00463.x

Smerdon JE et al (2004) Air-ground temperature coupling and subsurface propagation of annual temperature signals. J Geophys Res Atmosp 109(D21). https://doi.org/10.1029/2004jd005056

Smith L, Forster C, Woodbury A (1989) Numerical simulation techniques for modeling advectively-disturbed thermal regimes. In: Beck AE, Garven G, Stegena L (eds) Hydrogeological regimes and their subsurface thermal effects, pp 1–5. https://doi.org/10.1029/GM047p0001

Song Y et al (2012) A revised surface resistance parameterisation for estimating latent heat flux from remotely sensed data. Int J Appl Earth Obs Geoinf 17:76–84. https://doi.org/10.1016/j.jag.2011.10.011

Sophocleous M (2004) Climate change: why should water professionals care? Groundwater 42(5):637–637. https://doi.org/10.1111/j.1745-6584.2004.tb02715.x

Storey RG, Howard KWF, Williams DD (2003) Factors controlling riffle-scale hyporheic exchange flows and their seasonal changes in a gaining stream: a three-dimensional groundwater flow model. Water Resour Res 39(2). https://doi.org/10.1029/2002wr001367

Taniguchi M (1993) Evaluation of vertical groundwater fluxes and thermal properties of aquifers based on transient temperature-depth profiles. Water Resour Res 29(7):2021–2026. https://doi.org/10.1029/93wr00541

Taniguchi M (2000) Evaluations of the saltwater-groundwater interface from borehole temperature in a coastal region. Geophys Res Lett 27(5):713–716. https://doi.org/10.1029/1999gl002366

Taniguchi M et al (1999) Disturbances of temperature-depth profiles due to surface climate change and subsurface water flow: 1. An effect of linear increase in surface temperature caused by global warming and urbanization in the Tokyo Metropolitan Area, Japan. Water Resour Res 35(5):1507–1517. https://doi.org/10.1029/1999wr900009

Taniguchi M et al (2009) Anthropogenic effects on the subsurface thermal and groundwater environments in Osaka, Japan and Bangkok, Thailand. Sci Total Environ 407(9):3153–3164. https://doi.org/10.1016/j.scitotenv.2008.06.064

Thomas HR, Rees SW (1998) The thermal performance of ground floor slabs—a full scale in-situ experiment. Build Environ 34(2):139–164. https://doi.org/10.1016/S0360-1323(98)00001-8

Un-Habitat (2012) State of the World's Cities 2008/9: harmonious cities. Taylor and Francis

Van Vliet MTH, Ludwig F, Zwolsman JJG, Weedon GP, Kabat P (2011) Global river temperatures and sensitivity to atmospheric warming and changes in river flow. Water Resour Res 47(2):n/a–n/a. https://doi.org/10.1029/2010wr009198

Wankiewicz A (1984) Hydrothermal processes beneath arctic river channels. Water Resour Res 20(10):1417–1426. https://doi.org/10.1029/WR020i010p01417

Wehrli C (1985) Extra-terrestrial solar spectrum. World Radiation Centre, Davos Dorf

White DS, Elzinga CH, Hendricks SP (1987) Temperature patterns within the hyporheic zone of a Northern Michigan River. J N Am Benthol Soc 6(2):85–91. https://doi.org/10.2307/1467218

Zhan W et al (2014) Satellite-derived subsurface urban heat Island. Environ Sci Technol 48(20):12134–12140. https://doi.org/10.1021/es5021185

Zhu K, Blum P, Ferguson G, Balke K-D, Bayer P (2010) The geothermal potential of urban heat Islands. Environ Res Lett 5(4):044002. https://doi.org/10.1088/1748-9326/5/4/044002

Chapter 4
Geothermal Heat Pump

4.1 Thermal Installations

A thermal installation can be defined as a set of systems and instrumentation aimed at producing, distributing, storing and exchanging heat to satisfy the thermal energy demand of a building or infrastructure in general. The main technical systems that make up a thermal installation are described below.

4.1.1 External Heat Exchange Systems

An *external heat exchange system* is a device designed to maximise heat transfer between the thermal installation's operating distribution system and the external ambient heat reservoir (e.g., atmospheric air, river surface water, ground, etc.). A thermal installation will require heat to be exchanged both between its various constituent elements and with the environment in order to fulfil its function. There are different types of heat exchangers commonly classified according to the type of contact between the fluids exchanging heat. In *direct-contact heat exchangers* (DCHEs), two heat transfer fluids from open thermodynamic systems physically mix. More common are *indirect-contact heat exchangers* (ICHEs), where the fluids exchange heat through a thin metal wall that acts as an enclosure, thus avoiding contamination (closed systems). ICHEs are, in turn, classified into *surface heat exchangers* and *alternating heat exchangers*, depending on whether heat transfer fluids from different systems circulate through the same duct alternately or whether they circulate through adjacent ducts separated by a very thin metal wall. Depending on the morphology of the ducts through which the fluids circulate in surface heat exchangers, a distinction is made between *plate heat exchangers* and *shell and tube heat exchangers*. In many cases, the classification also refers to the state of aggregation of the fluids (liquid–liquid, liquid–gas, etc.) involved in the heat exchange.

A. García Gil et al., *Shallow Geothermal Energy*, Springer Hydrogeology,
https://doi.org/10.1007/978-3-030-92258-0_4

Geothermal heat exchangers are special external heat exchange systems, using a heat transfer fluid from the thermal installation's operating distribution system, with a multi-component medium consisting of a porous solid mineral phase and a fluid phase with varying degrees of groundwater saturation. The characteristics and properties of these exchangers are covered in detail in Chaps. 5 and 6 of this book.

4.1.2 Heat Production Systems

A heat production system is a thermal device or machine designed to generate one or more sources of heat, capable of acting as a source and/or sink. The heat sources generated are capable of producing heat, i.e., a transfer of thermal energy between thermodynamic systems.

The main heat production systems capable of generating a heat source are chemical combustion boilers (conventional, low-temperature and gas condensing boilers), heat pumps, cogeneration units and solar thermal panels, among others. Heat pumps are the main heat production system capable of generating a sink source. Non-reversible heat pumps are only designed for the production of a heat sink and are called chillers.

Taking into account the second law of thermodynamics, where energy tends to dissipate in the universe, generating a heat source or sink will require using a pre-existing heat source or using another form of energy for its transformation into heat. Transforming chemical, electrical, or mechanical energy into heat to produce a heat source is relatively straightforward through well-known processes such as chemical combustion or electrical resistance. Generating a heat sink is a more complicated or even impossible process if you take the universe as a thermodynamic system; however, it is possible to produce a heat sink in local isolated systems by performing mechanical work.

Conventional thermal installations usually require mechanical, electromagnetic or chemical energy to produce the heat sources needed to transfer thermal energy in a controlled manner. However, they can take advantage of natural heat sources, such as the sun or the earth's subsurface, in a constant process of heat transfer to improve the efficiency of heating and cooling. They can even use our environmental surroundings to harness heat sources such as the shallow geological environment and underground water reservoirs.

Solar thermal panels absorb the heat radiated by the sun and transform it into internal energy, generally by increasing the temperature of the absorber material of the solar panel. The absorber increases its internal energy and, consequently, its temperature. The absorber is in contact with a metal coil through which a heat transfer fluid circulates. Consequently, the absorber material transfers heat by conduction to the coil, and the coil transfers it by convection to the rest of the thermal installation. The absorbing material can be considered a heat production system as it is capable of transforming solar radiation into thermal energy, increasing its temperature and thus producing a heat source that can be used by a thermal installation.

4.1.3 *Heat Distribution Systems*

Technical distribution systems include all elements for transporting the heat carrier fluid and the fluids necessary for heat production. These elements of the distribution system include fuel, liquid and gaseous pipelines, pipelines designed to transport heat carrier fluid at different temperatures, pressures and states of aggregation (liquid and vapour), and water pipelines for thermal installation (DHW, surface water, groundwater).

The technical distribution systems also include all the necessary instrumentation such as essential accessories for controlling the operation of the installations. The most important *valves* are the gate valves, the shut-off valves, the ball or spherical valves, the butterfly valves for flow regulation and shut-off against backflow and isolation, the seat valves for regulation and control of small flows, and the check valves responsible for preventing the backflow of fluid in the pipes. Pipelines may have other types of accessory elements in addition to the valves, such as expansion compensators which are designed to absorb the thermal expansion of pipes.

Another important element of the distribution system is the circulating or delivery pumps. Although in some older installations the fluids in distribution systems are circulated by gravity or density differences induced by temperature changes in the different points of the pipe network, these days the section of the pipes is smaller. This reduces hydraulic system costs, and impulsion pumps are used to increase circulation speed and overcome the fluid-pipe friction associated with the reduction in diameter. Distribution systems in thermal installations almost exclusively use electric motor-driven centrifugal pumps because of their low noise emission. In-line pumps (pipe pumps) are also used, adapted to work at different pressures depending on the size of the length of the pipe network. The higher the pressure, the larger the distribution system, and they are also commonly used for DHW supply.

Distribution systems also use elements for heat exchange. The most common are liquid–liquid plate heat exchangers, which are considered to be elements of the distribution system when they are used for the purpose of conducting heat within the thermal installation.

Finally, in the distribution system of a thermal installation with a heat pump, the *operating distribution system* is distinguished as the part of the distribution system responsible for the operation of the heat reservoir by connecting the reservoir to the heat pump. The part of the distribution system of the thermal installation that transfers heat from the heat pump to the domestic system is called the *domestic distribution system.*

4.1.4 *Internal Heat Exchange Systems*

A technical internal heat exchange system is a device designed to maximise heat transfer between the distribution system of the thermal installation and the indoor environment with respect to the heat infrastructure. These systems are considered the terminal systems of a thermal installation, which satisfy the heat demand of the infrastructure containing them. The best-known internal heat exchangers are *fan coils*, air conditioning or air handling units (AHU), radiators and DHW preparation systems, among others.

Large *fan coil* units are called air handling units (AHUs). They represent one of the possible terminal systems of the heating system. They receive the heat transfer fluid from the heat distribution system and transfer it to the air via a series of *cooling* or *heating* coils, depending on demand. A return fan circulates air from inside the infrastructure and, after a variable mixture of outside air, it is passed through liquid–gas heat exchangers, commonly referred to as heat exchanger coils. The heat exchanger coils consist of coil pipes, in contact with thin metal plates called fins, through which the heat transfer fluid from the distribution system, treated by the heat production systems, circulates. Depending on the temperature difference of the heat transfer fluid and the temperature of the return air, the coil will dissipate or absorb heat from the surrounding air, in heating or cooling mode respectively. Finally, a fan will facilitate the discharge of the outgoing air from the heat exchange coils to the interior of the infrastructure, closing the cycle.

Another internal heat exchanger system is radiators, the terminal elements of a thermal installation. In essence, they are liquid–gas heat exchangers optimised for heat dissipation through the internal circulation of a heat carrier fluid, coming from the distribution system at a higher temperature than the air inside the infrastructure, thus constituting a heat source. Radiators consist of a metal body, a shut-off and/or regulation valve and an air vent for the internal heat transfer fluid circuit.

Finally, the terminal internal heat exchanger system is the DHW preparation system. It consists of a liquid–liquid heat exchanger, usually a plate or coil heat exchanger inside a storage tank, which dissipates heat from the high-temperature heat carrier fluid of the distribution system to a lower temperature drinking water circuit, until a temperature of approximately 39 °C is reached.

Table 4.1 summarises the potential different elements of a thermal installation and a geothermal installation (as a specific type of thermal installation).

4.2 Heat Pumps

A heat pump is a machine or device capable of applying electromechanical work to generate a transfer of thermal energy (heat) between two thermodynamic systems. The device is based on the reverse Carnot cycle (Sect. 2.1.3), which generates by work a thermal gradient capable of inducing such a transfer of thermal energy between two systems. A thermodynamic system in which it is desired to dissipate or absorb heat in a controlled manner is called a domestic thermodynamic system, *domestic system*. A thermodynamic system from which the heat demanded by the domestic system is to be absorbed or dissipated is called *heat reservoir*.

Table 4.1 Summary table of the potential main elements of a thermal installation

Thermal installation				
External exchange system	**Operating distribution system**	**Production system**	**Domestic distribution system**	**Internal exchange system**
General thermal installation				
Direct-contact heat exchangers **Indirect-contact heat exchangers** ***Surface heat exchangers*** Coil heat exchangers – Solar thermal panels (heat exchanger) Plate heat exchangers Shell and tube heat exchangers Geothermal heat exchangers ***Alternating heat exchangers***	**Pipelines** Fuel (liquid, gaseous) Heat carrier fluid – High/low pressure – High/low temperature – Gas (vapour, air, etc.)/liquid Water supply – Surface water – Ground water **Valves** **In-line pumps** **Plate heat exchangers** **Accessory elements**	**Heat source** – Chemical combustion boilers Conventional boilers Low temperature boilers Gas condensing boilers – Heat pumps – Cogenerators – Solar thermal panels **Heat sink** Heat pumps – Chillers – Heat pump (reversible)	**Pipelines** Heat transfer fluid – High/low pressure – High/low temperature – Gas-fired (steam, air, etc.) – Liquid Water supply – DHW **Valves** **In-line pumps** **Plate heat exchangers** **Accessory elements**	Fan coils Air handling units (AHU) Radiators DHW systems Floors/ceilings/heated walls
Geothermal installation				
	External secondary circuit	**Primary circuit**	**Internal secondary circuit**	
Geothermal heat exchangers **Closed-loop geothermal heat exchanger** – In vertical drilling – In horizontal boreholes or trenches – In thermoactive geostructures **Open-loop geothermal heat exchanger**	**Pipelines** Heat transfer fluid **Plate heat exchangers** **sediment filter** **Sludge settling tanks** **In-line pumps**	**Geothermal heat pump** – Direct expansion – Indirect expansion	**Pipelines** Heat carrier fluid Accumulator In-line pumps	Fan coils Air handling units (AHU) Radiators DHW systems Floors/ceilings/heated walls

As an example, for heating of the interior space of a building (domestic system), it is necessary to absorb heat from the outside environment (heat reservoir) and transfer it to the inside of the building. If both systems are at the same temperature (Fig. 4.1a), no such heat transfer will ever occur spontaneously. Therefore, the use of a heat pump is necessary, whereby, through the application of electromechanical work (reverse Carnot cycle), an artificial heat sink source (*cold* source) will be generated in the heat external reservoir, which induces heat flow in the heat reservoir at that point. Inside the domestic system (inside the building), the heat pump will generate by means of work (compression) a heat source (*hot* source) capable of inducing heat flow from the heat source to the rest of the domestic system, increasing its temperature (heating mode). In addition to the generation of the mentioned artificial heat source and sink, the heat pump also transports the heat between the source and sink by means of advection. In this way, thermal energy can be transferred by means of work from one thermodynamic system to another even if they have the same (or even higher) internal energy, without contradicting the second law of thermodynamics.

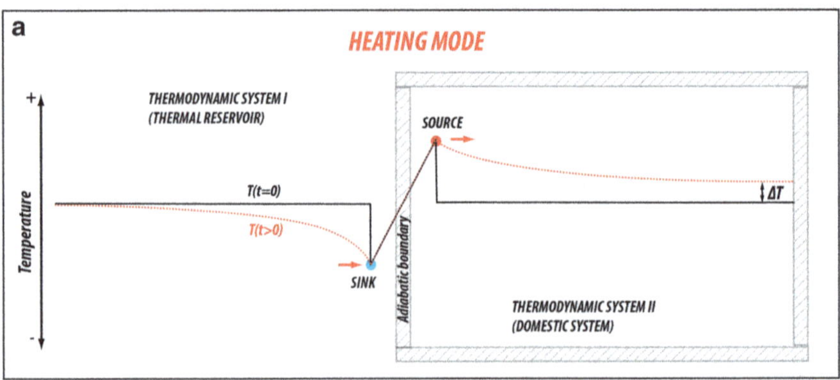

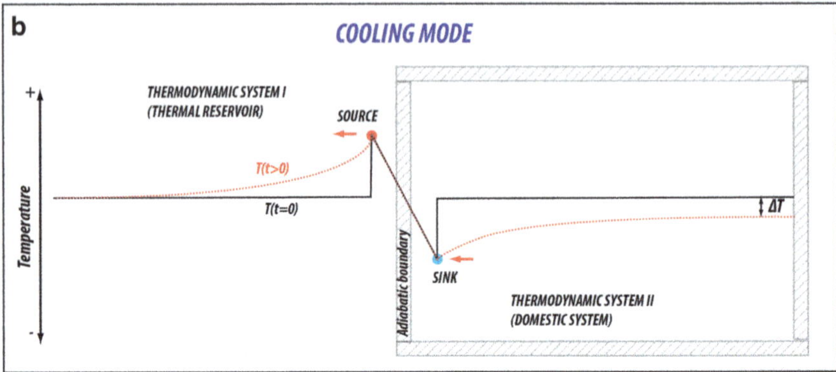

Fig. 4.1 Temperature profiles generated by the generation of artificial heat source and sink by means of a heat pump. **a** Heat pump in heating mode and **b** cooling mode. Temperature profile in black for the initial time and in red during the operation of a heat pump

If the cycle is reversed, the heat pump would switch from operating in heating mode (Fig. 4.1a) to cooling mode, reversing the heat source and sink (Fig. 4.1b). Historically, it is common for heat pumps designed only for cooling to be referred to as chillers. A heat pump limited to the refrigeration cycle will be called a chiller, while still being a heat pump in its general concept.

The amount of work a heat pump must do to achieve the heat energy transfer objective will obviously depend on the difference in thermal energy in the systems between which heat is to be transferred. Thus, if in the proposed example (Fig. 4.1) the temperature outside the building is much higher than inside the building, heat transfer will occur naturally even without the existence of a heat pump (*free cooling/heating*); activating the heat pump would accelerate, in a controlled manner, the heat energy transfer process, and it would require less mechanical energy compared with the first scenario.

An external heat reservoir can be constituted by a gaseous phase such as atmospheric air, *hot* air from a ventilation system or even *hot* gases from an industrial process. It can be liquid, coming from the environment, such as surface water (rivers, lakes, sea) and groundwater, or it can come from domestic wastewater, industrial wastewater or any other source. An external heat reservoir can be made up of solid materials such as any geological material (shallow geothermal reservoir) or anthropogenic material (building foundations, tunnels, etc.). The domestic system that will absorb or dissipate thermal energy by means of a heat pump may consist of the air inside a building or DHW for domestic use, recreational sports use in swimming pools or industrial processes that require it, among others.

The mechanical energy required to perform the compression work done by a heat pump can have different origins. The use of electrical energy in electrocompressors is the most common but, in very specific cases, it can originate from mechanical, thermomechanical, thermal or thermoelectric energy (*Peltier* effect).

4.3 Heat Transfer Through the Vapour Compression Cycle

Heat pumps generally use the vapour compression cycle for heat transfer (Sarbu and Sebarchievici 2015). The term vapour refers to a gas that under standard or ambient conditions of pressure and temperature is in a non-gaseous state. Gases are much more compressible than liquids and, therefore, it is convenient to use a fluid heat transfer that is in the gaseous phase during compression.

4.3.1 Ideal Vapour Compression Cycle

Heat pumps use the ideal vapour compression heat transfer cycle based on the reverse Carnot cycle, which makes use of the following basic components: a compressor, an evaporator, a condenser and an expansion system. The compressor is the main element of vapour compression heat pumps and is responsible for applying mechanical energy to generate artificial heat source and sink and, thus, enhance heat transfer between two systems. The mechanical energy required to perform compression work can have different origins, the most common being the use of electrical energy in compressors with electric motors. Almost anecdotally, there are very specific cases in which compression work is carried out using mechanical, thermo-mechanical, thermal or thermo-electric energy (Peltier effect).

Figure 4.2 shows the schematic diagram of a heat pump operating for refrigeration on the vapour compression cycle. The cycle takes place in a closed heat pump circuit, called the primary circuit. The primary circuit includes the compressor, where the compression of a gaseous heat carrier takes place. The compression of the gaseous heat carried causes its temperature to rise from approximately 3 °C at 1.7 bar to 73 °C at 13.5 bar, thus increasing its temperature significantly.

The *hot* compressed gas, resulting from compression, circulates through the primary circuit until it is introduced into a heat exchanger. The heat exchanger puts the primary circuit fluid in contact with the secondary circuit fluid, which is connected to an external heat reservoir, in this case at a lower temperature. The *hot* high-pressure gas from the primary circuit, during its transit through the heat

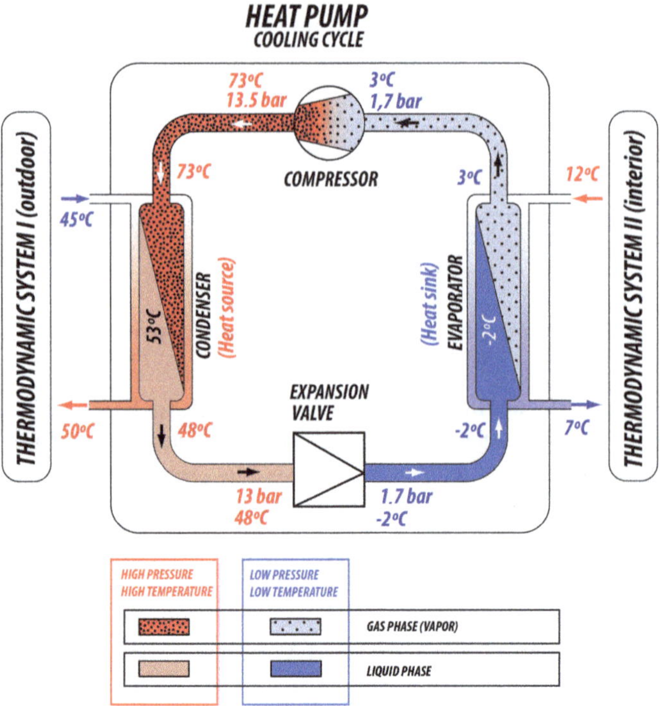

Fig. 4.2 Schematic diagram of a heat pump operating for refrigeration using the vapour compression cycle

exchanger, releases sensible heat to the secondary circuit and reduces its temperature. During heat dissipation and the consequent drop in temperature (at this constant pressure), condensation temperature is reached at around 53 °C, depending on the chemical composition of the heat carrier fluid, and the gas begins to condense, turning into a liquid phase. When it starts to condense inside the heat exchanger, the sensible heat transferred until then becomes latent heat and all the latent heat transferred to the secondary circuit will be used to change phase (from gas to liquid). Consequently, the temperature during the condensation process will be constant until all the gas condenses (isothermal process). Once all the gas has condensed, if heat continues to be transferred to the secondary circuit, the heat transferred will again be sensible heat and the temperature of the liquid (formerly gas) in the primary circuit will continue to decrease to approximately 48 °C, maintaining the pressure at 13.5 bar.

The heat exchanger where gas condensation occurs during heat dissipation with the secondary circuit is called a condenser. This acts as a heat sink for the primary circuit, but also as a heat source for the secondary circuit. The secondary circuit introduces a heat carrier fluid at approximately 45 °C and leaves at 50 °C. The gas–liquid phase change (condensation), with the consequent latent heat transferred, allows the temperature in the condenser to remain constant at the condensing temperature (at 13.5 bar), thus maintaining a higher thermal gradient between the primary and secondary circuits during heat transfer. If no phase change occurrs, the temperature in the primary circuit in the heat exchanger continues to decrease even more and, consequently, would decrease the heat transfer efficiency between the two circuits. Note that as the temperature of the heat transfer fluid in the primary circuit decreases and the temperature of the heat transfer fluid in the secondary circuit increases, the temperature difference required for heat transfer between the two circuits within the heat exchanger becomes smaller.

The liquid heat carried from the primary circuit at the condenser outlet, although at a lower temperature than when entering, still has a considerable temperature and maintains the pressure of 13.5 bar. Under these conditions, the liquid passes to the expansion valve and the fluid, liquid at 13.5 bar and 48 °C, undergoes an expansion (negative work) and passes from 1.7 bar to −2 °C, also in a liquid phase.

The liquid heat carrier fluid, which leaves the expansion valve at low pressure and temperature, can be configured as a heat sink with respect to another secondary circuit connected to the indoor domestic thermodynamic system. Therefore, this low-temperature fluid is driven through the primary circuit to another heat exchanger where, in this case, the fluid absorbs heat from the domestic secondary circuit. The fluid from the primary circuit in the heat exchanger absorbs heat and increases its temperature to its boiling point at 1.7 bar. This heat exchanger is called an evaporator. During boiling, the liquid heat transfer fluid absorbs latent heat and the process is kept at a constant temperature. The secondary circuit coming from the internal system enters the evaporator at a temperature of 12 °C and leaves at 7 °C. The evaporator acts as a heat source for the primary circuit but as a heat sink for the secondary circuit.

Finally, the fluid gas leaving the evaporator has a temperature of approximately 3 °C, maintains a pressure of 1.7 bar and is directed through the primary circuit to the compressor, where the cycle is closed.

From the compressor outlet to the expansion valve inlet, the entire primary circuit has been maintained at a constant high pressure of 13.5 bar and at a high temperature, although it has varied. In the rest of the primary system, a low pressure of 1.7 bar has been maintained at a low temperature, although it also has varied. These domains within the primary circuit are represented with red and blue colours respectively in Fig. 4.2.

The heat transferred between the heat reservoir and the indoor system is not direct but requires the primary circuit, where the inverse Carnot cycle takes place and two secondary circuits transport the heat by advection to the systems. Therefore, the vapour compression heat transfer process is considered an indirect process. That is, three thermodynamic subsystems are needed to transfer heat between two thermodynamic systems if a heat pump with vapour compression heat transfer is used.

The basic vapour compression cycle (Fig. 4.3a) includes an isentropic compression, process 1-2 Fig. 4.3b, from saturated vapour at evaporating pressure to condensing pressure, without overheating of the vapour or undercooling of the liquid. The compression process is designed as a function of the heat transfer compound to go from pressure (p_{ev}) and temperature (T_{ev}) of evaporation to pressure (p_{cnd}) and temperature (T_{cnd}) of condensation. According to the second law of thermodynamics (Eq. 2.45) a process is isentropic when it is reversible adiabatic. Obviously, this is an idealised process where there is no heat loss during compression, no friction, no inelastic deformation, etc. The isentropic compression process 1-2 is represented as a fully vertical path in the T-S diagram (Fig. 4.3b).

The *hot* vapour from the compressor is fed into the condenser (Fig. 4.4) where the 2-2′ isobaric vapour cooling process takes place, giving off sensible heat from the compressor outlet temperature (T_2) to the condensing temperature ($T_{2'} = T_{cnd}$). When the condensing temperature is reached, the 2′-3 process of isobaric and isothermal condensation takes place (Fig. 4.3c). During the phase change latent heat transfer the process remains isothermal. Figure 4.4 represents a steam-liquid shell and tube heat exchanger. High-pressure and high-temperature steam is introduced and comes into contact with tubes through which a lower temperature cooling fluid circulates. The vapour is cooled to condensing temperature (process 2-2′) and condenses (process 2′-3) on the tube surface. As liquid accumulates in the tube, as vapour condenses, it drips into a pool of liquid heat carried fluid at the base of the condenser. The liquid from the base is pumped and continues through the primary circuit to the expansion valve.

The liquid heat transfer fluid leaving the condenser and is fed back to an expansion valve where the isentropic expansion process (3-4) takes place. The isentropic process is represented as a fully vertical path in the p-H diagram (Fig. 4.3c). According to Eq. 2.18 it follows that:

$$H_2 - H_1 = U_2 - U_1 + p(V_2 - V_1) \tag{4.1}$$

A priori, from this equation it is recognised that the pressure is constant during a process where the enthalpy (H) is calculated. In the case of an ideal gas flowing

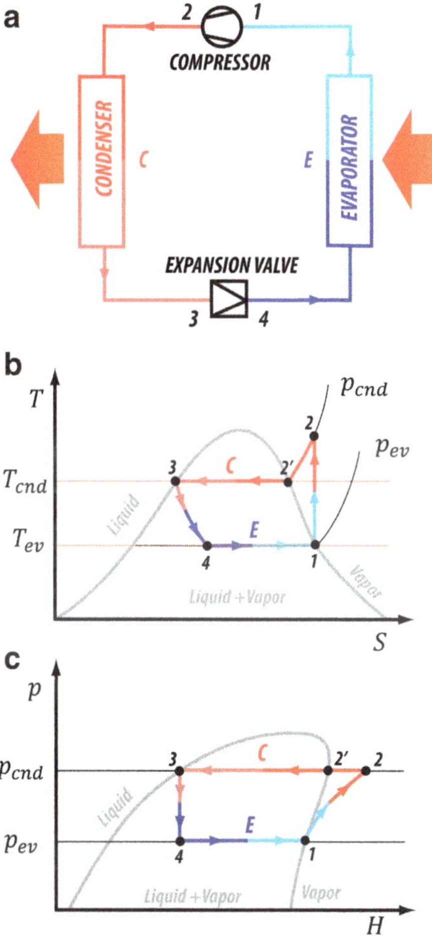

Fig. 4.3 **a** Simplified schematic of the vapour compression heat transfer cycle process along the primary circuit of a heat pump and its representation in the **b** temperature-entropy diagram and **c** pressure-enthalpy diagram

through a valve, it can be seen that the inlet pressure is greater than the outlet pressure. In general, when a fluid expands from a high-pressure region to a low-pressure region, work is done and changes in the potential and kinetic energy of the gas occur. When a throttling valve is used, a restriction is imposed on the flow, which causes a decrease in pressure, no work is done and the heat transfer is generally small. The throttling process and the temperature and pressure drop is governed by the so-called *Joule–Thomson effect* equation (Wark 1994). During the throttling process, under adiabatic conditions, the decrease in cross section causes an increase in fluid velocity (*Venturi* effect) so that mass conservation is maintained and the temperature of the system decreases by allowing it to expand freely while keeping the enthalpy constant. The

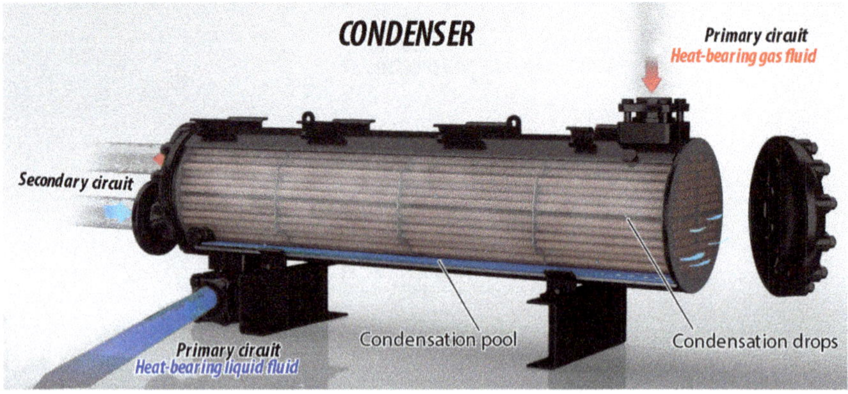

Fig. 4.4 Schematic of a shell and tube heat exchanger in condenser mode operation. Modified from https://www.bitzer.de/

pressure loss is compensated for by the increase in volume, producing the effect of a free expansion at constant pressure where the internal energy of the system remains unchanged. The pressure loss, from around 11 bar down to 1.7 bar, drops the temperature of the heat transfer fluid down to −2 °C. The low-temperature and low-pressure fluid is finally subjected to a new heat exchange with another secondary circuit. This uses the same heat exchanger as seen in the condensation process, but the heat flow is reversed. In process 4-1, the fluid is introduced into the heat exchanger at a low temperature and absorbs heat isobarically, resulting in boiling at the condensation temperature and low pressure. The latent heat absorbed makes the process 4-1 isothermal as well as isobaric (Fig. 4.3b). The fact that boiling of the liquid takes place makes the heat exchanger an evaporator (although the boiling phenomenon takes place). The liquid is converted to vapour by absorbing latent heat from a second secondary circuit and the low-pressure gas phase vapour (isobaric process) leaving the evaporator is passed to the compressor to close the cycle (Fig. 4.3) (Cengel and Boles 2006).

During reversible adiabatic (isentropic) compression, the first law of thermodynamics (Eq. 2.9) indicates that the change in internal energy expressed as enthalpy in the p-H diagram (Fig. 4.3b) will be equal to the work of compression:

$$W' = H_2' - H_1' \tag{4.2}$$

where W' [J kg^{-1}] is the specific work of compression, H_2' [J kg^{-1}] is the specific enthalpy at point 2, H_1' [J kg^{-1}] is the specific enthalpy at point 1. The heat per unit mass of heat transfer fluid to be transferred in the condenser Q_{cnd}' [J kg^{-1}] and in the evaporator Q_{ev}' [J kg^{-1}] is given by:

$$Q_{cnd}' = H_2' - H_3' \tag{4.3}$$

$$Q'_{ev} = H'_1 - H'_4 = H'_1 - H'_3 \tag{4.4}$$

The coefficient of performance (COP) as the ratio between the thermal energy transferred and the mechanical work done by the heat pump is given by:

$$COP = \frac{Thermal\ energy\ transferred}{Mechanical\ work} = \frac{Q'_{cnd}}{W'} = \frac{H'_2 - H'_3}{H'_2 - H'_1} \tag{4.5}$$

The thermal power for cooling (P_{RF}) [W] and heating (P_{CF}) [W] of the heat pump is:

$$P_{RF} = m_{fm} Q'_{cnd} \tag{4.6}$$

$$P_{CF} = m_{fm} Q'_{cnd} \tag{4.7}$$

where m_{fm} [kg s^{-1}] is the mass flow rate of heat carrier fluid. The thermal power to realise the isentropic compression process P_{ie} [kW] can be expressed as:

$$P_{ie} = m_{fm} W' \tag{4.8}$$

Since there are always energy losses in the process, a dimensionless efficiency factor η_{ie} is introduced to determine the thermal power to realise the effective isentropic compression process (P_{ef}) [kW]:

$$P_{ef} = \frac{1}{\eta_{ie}} P_{ie} \tag{4.9}$$

4.3.2 Real Vapour Compression Cycle

In the real vapour compression cycle, during vapour compression, friction and other irreversible processes take place which, together, make this process irreversible ($\Delta S > 0$). The inefficiently applied energy during compression means that the internal energy of the compressed gas does not reach point 2 in the diagram above (Fig. 4.3c) but only reaches point 2sc (Fig. 4.5), so that condensing pressure and temperature design conditions are not achieved. To reach these design conditions, additional work is required ΔW. However, the additional work will allow to reach a higher internal energy at point 2 (Fig. 4.5), which means that an enthalpy increase in the condensation process equal to ΔQ_{cnd} (Fig. 4.5) will be produced.

The heat loss during compression is sufficiently low that the actual compression is still considered adiabatic. In any case, a real total work W^r is required to achieve the same pressure and temperature conditions as in the ideal cycle. Therefore, the

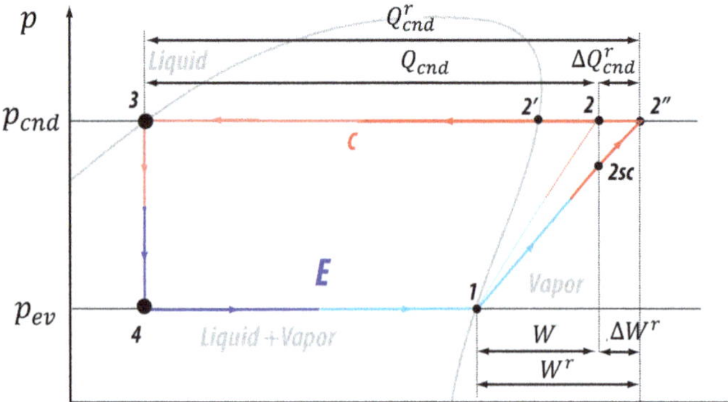

Fig. 4.5 Pressure–temperature diagram of the real vapour compression heat transfer cycle along the primary circuit of a heat pump

reversible work of the ideal cycle W is always less than W^r since the real process is always less efficient. The ideal isentropic compression cycle represents maximum efficiency and this is the reference used to evaluate the efficiency of the real irreversible compression process. The adiabatic compressor efficiency coefficient η_{ad} can be defined as:

$$\eta_{ad} = \frac{W}{W^r} = \frac{W^r - \Delta Q_{cnd}}{W^r} = 1 - \frac{\Delta Q_{cnd}}{W^r} \tag{4.10}$$

The real work will always be greater than the ideal ($W^r > W$) to compensate for the loss of energy used in unavoidable irreversible processes, so that $\eta_{ad} < 1$. η_{ad} will tend to unity as the more efficient the actual compression process is. The existence of irreversible processes makes it necessary to correct the coefficient of efficiency defined above in order to obtain the real coefficient of efficiency η_r:

$$\eta_r = \frac{Q^r_{cnd}}{W^r} = \frac{Q_{cnd} + \Delta Q_{cnd}}{W + \Delta W} = \frac{Q_{cnd} + \Delta Q_{cnd}}{W + \Delta Q_{cnd}} \tag{4.11}$$

By subtracting ΔQ_{cnd} from Eq. 4.10, taking the effective work to be $W^r = W + \Delta Q_{cnd}$ and considering the resulting expression into Eq. 4.11, gives

$$\eta_r = \frac{Q^r_{cnd}}{W^r} = \frac{Q_{cnd} + W\left(\frac{1}{\eta_{ad}} - 1\right)}{W + W\left(\frac{1}{\eta_{ad}} - 1\right)} \tag{4.12}$$

Combined with Eq. 4.5:

$$\eta_r = (COP - 1)\eta_{ad} + 1 \tag{4.13}$$

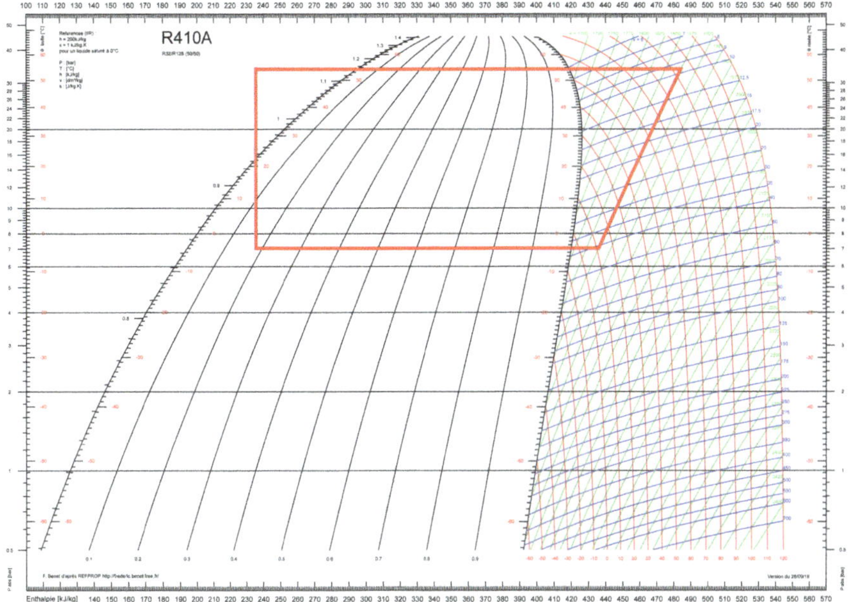

Fig. 4.6 P-H diagram for heat transfer fluid R410A. Red lines represent isotherms, green isentropic lines and blue isochoric lines. *Source* http://frederic.benet.free.fr

Note that $\varepsilon_r < COP$. A further coefficient η_{el} is added to the thermodynamic efficiency losses for losses inherent to the electric motor (electrical and mechanical resistivities) which supplies the mechanical energy, thus allowing calculation of the COP_{el}:

$$COP_{el} = \eta_{el}\eta_r \tag{4.14}$$

Each heat transfer fluid will have different thermodynamic properties and, therefore, different p-H and T-S plots. A real example for heat transfer fluid R410A is shown in Fig. 4.6.

4.4 Reversibility

Heat pumps have historically been used in cooling mode as the main component of chillers and air conditioners. Today, the heat pump concept is considered for cooling or heating, depending on the direction of flow of the heat carrier fluid in the primary circuit. By incorporating a four-way valve, heat pumps can transfer heat in both directions without altering the operation of the compressor. Note that the compressor not only compresses the fluid but also provides the pressure difference

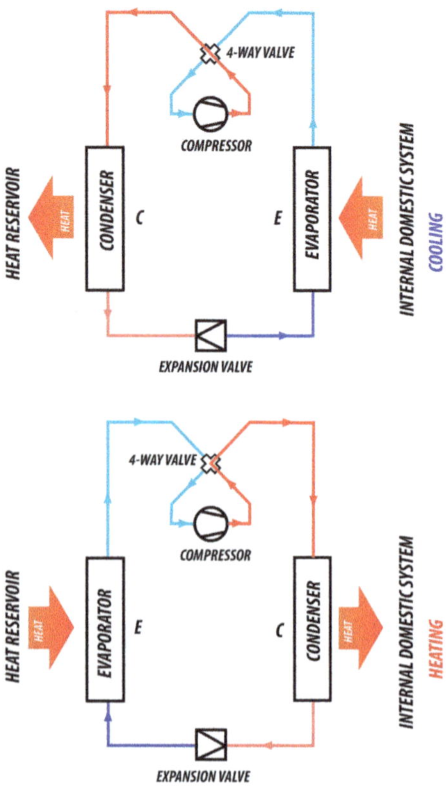

Fig. 4.7 Reversible cooling and heating cycles using a four-way valve

to circulate the fluid throughout the primary circuit. Figure 4.7 shows a schematic of the heating and cooling cycles using a four-way valve. The two heat exchangers in a heat pump's primary circuit are structurally the same and, depending on the direction of flow of the heat carrier fluid, act as condenser or evaporator.

4.5 Operating Mode of Heat Pumps

A heat pump operates according to the heat dissipation or heat absorption demand of a thermal installation. There are infinite temporal distributions of heat demand at different scales, from a minutal to an annual scale. It is common for the temporal distribution of demand to be cyclical in nature, with frequencies and amplitudes that will depend largely on the type of activity of the thermal installation. For example, supermarkets, hospital operating theatres, or data processing centres may have a large cooling demand throughout the year on a constant basis, while a holiday centre may have a large seasonal cooling demand concentrated over a few months of the year.

A heat pump will be able to satisfy any demand of any thermal installation, both cooling and heating, as long as the required supply temperatures are within the operating range of a heat pump. As seen in the vapour compression cycle, the condensing temperature has a limit (critical point) beyond which vapour condensation is not possible. It is common for a conventional heat pump to have a maximum heat transfer fluid production temperature of 55–65 °C. This temperature is sufficient for air conditioning and DHW supply in most common buildings and infrastructures, especially if there is adequate thermal insulation.

In thermal installations where a supply temperature range outside the range provided by a heat pump is required, the combined use of heat pumps with other complementary heat production systems is possible. It is common for large-scale thermal installations, which use heat pumps as a heat production system, to be designed in such a way that they cover the base demand and there is a complementary production system to meet peak demand. Depending on this need for supplementary production systems, the following operating modes of a heat pump are defined: monovalent mode, bivalent mode (parallel or alternating) and mono-energy mode.

When a heat pump is the only type of production system in a thermal installation, where the existing demand is completely satisfied throughout the year, it is considered to be operating in a *monovalent regime*; this is a situation where the temperature demand is approximately in the range of −2 to 65 °C. Most geothermal heat pumps operate in a monovalent regime.

A heat pump operates in a *bivalent mode* if it is part of a thermal installation with two or more heat production systems. One of the heat production systems is the heat pump itself and at least one of the other heat production systems is based on combustion methods. In addition, if both systems operate simultaneously, the bivalent regime is called parallel. If both production systems work alternately, the bivalent regime is called alternating.

Heat pumps operating in a *mono-energy mode* are heat pumps that operate together with another heat production system in the thermal installation, where the second heat production system is also an electrically driven heat pump.

Heat pumps make it possible to build up the safest and most economical thermal installations, when they are properly designed by adapting the available thermal reservoir and the desired distribution system to the thermal energy demand. They constitute the most energy efficient option, consuming minimal energy during operation, and they are the optimal solution in the energy design of buildings and facilities (Sarbu and Sebarchievici 2015). Thermal installations with reversible heat pumps, capable of providing both cooling and heating, are more effective than thermal installations with traditional production systems. In the case of renovation of production systems with pre-existing traditional systems, the cost effectiveness is not trivial and needs to be assessed on a case-by-case basis. The biggest challenge for heat pumps today is to achieve production temperatures above 55 °C, and to reach the 70–90 °C needed in some thermal installations with high heating demand, or even to meet sanitary requirements for DHW production.

4.6 Performance

The COP defined in Eq. 4.5 relates the thermal energy transferred by a heat pump to the mechanical (electrical) work done by the compressor of the heat pump itself. As deduced in Sect. 2.1.4 (Eq. 2.43) the efficiency of the Carnot cycle, the thermodynamic basis of the vapour compression cycle in heat pumps, ultimately depends on the temperature difference between the systems in which heat is transferred. In the case of heat pumps, this refers to the difference between the temperature of the heat reservoir outside the installation and the temperature of the indoor domestic system (Fig. 4.8).

Thermal installations operating with an external heat reservoir, in which temperature varies over time, and an internal domestic system, with a temperature that remains constant, will experience varying efficiency over time as a function of the temperature in the external heat reservoir. If the internal system temperature tends to decrease, the heat pump will operate in heating mode, and if it tends to increase, it will operate in cooling mode. This happens in the air-conditioning of buildings using air-to-water or air-to-air heat pumps, making use of the atmosphere as an external heat reservoir where, as an example, its temperature changes from 3 °C in winter to 35 °C in summer.

It is therefore appropriate to specify whether the COP is an annual average or whether it corresponds to a period of heating or cooling mode operation.

A representative energy efficiency indicator for the entire heating season can be calculated using the *Seasonal Coefficient of Performance* (SCOP) defined as:

$$SCOP = \frac{E_{CF}}{W} \tag{4.15}$$

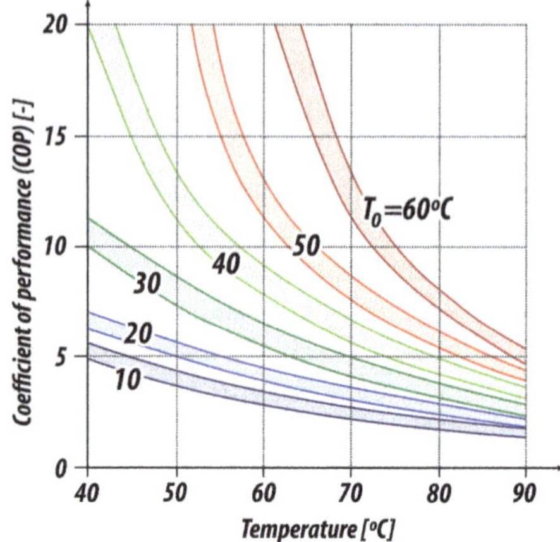

Fig. 4.8 Variation of heat pump efficiency (COP) as a function of the temperature of the heat reservoir (T_0) and the temperature of the internal domestic system of the thermal installation (T). Data: Sârbu and Sebarchievici (2010)

where E_{CF} [J] is the heat energy transferred by the heat pump in heating mode, i.e., the energy transferred from the external heat reservoir to the domestic system during the heating season throughout a year of operation. W [J] is the mechanical work done by the compressor to transfer that thermal energy in that period. The SCOP index is the reference value used for energy certification and legislation (European Regulation 626/2011). The index takes into account that the operating regime of a thermal installation is different during the year. The industry sector makes an approximation, taking into account an operation at different loads of 100, 74, 47 and 21% of its capacity. The calculation method divides the *cold season* into a number of hours, with different temperatures of the external heat reservoir. The calculation method can be found in the UNE 14825-2014 standard.

For the cooling season, the *Seasonal Energy Efficiency Ratio* (SEER) for cooling is used, defined as (Purushothama 2009):

$$SEER = \frac{E_{RF}}{W} \tag{4.16}$$

where E_{RF} [J] is the thermal energy transferred by the heat pump operating under cooling mode, i.e., the thermal energy transferred from the domestic system to the external heat reservoir during the cooling season throughout a year of operation. W [J] is the mechanical work done by the compressor to transfer that thermal energy in that period. Since each system has a different operating regime during the year, the industry sector makes an approximation, taking into account operation at different loads of 100, 74, 47 and 21% of its capacity for outdoor heat reservoir temperatures of 35, 30, 25 and 20 °C respectively, with an indoor temperature of the domestic system always at 27 °C. In addition, the SEER of a heat pump determines its cooling energy rating. Systems with SEER > 8.5 have the highest energy rating, while a SEER < 3.6 has the lowest energy rating, with other intermediate ratings for intermediate ranges. The UNE EN 14825 standard establishes the specific calculation methodology for SEER, weighted by partial loads.

4.7 CO_2 Emissions

Considering E_T [kWh] as the total thermal energy transferred by a heat pump over a period of time, typically one year, and W_T [kWh] is the total electrical energy consumed by the heat pump as work to perform the heat energy transfer under consideration. The CO_2 emissions [kg] emitted (G_{HP}) during the operation of a heat pump shall only be those linked to the production of thermal energy consumed. They can be simply calculated by:

$$G_{HP} = W_T \xi \tag{4.17}$$

where ξ [kg kWh^{-1}] is the *emission factor* linked to the production of electrical energy, indicating the kg of CO_2 emitted into the atmosphere when producing one kWh of electrical energy, for example, the kg of CO_2 emitted by a coal-fired power plant to produce one kWh of electricity.

Note that this approach neglects other CO_2 emissions linked to the life cycle of a heat pump, such as CO_2 emissions from the production of heat carrier fluids used in the vapour compression cycle, installation of the pump, or drilling of geothermal boreholes. As there are more GHGs in addition to CO_2, the emission factor can be expressed as kgCO_2 kWh^{-1} specifically or by expressing all GHGs as kilogram CO_2 equivalents.

It is common for heat pumps to use electrical energy from the general grid, produced from various sources (electrical *mix*), of n sources, where each source has an emission factor. Therefore, the emissions from heat pumps (G_{HP}) using this type of network can be calculated as follows:

$$G_{HP} = \gamma_1 W_T \xi_1 + \gamma_2 W_T \xi_2 + \cdots + \gamma_n W_T \xi_n = \sum_{i=1}^{n} \gamma_i \cdot W_T \xi_i \tag{4.18}$$

where γ_i and ξ_i are the fraction and emission factor of each source of electricity production, respectively. With $\sum_{i=1}^{n} \gamma_i = 1$. By taking the W_T as common factor:

$$G_{HP} = W_T(\gamma_1 \xi_1 + \gamma_2 \xi_2 + \cdots + \gamma_n \xi_n) = W_T \sum_{i=1}^{n} \gamma_i \xi_i \tag{4.19}$$

The effective emission factor ξ_{ef} [kgCO_2 kWh^{-1}] can be defined as:

$$\sum_{i=1}^{n} \gamma_i \xi_i = \xi_{ef} \tag{4.20}$$

$$G_{HP} = W_T \xi_{ef} \tag{4.21}$$

The average effective CO_2 emission factor in Europe ranges from 0.04 kgCO_2 kWh^{-1} in Sweden to 1.08 kgCO_2 kWh^{-1} in Luxembourg, and the European average is 0.486 kgCO_2 kWh^{-1} (Castells 2012). This average almost matches the Spanish emission factor of 0.480 kgCO_2 kWh^{-1} (IEA 2015). If the heat pump replaces a production system of a thermal installation, the savings in CO_2 emissions can be calculated as (Bayer et al. 2012):

$$G_{sav} = G_{sps} - G_{HP} \tag{4.22}$$

where G_{sps} [kgCO_2 kWh^{-1}] is the CO_2 emissions of the substituted production system.

4.8 Types of Heat Pumps

Heat pumps can be classified according to their use as heating heat pumps, DHW heat pumps, air conditioners, chillers, ventilation machines, heat pump dryers, heat recovery units, etc. They can also be classified according to the heat reservoir they exploit in order to transfer heat to the domestic system, distinguishing between air heat pumps, exhaust air heat pumps (EAHP), surface water heat pumps, groundwater heat pumps (GWHP), ground source heat pumps (GSHP), among other types (Ochsner 2012). It is also very common to classify heat pumps according to the type of heat transfer fluids used, both at source in the thermal reservoir exploited and in the distribution system of the thermal installation, distinguishing between water-water heat pumps, brine-water, direct expansion-water, air-water, air-air, etc. Finally, in the commercial field, heat pumps are classified according to their construction characteristics: compact heat pumps, indoor, outdoor, number of compression stages, etc.

The following text provides an overview of the most common types of vapour compression heat pumps on the market for residential, industrial, commercial and community buildings.

Air-to-air heat pumps: These are the most popular and mass-produced heat pumps in the industry in compact form. Both the evaporator and the condenser of a heat pump are of the *thin fin* type, designed to optimise metal-to-air surface heat transfer. They are designed to exploit a gaseous reservoir, usually atmospheric air, and transfer conditioned air to an interior space.

It is important to note that the outdoor heat reservoir of atmospheric air fluctuates seasonally and inversely to the demand of buildings. If the outdoor temperature rises in summer, the cooling demand increases and, consequently, the heat transfer capacity is significantly reduced due to reduced energy efficiency.

Air-to-air heat pumps can also be commonly found operating as waste heat recovery unit (WHRU) from air handling units (AHUs) in large thermal installations.

Air-to-water heat pumps: These are used in thermal installations where atmospheric air is to be used as an external heat reservoir, but the heat is distributed through the thermal installation as a liquid fluid, improving efficiency and reducing the noise produced by fans in overhead distribution circuits. Generally, one of the heat pump heat exchangers is installed outside and is connected via heat transfer fluid piping to the central structure of the thermal installation. This type of heat pump is mainly found working in bivalent operations together with heating, cooling, DHW production and heat recovery systems. The heat pump covers 75–80% of the demand and the peaks of demand are produced by more traditional means such as electric heaters or diesel or natural gas boilers.

Water-to-water (brine-to-water) heat pumps: water-to-water heat pumps are those where both the heat carrier fluid of the operating distribution system, which interacts with one of the heat exchangers of the heat pump (evaporator or condenser), and the fluid used to distribute the heat through the domestic system, which interacts with the other heat exchanger of the heat pump (evaporator or condenser), are liquid water. These pumps can circulate surface water (rivers, lakes, sea, etc.) or groundwater directly, through one of the exchangers (evaporator or condenser), or indirectly, by adding a water-water plate heat exchanger to avoid heat pump deterioration and environmental contamination by the heat pump refrigerant (which can be polluting). Brine-to-water heat pumps refer to the same type of pump as water-to-water heat pumps but, unlike water-to-water heat pumps, they are coupled to the ground via a geothermal heat exchanger (external heat exchanger system), where a U-shaped pipe goes into the ground and connects to the heat pump. Brine refers to the fact that brine was previously used at low temperature to absorb heat from the ground. These days, a mixture of water and antifreeze is used. Both types of heat pumps are used for monovalent heating and cooling, heat recovery and DHW production.

Several water-to-water heat pumps can be used together in the same thermal installation, supplying the heat exchange coils of an AHU. This type of heat pump application has the advantages of greater control over heat production, efficient centralised maintenance, and it provides redundancy and flexibility when dealing with a given operating problem.

Direct expansion heat pump-water: one of the heat exchangers of the heat pump itself is introduced into the heat reservoir. In other words, part of the primary circuit is introduced into the thermodynamic system to be thermally exploited. In this way, the efficiency losses of having an additional heat exchanger are removed. This type of heat pump was initially used in shallow geothermal systems with closed-loop geothermal heat exchangers. The fact that both the heat carrier fluids used in the compression cycle and the compressor lubrication oil can be considered pollutants led to their obsolescence.

High temperature heat pump: In newer thermal installations, with efficient thermal insulation and low temperature internal heat exchange systems, single stage heat pumps are sufficient and suitable, ideal for radiant floors/ceilings and walls (35 °C). Such pumps deliver heat transfer fluid at temperatures up to approximately 55 °C, and some medium temperature pumps (also single stage) can produce temperatures up to approximately 65 °C. When pushed to the limit, they can provide temperatures usable in most existing radiators; however, at the limit of operating conditions, their efficiency is questionable compared to condensing boilers. This also explains the efficiency losses when producing DHW from the heat pump alone. In case an old heating installations (high temperature radiators, poor thermal insulation, etc.) or industrial heating installations demands higher temperatures (70–90 °C) high temperature heat pump are required. There are a number of new generation heat pumps capable of supplying up to 80–90 °C, and −25 °C in cooling mode operation (Aikins et al. 2013). The technology developed uses two cascade two-stage vapour compression cycles with different heat transfer fluids (R410A and R134a).

4.9 Geothermal Heat Pumps

These days, ground source heat pumps are becoming increasingly important in the thermal sector of air conditioning of buildings and industrial heat production (heating and cooling), as well as other applications. This is due to their optimal characteristics in terms of energy efficiency and environmental quality compared to other alternatives (Egg and Howard 2010). They use high-efficiency technology based on the vapour compression cycle, coupled to a shallow geothermal reservoir.

The shallow geothermal reservoir (<400 m) behaves approximately like an ideal reservoir. This is defined as a thermodynamic system, with sufficient thermal capacity so that when finite heat transfer occurs the temperature effectively remains constant, and it can therefore act as both a heat source and a heat sink (Cengel and Boles 2006). As explained in Chap. 3, the thermal regime of the soil deeper than 10–15 m remains constant, with values close to two degrees Celsius above the annual average air temperature of the locality under consideration. In other words, the subsurface from this depth is in thermal equilibrium with the climate (Oke 1987), with a small constant effect of two degrees Celsius, as a result of the geothermal gradient. This situation does not occur in the ambient air thermal reservoir. When cooling demand occurs in the air conditioning of a building in summer, the temperature of the ambient air reservoir is increased. The same problem occurs in winter. If it is taken into account that the efficiency of a heat pump depends on the temperature of the thermal reservoir and the target temperature of the internal system to be heated or cooled (Fig. 4.8), it is easy to understand how heat pumps coupled to a stable thermal reservoir all year round will give better performance than if coupled to an unstable reservoir with variations of temperature. The most direct result of this thermal stability is that in summer the ground is *cooler* and in winter it is *warmer* than the air outside a building or infrastructure.

The general concept of a ground source heat pump refers to the coupling of a water-to-water heat pump to the ground. It is a standard heat pump based on the vapour compression cycle and, although it may include minor modifications such as adaptations introduced by manufacturers, the essential thermodynamic cycle described in previous sections of this chapter remains unchanged. The mode of ground coupling refers to the type of heat exchanger used to transfer heat with the ground. There are two main types of geothermal heat exchangers (ASHRAE 2016):

- *Closed-loop geothermal heat exchangers*: those based on a closed-loop (secondary circuit of the heat pump) where a controlled heat transfer fluid is recirculated by means of an impulsion pump. Most of the closed circuit is introduced into the ground through a vertical borehole, which enables heat transfer with the ground.
- *Open-loop geothermal heat exchangers*: those based on an open-loop where the heat carrier fluid is the groundwater itself. The circuit is called open-loop because it exchanges mass. The groundwater in equilibrium with the ground (Jacob 1950) is circulated through one of the heat exchangers of the heat pump (condenser or evaporator) and is then injected into the source aquifer. It is common to find a

plate heat exchanger in the operating distribution system to prevent groundwater from entering the heat exchanger of the pump.

Open-loop geothermal heat pumps have a SCOP for heating of approximately 3.0–4.0 and EER for cooling of 11.0–17.0. For closed system heat pumps, the SCOP for heating ranges from 2.5 to 4.0, and the EER for cooling ranges from 10.5 to 20.0 (Heinonen et al. 1997).

References

Aikins KA, Lee S-H, Choi JM (2013) Technology review of two-stage vapor compression heat pump system. Int J Air-Conditioning Refrig 21(03):1330002. https://doi.org/10.1142/s2010132513300024

ASHRAE (2016) Ashrae handbook 2016: HVAC systems and equipment, SI edn. ASHRAE

Bayer P, Saner D, Bolay S, Rybach L, Blum P (2012) Greenhouse gas emission savings of ground source heat pump systems in Europe: a review. Renew Sustain Energy Rev 16(2):1256–1267. https://doi.org/10.1016/j.rser.2011.09.027

Castells XE (2012) Energía, Agua, Medioambiente, territorialidad y Sostenbilidad. Editorial Díaz de Santos, S.A.

Cengel YA, Boles MA (2006) Thermodynamics: an engineering approach. McGraw-Hill Higher Education, New York

Egg J, Howard BC (2010) Geothermal HVAC. McGraw-Hill Education, New York

Heinonen EW, Tapscott RE, Wildin MW, Beall AN (1997) Assessment of antifreeze solutions for ground-source heat pump systems. In: American Society of Heating, Refrigerating and Air-Conditioning Engineers (ASHRAE) annual meeting. American Society of Heating, Refrigerating and Air-Conditioning Engineers, Inc., Boston, MA (United States), p 1072

IEA (2015) Energy policies of IEA countries: Spain 2015 review. International Energy Agency, Paris. 978-92-64-23924-1

Jacob CE (1950) Flow of groundwater in engineering hydraulic. In: Rouse H (ed) Engineering hydraulics. Wiley, New York, pp 321–386

Ochsner K (2012) Geothermal heat pumps: a guide for planning and installing. Taylor & Francis, London

Oke TR (1987) Boundary layer climates. Routledge, London

Purushothama B (2009) Maintenance of humidity. In: Purushothama B (ed) Humidification and ventilation management in textile industry. Woodhead Publishing India, pp 121–161. http://doi.org/10.1533/9780857092847.121

Sârbu I, Sebarchievici C (2010) Heat pumps—efficient heating and cooling solution for buildings. WSEAS Trans Heat Mass Transf 5:31–40

Sarbu I, Sebarchievici C (2015) Ground-source heat pumps: fundamentals, experiments and applications. http://doi.org/10.1016/b978-0-12-804220-5.0000

Wark K (1994) Advanced thermodynamics for engineers: solutions manual. McGraw-Hill Companies, New York

Chapter 5
Shallow Geothermal Systems with Closed-Loop Geothermal Heat Exchangers

Shallow geothermal installations that use a geothermal heat exchanger with a closed loop in which a heat carrier fluid is recirculated (Fig. 5.1) are referred to as *ground source heat pump* (GSHP) systems. A general characteristic of these systems is their ubiquity. They are designed to operate in any terrain, independent of the geology and/or hydrogeology of the surrounding area. Closed systems can be located in almost any environment, from hard rocks (granite, quartzite, etc.) to unconsolidated sediment, even in artificial terrain of an anthropogenic nature.

5.1 General Characteristics

There are two main types of closed-loop systems, depending on the characteristics of the geothermal heat exchanger used: *direct expansion* (DX) and *indirect expansion* systems (Fig. 5.2). The former are pioneering systems that were widely used in the past, while the latter currently dominate the European markets. Systems with direct expansion geothermal heat exchangers circulate the heat pump's own heat carrier (primary circuit) into the ground via a closed loop. In these systems, the closed loop is buried in the ground, acting as the heat pump's evaporator/condenser and consists of a metal pipe, usually made of copper. The heat carrier fluid used is usually a fluorocarbon refrigerant (e.g., R407c). Installations with indirect expansion geothermal heat exchangers add a second closed circuit external to the heat pump (secondary circuit), which is buried in the ground through which an aqueous heat carrier fluid is recirculated.

In both types of installations, there is a part of the circuit (primary or secondary) buried in the subsoil. The closed-loop circuit, when introduced in vertical boreholes, conform the so called *borehole heat exchanger* (BHE), which consists of a vertical borehole where a pipe is inserted, generally in the shape of a U, and filled with thermally enhanced grout.

A. García Gil et al., *Shallow Geothermal Energy*, Springer Hydrogeology,
https://doi.org/10.1007/978-3-030-92258-0_5

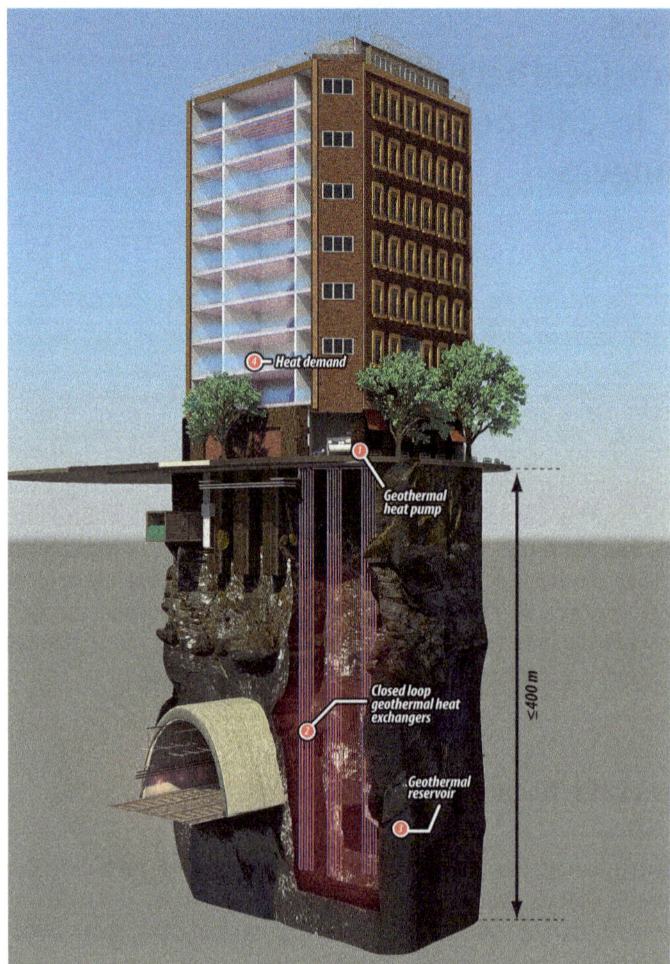

Fig. 5.1 Shallow geothermal system with closed-loop geothermal heat exchangers

DX-GSHP systems (Soni et al. 2016) tend to have a higher efficiency than closed indirect expansion systems due to the extreme temperatures of the operating fluid in the primary circuit, which can reach values below −10 °C in heating mode and more than 60 °C in cooling mode. Another advantage of these systems is that there is a single heat exchange process; the heat from the ground is transferred with the heat pump in a single step, thus avoiding secondary heat losses. In addition, the metal pipe of the closed circuit has a higher thermal conductivity than the PVC of indirect expansion systems. A disadvantage is the possible loss of oil from the compressor when oil enters the circuit, especially in deep vertical boreholes where the oil accumulation causes additional frictional forces in the refrigerant flow, which hinders the circulation of the refrigerant. Fluorocarbons are considered to be persistent organic

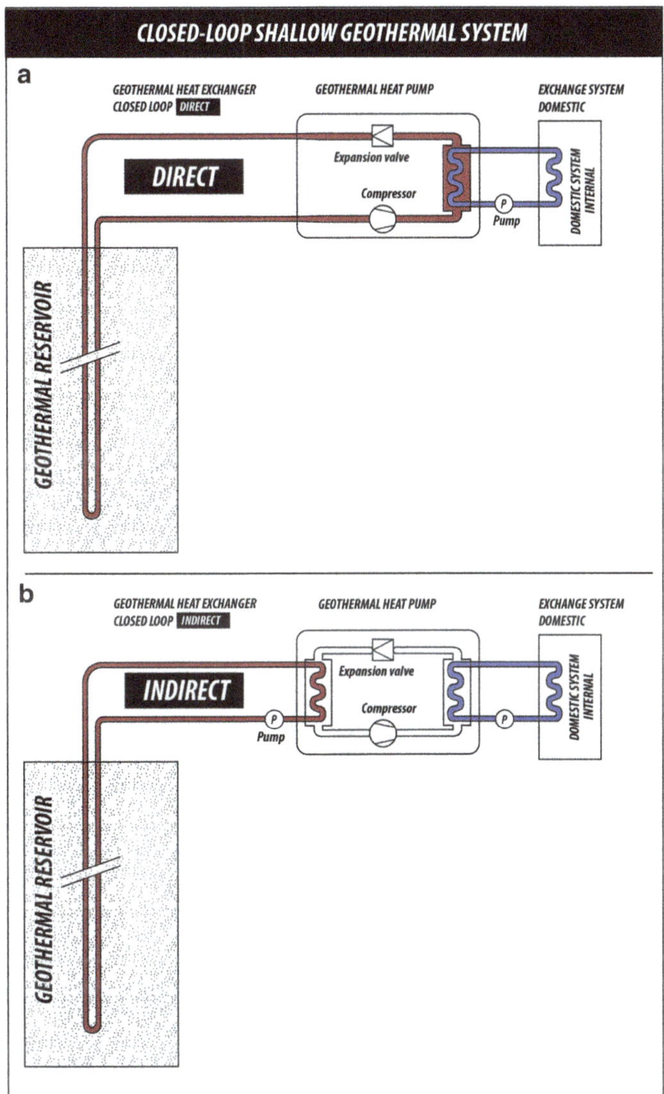

Fig. 5.2 Typology of shallow geothermal installations with closed loop geothermal heat exchangers: direct expansion or DX (**a**) and indirect expansion (**b**)

pollutants in the environment. Their recirculation in underground metallic pipes poses a potential contamination risk. Mechanical stresses from the ground on the structural integrity of the pipes can cause leakage of the circuit and, therefore, a loss of pollutant. Also, in certain redox, acid-base or salinity environments, corrosion phenomena also threaten the tightness of the DX-GSHP closed circuit. Although DX-GSHP systems are the most efficient, this environmental disadvantage has caused them to have fallen out of use compared to indirect systems in some countries. In other countries, such as Austria, direct expansion closed-loop systems are very popular. To overcome the rejection of the use of polluting refrigerants, new systems are being developed that experiment with the use of CO_2 as heat carrier fluid (CO_2-DX-GSHP), combined with stainless steel piping (Gao et al. 2017b).

Indirect expansion systems use alcohols and glycols as antifreeze agents. These compounds biodegrade rapidly compared to fluorocarbons and avoid the above-mentioned problems of possible environmental contamination.

5.2 Closed-Loop Geothermal Heat Exchangers

Closed-loop shallow geothermal systems use one or more geothermal heat exchangers to transfer thermal energy with the ground. The term closed-loop geothermal heat exchanger refers to an assembly of PVC pipes constituting a closed loop that is installed, usually, within a vertical borehole in the ground for heat exchange (Fig. 5.3). The vertical borehole consists of a borehole approximately 150 mm in diameter and 100–200 m deep. A pipe assembly of approximately 30 mm inside diameter is inserted into the borehole space in a U-shape (Fig. 5.3a). In most

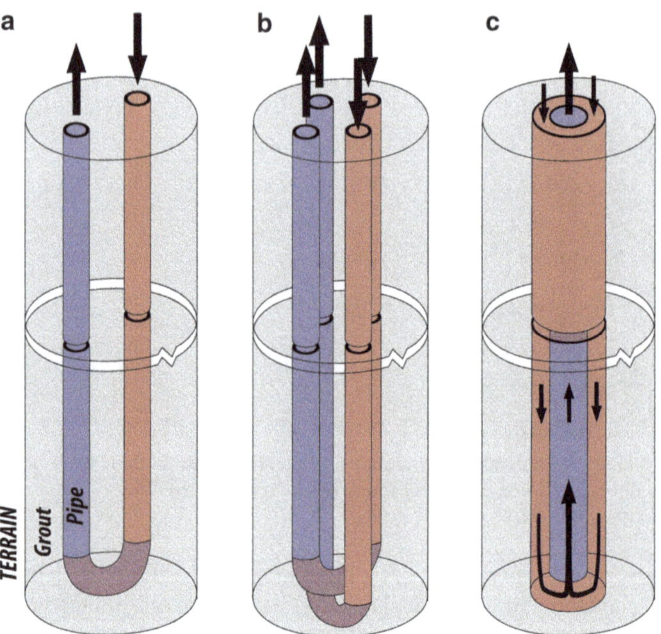

Fig. 5.3 Different pipe assembly designs of closed-loop geothermal heat exchangers in vertical boreholes. **a** Single U-exchanger, **b** double U-exchanger and **c** annular coaxial exchanger. The arrows show the flow direction of the recirculated heat carrier fluid

cases it is filled with thermally enhanced grout, which consists of a bentonite-cement mixture. When the grout solidifies, it forms a thermoactive material that facilitates the sealing of the exchanger, preventing groundwater interactions, and improves the contact between the pipes and the ground. However, in hard rock soils, the grout may not be introduced and the annular space between the rock and the pipe assembly is left free, filled only by existent groundwater. Once the borehole has been completed, the pipe assembly is inserted and, by adding the grout, the heat exchanger is constructed. It is not recommended to use the drilling cuttings to fill the annular space; it is recommended to first review local regulations and guidelines for the construction of geothermal heat exchangers (Hiller et al. 2000; IGSHPA 2009; UNE 2014). It is advisable to use an impermeable material that hydraulically isolates the pipe circuit from the ground in order to avoid environmental problems including: (1) seepage of contaminants from the surface, (2) uncontrolled transfer of water between different aquifer levels, (3) uncontrolled transfer of contaminants between different aquifer levels and (4) elimination of artesian conditions of confined aquifers and vertical circulation of water along the geothermal heat exchanger.

In addition, drill cuttings are not a material designed to optimise heat transfer, as thermally enhanced grout presents: (1) relatively high thermal conductivity, (2) relatively low viscosity for pumping to the bottom of the borehole and filling the annular space during installation, (3) complete filling of the annular space and prevention of voids and air bubbles, (4) firm adherence to HDPE pipe, (5) little or no settling or shrinkage after placement, and (6) maintenance of its thermal properties over the long term (Chiasson 2016). The grout is introduced into the borehole using the *Tremie* method, which involves introducing an auxiliary pipe (Fig. 5.4a), at the same time as the heat exchanger pipe assembly is introduced, to pump the grout to the bottom of the borehole and backfill from the bottom up. The grout slurry is pumped in as the Tremie pipe is withdrawn. This ensures that the grout is pumped in from the bottom, displacing any mud particles or debris from the borehole wall. Exchangers where proper grouting is not done could result in a loss of heat transfer efficiency of the constructed geothermal heat exchanger, possible groundwater contamination processes and development of subsidence phenomena.

5.2.1 Types of Geothermal Heat Exchangers

There are different types of pipe assembly configurations within the borehole that result in different types of closed-loop geothermal heat exchangers. Figure 5.3 shows the most commonly used assemblies.

The first major and most commonly used group is the single U-shaped assemblies. In these, a single internal U-shaped pipe is placed in the centre of the borehole (Fig. 5.3a). The heat carrier fluid is introduced at one end of the pipe and simply exits at a different temperature at the other end. All the remaining borehole space is filled with thermally enhanced grout.

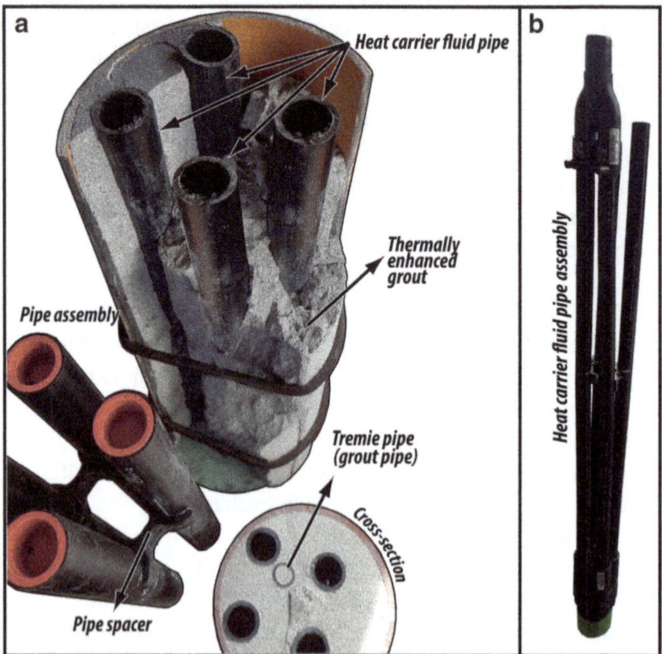

Fig. 5.4 Middle section of a geothermal double U-shaped geothermal heat exchanger (**a**). Example of bottom and top termination of the heat carrier fluid circulation pipe assembly for a double U-shaped geothermal heat exchanger design (**b**)

Another group is known as double U-shaped assemblies (Fig. 5.3b). In this case, the assembly is made up of a double pipe in parallel forming a U, thus obtaining two ends of pipe where the heat carrier fluid is introduced and two ends where it leaves the assembly (Figs. 5.3b and 5.4). This type of exchanger provides an advantage in thermal performance due to the fact that there is a second circuit through which the heat circulates efficiently by forced convection. It is much more efficient than the thermally enhanced grout that it replaces in the borehole. Double U-shaped geothermal heat exchangers predominate in Switzerland where they are the most widely used assembly (Rybach and Sanner 2000). The application of these exchangers improves their efficiency by reducing the thermal resistance of the exchanger, allowing the drilling depth of boreholes to be reduced by 22% compared to the use of single U-exchangers (Yavuzturk and Chiasson 2002).

A third group of assemblies use coaxial piping. Coaxial heat exchangers are called *annular* when the heat carrier fluid is introduced into the annular space and extracted through the central pipe (Fig. 5.3c). They are called *central* when the direction is reversed, the heat carrier fluid is introduced through the central pipe and extracted through the annular space of the pipe (Diersch et al. 2011).

Geothermal heat exchangers for vertical boreholes can be used to dissipate thermal energy when the fluid entering the heat exchanger is at a higher temperature than the ground (cooling mode). They can also be used to absorb thermal energy from the ground if the inlet fluid is at a lower temperature (heating mode).

At present, it is recommended that single U-shaped assemblies use only *high-density polyethylene* pipe (HDPE) due to its high flexibility, low cost, and relatively acceptable thermal properties. It is very resistant to wear and tear and has a longer durability (>30 years) than metallic pipes, as it is resistant to corrosion. In addition, HPDE pipes allow thermofusion welding, resulting in a very strong joint. The first geothermal heat exchangers used copper piping but, after experimentation with other materials, it was concluded that the thermal properties of the underground heat exchanger pipe were a minor factor in determining its overall efficiency in heat transfer with the ground. The heat carrier fluid in the geothermal heat exchanger is a mixture of water and 20–25% antifreeze in order to operate at temperatures below the freezing point. Commonly used antifreezes mixed with water are ethylene glycol, propylene glycol, methanol or salt. Increasing the mixing ratio will result in freezing points at lower temperatures.

The most commonly used polyethylene pipe has a 26–40 mm external diameter. The heat carrier fluid is pressurised to 2–3 bar during operation of the heat exchanger, but pressurisation tests are usually carried out at 10–16 bar. The sizing of the pipe diameter is a process of optimising the flow rate, friction losses and the power of the heat pump to which the heat exchanger is connected. The optimal pipe diameter of the geothermal heat exchanger will generate a turbulent flow regime, a condition that improves the heat exchange between the pipe wall (ground) and the heat carrier fluid in the closed circuit. However, turbulent flow increases frictional pressure losses compared to laminar flow, and a turbulent regime with minimum pressure losses is desirable. It must be taken into account that, when using antifreeze at different mixing ratios, the density and viscosity of the resulting heat carrier fluid in the circuit will change. The Reynolds number (Re) is a dimensionless parameter relating the inertial forces to the viscous forces present in fluid flow. For a fluid flowing in a pipe, the Re is given by Eq. 2.143. For values of $Re > 4000$, the fluid will have turbulent behaviour. In addition, both the density and dynamic viscosity of the fluid will change with temperature, with density and viscosity generally tending to decrease with increasing temperature (Fig. 5.5). Dynamic viscosity is particularly sensitive to temperature; in the specific case of water, it is approximately twice as viscous at 0 °C as it is at 25 °C.

The turbulent flow and head loss must be optimized in order to have a sufficient flow rate to carry the heat for which the heat pump exchanger is designed. A typical flow rate is ~3 L min^{-1} kW t^{-1}.

5.2.2 *Grids of Closed-Loop Geothermal Heat Exchangers (BHEs)*

In large installations (>12 kW) it is common to drill several boreholes to install several geothermal heat exchangers (BHEs). In installations where it is necessary

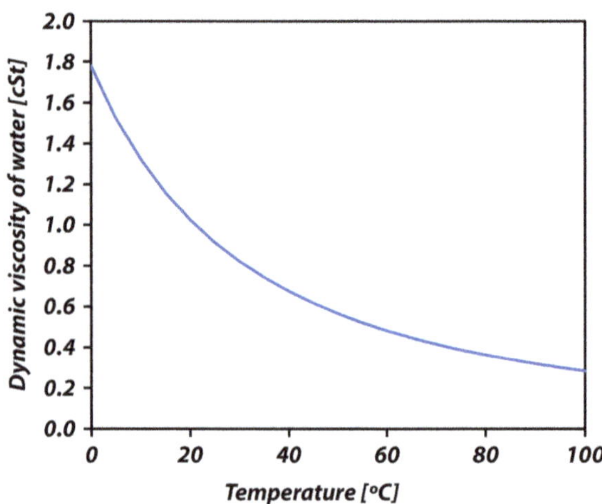

Fig. 5.5 Dynamic viscosity of water as a function of temperature. Data from Hellstrom (1991)

to build a large number of BHEs, grids of vertical boreholes of a given depth are constructed, allowing heat transfer to be maximised in the available land area.

With more and deeper BHEs, the more heat exchange/storage capacity is available for a given installation. When the land area is limited, the geothermal heat exchangers tend to be deeper. To determine the number and depth of boreholes required for a given area and heat demand, the following factors must be taken into account: (1) the thermal-hydraulic properties of the ground, (2) the length of the heating and cooling periods, (3) the degree of imbalance between cooling and heating demand over the year, (4) the spacing between BHEs, (5) the temperature change over which the geothermal exchangers are designed to operate, and (6) the temporal distribution of heat demand. These factors, each of considerable complexity, require case-to-case optimization. To make an initial approximation, a rule of thumb that is wide-spread in the shallow geothermal industry in Europe considers 50 W m^{-1} as peak power per linear metre of borehole for heat absorption and 70 W m^{-1} for heat dissipation. This means that a heat exchanger, which typically has a depth of between 40 and 200 m, would have between 2 and 10 kW for heating and 2.8–14 kW for cooling. Therefore, an initial estimate could be made of the number of geothermal heat exchangers needed to achieve the total thermal output of a given system. This rule of thumb is broadly consistent with the average values observed in closed-loop systems in the UK (Fig. 5.6) and the USA, whose ranges of peak heat transfer rates per linear metre are between 40–100 kW m^{-1} and 68–82 kW m^{-1} respectively.

For example, Figs. 5.7 and 5.8 show four systems with a heat exchange capacity of 100 kW, where two systems reasonably match the basic rule of 50–70 kW m^{-1} using 15 heat exchangers (1400–1600 m boreholes in total). However, a third system will use 20 exchangers (2100 m drilled in total) and a fourth system only 10 exchangers (850 m drilled in total), meaning that the third system, assuming the simple rule, would have built boreholes of 500–700 m less depth to meet its needs and the fourth

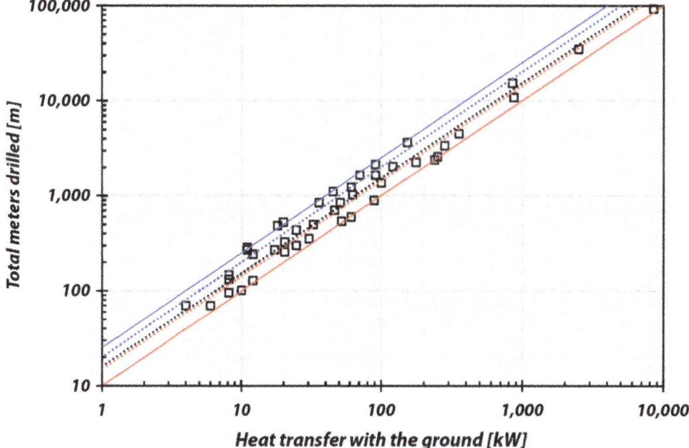

Fig. 5.6 Relationship between heat transfer rates and total drilled metres of geothermal heat exchangers of GSHP systems in the UK. The thermal power per linear metre of minimum (40 kW m^{-1}), average (65 kW m^{-1}) and maximum (100 kW m^{-1}) borehole heat output observed in the systems is shown as a linear relationship. Lines corresponding to the European standard of 50 kW m^{-1} and 70 kW m^{-1} for heating and cooling, respectively, have been added. Modified from Banks (2012)

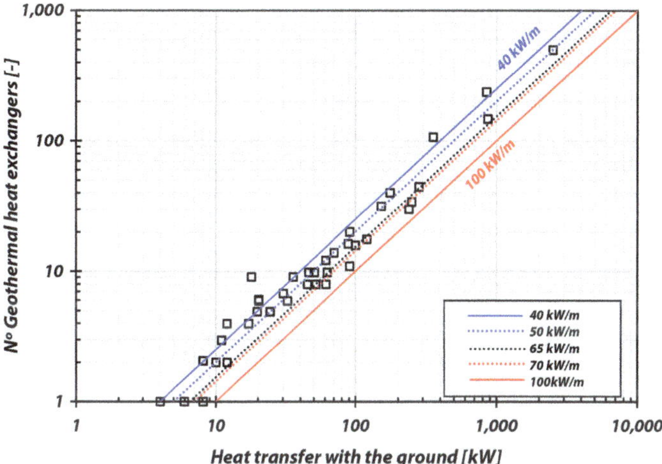

Fig. 5.7 Relationship between heat transfer rates and total number of geothermal heat exchangers of GSHP systems in the UK. The minimum (40 kW m^{-1}), average (65 kW m^{-1}) and maximum (100 kW m^{-1}) heat output per linear metre of borehole observed in the systems is shown as a linear relationship. Lines have been added for the European standard of 50 kW m^{-1} and 70 kW m^{-1} for heating and cooling, respectively, for 100 m deep heat exchangers (average depth). Modified from Banks (2012)

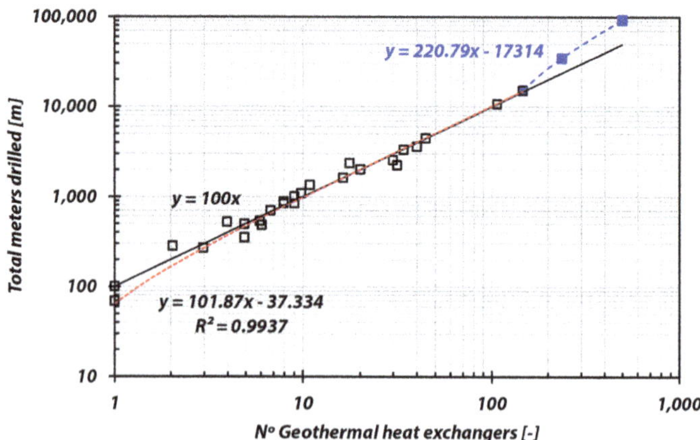

Fig. 5.8 Ratio between the number of heat exchangers and the total drilled metres of geothermal heat exchangers of GSHP systems in the UK. Modified from Banks (2012)

system would have built exchangers of 550–750 m (almost twice as much as built). It is true that the error is a constant percentage and, therefore, that error in a small domestic 12 kW system, adding a small safety factor, drilling a few extra metres is more cost effective than doing a specific design study. According to the UK experience (Fig. 5.6), drilling three 100 m deep BHEs could meet the peak demand of a 12 kW system.

This is the simplistic approach that some installers of closed-loop geothermal heat exchangers in domestic installations may take. However, these rules arise from statistical calculations of previous experiences and are, therefore, subject to the specific conditions of the geological, hydrogeological and climatic environment from which this information is extracted.

5.2.3 *Drilling Systems in the Construction of Geothermal Heat Exchangers*

Knowledge of the lithology of the site where a geothermal heat exchanger is to be located has implications for the drilling system chosen. Very briefly, two main groups of lithology can be distinguished: hard rocks (granites, basalts, etc.) and unconsolidated sediment. It is necessary to consider both the lithology outcropping on the surface and the lithologies to be traversed, as it is common to find several geological formations that require different instrumentation, depending on the sections to be drilled. In unconsolidated soils, it is necessary to case the borehole through the use of auxiliary drill pipes (casing), which prevents the borehole from collapsing. The casing can sometimes complicate the operation, in some cases requiring reduction in diameter as the borehole is deepened. Although the casing can be recovered on completion of the geothermal heat exchanger, in some cases this is not possible. When the casing is removed, the borehole collapses and the ground material comes into contact with the HPDE pipe directly. The increased number of manoeuvres involved in laying the casing increases the cost of boreholes in unconsolidated sediment.

There are two main types of drilling methods: rotary drilling (Auger, mud rotary, sonic, among others) and percussion drilling (with cable tool rigs and rotary-percussion drilling) (Table 5.1). The characteristics and rigs of each of the aforementioned drilling methods can be found in the specialised literature.

Table 5.1 Main characteristics of commonly used drilling methods

Property	Percussion drilling	Auger drilling	Rotary drilling	Air rotary drilling	Rotary-percussion drilling	Sonic drilling
High feed rate (m/day)	No	Yes	Yes	Yes	Yes	Yes
Suitable for sampling geological material	Yes	Yes	Yes	No	No	Yes
Capital investment, high maintenance and operating costs	No	No	Yes	Yes	Yes	Yes
Significant water supply needs	No	No	Yes	No		
Working capacity in hard rock	No	No		Yes	Yes	Yes
Working capacity in unconsolidated sediment without cobbles	Yes	Yes	Yes	No	No	Yes
Working capacity in unconsolidated sediment with cobbles	Yes	No	Yes	No	No	Yes
Limited to shallow depths (< 30 m)	No	Yes	No	No	No	No
Risk of aquifer contamination	No	No	Yes	No	No	
Cake formation in borehole (negative skin effect)	No	No	Yes	No		
Market experience	Yes	Yes	Yes	Yes	Yes	No
Reliability	Yes					
High number of operators and specialisation	No		Yes	Yes		
Working in remote locations	Yes					
Minimum diameter		Yes				
Stability of borehole walls			Yes	Yes		

5.3 Heat Transfer in Closed Geothermal Heat Exchangers

Closed systems transfer thermal energy with the ground by means of closed-loop geothermal heat exchangers, described in Sect. 5.2.1. The heat transfer mechanism is complex, as it involves both heat conduction and advection (forced convection) of heat in a porous media with several components. These components include both intrinsic components of the porous media and components of the heat exchangers, including piping, moving heat carrier fluid and thermally enhanced grout. The exchanger geometry can become complex (Fig. 5.1), where each component interacts with the others in a non-trivial way in a transient regime. Each component is in direct contact with one or more components of the heat exchanger-terrain system and, indirectly, with the others. Broadly speaking, the most important factors that determine the efficiency of heat transfer between the heat exchanger itself and the ground are the size of the borehole and the spacing between pipes (shank spacing). The inherent complexity of heat transport within a geothermal heat exchanger results in different mathematical models of varying complexity, depending on the simplifications adopted. In general, it is accepted that it is a heat flow problem in a multicomponent system, made up of components between which there is no thermal equilibrium (Aifantis 1979; Nield and Bejan 1999) analogous to the one presented in Sect. 2.3.1 (see Fig. 2.9).

To understand the geothermal exchanger models, it is necessary to describe the thermal regime in multi-component systems. To do that, the heat flow model in a single-component system will be described, and then others will be introduced.

The differential equation of heat conduction for a stationary, homogeneous, isotropic, single-component system where no work is applied, and there is no internal heat generation, can be obtained from the energy balance in a control volume (Fig. 5.9a). Applying the first law of thermodynamics, the net heat transfer will be equal to the internal energy change:

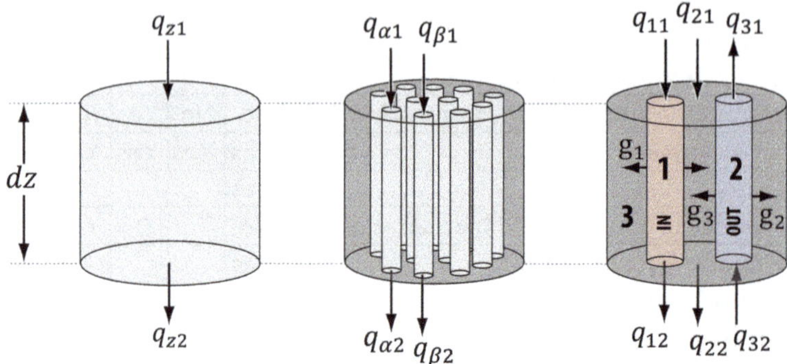

Fig. 5.9 Differential control volume for the one-dimensional flow in a single-component system (**a**), a multi-component system of *n* components (**b**) and a three-component system (**c**)

$$\frac{dU}{dt} = \frac{dQ}{dt} \tag{5.1}$$

where dU/dt [W] is the change in internal energy over time and dQ/dt [W] is the net heat flux transferred into the control volume. The net heat transferred to the control volume will be the heat flow out minus the heat flow over a given period of time:

$$\frac{dQ}{dt} = (q_{z1} - q_{z2}) \tag{5.2}$$

Taking into account Fourier's Law (Eq. 2.50), the incoming conduction heat flow in the direction of z (q_{z1}) to the control volume will be:

$$q_{z1} = -\lambda_z A \frac{\partial T}{\partial z} \tag{5.3}$$

where A [m] is the cross section, λ_z [W m^{-1} K^{-1}] is the thermal conductivity in the z direction and $\partial T/dz$ [K m^{-1}] is the temperature gradient in the z direction. The heat flow out of the control volume q_{z2} [W m^{-1} K^{-1}] in the z direction will depend on the heat flow at the point $z = dz$, at the outlet of the control volume. As λ_z and A are constant, a heat flux change along dz would be a consequence of a change in the temperature gradient at the point $z = dz$. To estimate q_{z1} at the point $z = dz$ using Taylor series, the polynomial $P(z)$ very similar to q_{z2} is obtained:

$$q_{z2} \cong P(z) = q_z(dz) + \frac{\partial q_z(dz)}{\partial z}(z - dz) + \cdots + \frac{\partial^{(n)} q_z(dz)}{\partial z}\frac{1}{n!}(z - dz)^n \tag{5.4}$$

Taking the first two terms:

$$q_{z2} \cong q_z(dz) + \frac{\partial q_z(dz)}{\partial z}(z - dz) \cong q_z(0) + \frac{\partial q_z(0)}{\partial z}(0 - dz) \tag{5.5}$$

That is to say:

$$q_{z2} \cong q_{z1} + \frac{\partial q_{z1}}{\partial z} dz \tag{5.6}$$

$$q_{z2} \cong -\lambda_z \frac{\partial T}{\partial z} A + \frac{\partial}{\partial z}\left(-\lambda_z \frac{\partial T}{\partial z} A\right) dz \tag{5.7}$$

The net heat transferred to the control volume in the z direction:

$$q_{z1} - q_{z2} \cong \frac{\partial}{\partial z}\left(-\lambda_z \frac{\partial T}{\partial z}\right) A dz = -\lambda_z \frac{\partial^2 T}{\partial z^2} A dz \tag{5.8}$$

The net heat transferred to the control volume is:

$$\frac{dQ}{dt} = -\lambda_z \frac{\partial^2 T}{\partial z^2} A dz \tag{5.9}$$

If thermal energy storage occurs, the temperature of the system will change accordingly. The quantity relating the energy required to increase by 1 K is the specific heat capacity c [J kg^{-1} K^{-1}]. When the change of internal energy occurs at constant volume, the specific heat at constant volume c_V is defined as:

$$c_V = \left.\frac{\partial u}{\partial T}\right|_V \Rightarrow u = c_V \partial T + u_{ref} \tag{5.10}$$

where u_{ref} is the arbitrary reference internal energy, which may be zero for simplicity. For an incompressible solid, the heat capacity at constant volume and pressure is equal to the heat capacity at constant pressure $c_V = c_P = c$. Note that u [J kg^{-1}] and U [J], therefore $U = u\, dm$. The control volume is $dV = A\, dz$. Taking into account the density ρ [kg m^{-3}], dm can be expressed as:

$$\rho = \frac{dm}{dV} \Rightarrow dm = \rho\, A\, dz \tag{5.11}$$

And therefore:

$$\frac{dU}{dt} = \frac{d(u\, dm)}{dt} = \frac{\partial T}{\partial t} c \rho A dz \tag{5.12}$$

Substituting Eqs. 5.9 and 5.12 into Eq. 5.1 gives the general equation for heat conduction in one dimension:

$$c\rho \frac{\partial T}{\partial t} = -\lambda_z \frac{\partial^2 T}{\partial z^2} \tag{5.13}$$

In addition, heat transfer by the convection mechanism can be considered (see Sect. 2.2.4) and the convection term (Eq. 2.88) is added to Eq. 5.13:

$$c\rho \frac{\partial T}{\partial t} = -\lambda_z \frac{\partial^2 T}{\partial z^2} + \frac{\partial T}{\partial z} v_z \rho_f c_f \tag{5.14}$$

where v_z is the fluid velocity and the subscript f refers to the fluid carrying thermal energy. Considering that q is the heat flux transported by both conduction and convection, then:

$$q = -\lambda_z \frac{\partial T}{\partial z} + v_z \rho_f c_f T \tag{5.15}$$

Equation 5.14 can be rewritten in a more compact form by taking Eq. 5.15 into account as follows:

$$c\rho \frac{\partial T}{\partial t} = \frac{\partial q}{\partial z} \tag{5.16}$$

With Eq. 5.16 the equation for heat conduction and convention is obtained for a simple system with a single component and in one dimension. Including a source term g [W m^3] the equation is:

$$c\rho \frac{\partial T}{\partial t} + \mathrm{g} = \frac{\partial q}{\partial z} \tag{5.17}$$

5.3.1 Heat Transfer Equation for Multicomponent Systems

If other components such as the simple U-shaped pipe of a BHE are to be taken into account, it is necessary to expand Eq. 5.17 for multi-component systems. The energy balance of a control volume with n components (Fig. 5.9b) can be studied proceeding analogously to the one-component system. The multicomponent system will have n components at temperature $T_1, T_2, \ldots, T_n$ with net heat fluxes transferred in each component of $q_1, q_2, \ldots, q_n$ and their respective source/sink terms $\mathrm{g}_1, \mathrm{g}_2, \ldots, \mathrm{g}_n$.

Equation 5.17 only takes into account heat transport in the z dimension. In order to take into account heat transfer within the control volume between the different components, heat transfer is introduced as a source term. Thus, e.g., q_1 will refer to the net heat transferred to the component $\alpha = 1$ in the direction z and the source/sink term g_1 will refer to the net heat transferred between the component $\alpha = 1$ and the other components within the control volume. Regardless of the specific interactions between components in each component, Eq. 5.16 must be fulfilled. Assuming the α types of pipes with n components with $\alpha = 1, 2, \ldots, n$. For the n pipes of a system:

$$\frac{\partial T_\alpha}{\partial t} c_\alpha \rho_\alpha + \mathrm{g}_\alpha = \frac{\partial q_\alpha}{\partial z} \tag{5.18}$$

The energy balance for a system of n components of type α will be given by the sum of the energy balances of the individual components. Taking into account Eq. 5.16:

$$\frac{\partial T_1}{\partial t} c_1 \rho_1 + \mathrm{g}_1 + \frac{\partial q_1}{\partial z} + \frac{\partial T_2}{\partial t} c_2 \rho_2 + \mathrm{g}_2 + \frac{\partial q_2}{\partial z} + \cdots + \frac{\partial T_n}{\partial t} c_n \rho_n + \mathrm{g}_n + \frac{\partial q_\alpha}{\partial z} = 0 \tag{5.19}$$

$$\sum_{\alpha=1}^{\alpha=n}\left(\frac{\partial T_\alpha}{\partial t}c_\alpha\rho_\alpha + g_\alpha + \frac{\partial q_\alpha}{\partial z}\right) = 0 \tag{5.20}$$

If all pipes have the same thermal properties, they can be taken out as a common factor:

$$\begin{aligned} c_1\rho_1\frac{\partial T_1}{\partial t} + c_2\rho_2\frac{\partial T_2}{\partial t} + \cdots + c_n\rho_n\frac{\partial T_n}{\partial t} + g_1 + g_2 + \cdots + g_n \\ + \frac{\partial q_1}{\partial z} + \frac{\partial q_2}{\partial z} + \cdots + \frac{\partial q_\alpha}{\partial z} = 0 \end{aligned} \tag{5.21}$$

$$\begin{aligned} c\rho\left(\frac{\partial(T_1 + T_2 + \cdots + T_n)}{\partial t}\right) + g_1 + g_2 + \cdots + g_n \\ + \frac{\partial(q_1 + q_2 + \cdots + q_n)}{\partial z} = 0 \end{aligned} \tag{5.22}$$

$$c\rho\frac{\partial\left(\sum_1^n T_\alpha\right)}{\partial t} + \sum_{\alpha=1}^{n} g_\alpha + \frac{\partial\left(\sum_1^n q_\alpha\right)}{\partial z} = 0 \tag{5.23}$$

Taking into account Eq. 5.15, q_α is the heat flux transported both by conduction and convection in the z direction for each of the components α:

$$q_\alpha = -\sum_{\beta=1}^{n}\lambda_{\alpha\beta}\frac{\partial T_{\alpha\beta}}{\partial z} + \sum_{\beta=1}^{n} v_{\alpha\beta}\rho_{\alpha\beta}c_{\alpha\beta}T_{\alpha\beta} \tag{5.24}$$

where the subscript $\alpha\beta$ refers to the physical property of the material at the contact of these phases (see example $n = 3$ presented below). It should be noted that convective flow does not necessarily always exist and, therefore, this term should be eliminated. For example, if a component represents a pipe and is in contact with the grout, it is obvious that convective heat flow will not occur. There would not even be conduction because, by limiting the transport to the heat flow in the z direction, there will only be a vertical convective heat transfer along the pipes and horizontal conduction between the pipes and the grout. To define the heat transferred by conduction and convection between the elements, a source/sink term is introduced. It is interpreted as that component experiencing a heat gain or loss, not to other systems in the environment of the control volume, but to the other components. The heat transferred between components can be estimated according to *Newton's Law of Cooling*, which states that the rate of heat transfer of a body is proportional to the temperature difference between a body and its surroundings according to:

$$\frac{dQ}{dt} = -bA\Delta T \tag{5.25}$$

where b [W m^{-2} K^{-1}] is the heat transfer coefficient, A [m^2] is the area where the heat transfer takes place and ΔT [K] is the temperature difference between the body and the surroundings. In general, this law is considered to be valid with small temperature gradients, especially in free convection processes. Nevertheless, it is considered as a reasonable approximation for forced fluid flows in contact with a solid, i.e., in advection processes.

The heat absorbed or dissipated internally by each of the components is determined by the sum of the heat transferred with the other components according to Newton's cooling law per unit area of heat exchange:

$$g_\alpha = \sum_{\beta=1}^{n} b_{\alpha\beta}\left(T_{\alpha\alpha} - T_{\alpha\beta}\right) \tag{5.26}$$

Introducing Eq. 5.23 the expressions obtained in Eqs. 5.24 and 5.26, the multicomponent equation is obtained:

$$c\rho\frac{\partial\left(\sum_{\alpha=1}^{\alpha=n} T_\alpha\right)}{\partial t} + \sum_{\alpha=1}^{n}\sum_{\beta=1}^{n} b_{\alpha\beta}\left(T_{\alpha\alpha} - T_{\alpha\beta}\right) + \frac{\partial\left(\sum_{\alpha=1}^{\alpha=n}\left(-\sum_{\beta=1}^{\beta=n}\lambda_{\alpha\beta}\frac{\partial T_{\alpha\beta}}{\partial z} + \sum_{\beta=1}^{\beta=n} v_{\alpha\beta}\rho_{\alpha\beta}c_{\alpha\beta}T_{\alpha\beta}\right)\right)}{\partial z} = 0 \tag{5.27}$$

5.3.2 *General Heat Transfer Equation for Closed Geothermal Heat Exchangers*

Applying Eq. 5.23 for a three-component system gives the heat balance for a single U-shaped BHE (Fig. 5.3a). Where the first component ($\alpha = 1$) would represent the inlet pipe, the second component ($\alpha = 2$) would represent the outlet pipe, and the third component ($\alpha = 3$) would represent the grout. The heat flow q_α in each component α for a system with three components ($n = 3$) according to Eq. 5.23 will depend on the existing heat transfer with the rest of the components $\beta = 1, 2, 3$ in the z direction:

$$\begin{aligned} q_1 = &-\lambda_{11}\frac{\partial T_{11}}{\partial z} + v_{11}\rho_{11}c_{11}T_{11} - \lambda_{12}\frac{\partial T_{12}}{\partial z} + v_{12}\rho_{12}c_{12}T_{12} \\ &- \lambda_{13}\frac{\partial T_{13}}{\partial z} + v_{13}\rho_{13}c_{13}T_{13} \\ q_2 = &-\lambda_{21}\frac{\partial T_{21}}{\partial z} + v_{21}\rho_{21}c_{21}T_{21} - \lambda_{22}\frac{\partial T_{22}}{\partial z} + v_{22}\rho_{22}c_{22}T_{22} \\ &- \lambda_{23}\frac{\partial T_{23}}{\partial z} + v_{23}\rho_{23}c_{23}T_{23} \end{aligned}$$

$$q_3 = -\lambda_{31}\frac{\partial T_{31}}{\partial z} + v_{31}\rho_{31}c_{31}T_{31} - \lambda_{32}\frac{\partial T_{32}}{\partial z} + v_{32}\rho_{32}c_{12}T_{32} - \lambda_{33}\frac{\partial T_{33}}{\partial z} + v_{33}\rho_{33}c_{33}T_{33} \quad (5.28)$$

Equation 5.28 corresponds to the application of the general equation (Eq. 5.24) for three components $n = 3$. The first component q_1 can be read as the heat balance in the inlet pipe (component $\alpha = 1$) will be the sum of heat conduction and heat advection between the inlet pipe and the inlet pipe itself ($\beta = 1$). That is, the heat transfer along the pipe itself. Plus the heat that is transferred by conduction or advection between the inlet pipe ($\alpha = 1$) and the outlet pipe ($\beta = 2$). Plus the heat transferred by conduction or convection between the inlet pipe ($\alpha = 1$) with the thermally enhanced grout ($\beta = 2$). Note that the defined control volume has dimension only in the z direction, and that all heat transfer between components other than itself corresponds to heat transport in the horizontal plane perpendicular to the direction of the pipes. Consequently, the partial derivatives of T_{12}, T_{13}, T_{21}, T_{23}, T_{31}, T_{32} with respect to z are hereby cancelled. In addition, the velocity of a fluid in a one-dimensional domain defined only as v_{11}, v_{22} and v_{33} exists, representing the velocity of fluid flowing through each of the components in the vertical z direction. The component $\alpha = 3$ represents the thermally enhanced grout with very low hydraulic conductivity, and. v_{33} can be neglected. Therefore, Eq. 5.28 can be rewritten as:

$$\begin{aligned} q_1 &= -\lambda_{11}\frac{\partial T_{11}}{\partial z} + v_{11}\rho_{11}c_{11}T_{11} \\ q_2 &= -\lambda_{22}\frac{\partial T_{22}}{\partial z} + v_{22}\rho_{22}c_{22}T_{22} \\ q_3 &= -\lambda_{33}\frac{\partial T_{33}}{\partial z} \end{aligned} \quad (5.29)$$

Proceeding in the same way, the source/sink terms representing the internal interactions within the system between the different components will be determined by the sum of the heat transferred with the other components, according to Newton's cooling law described in Eq. 5.26. Applying it to the specific case of a single U-shaped BHE:

$$\begin{aligned} g_1 &= b_{11}(T_{11} - T_{11}) + b_{12}(T_{11} - T_{12}) + b_{13}(T_{11} - T_{13}) \\ g_2 &= b_{21}(T_{22} - T_{21}) + b_{22}(T_{22} - T_{22}) + b_{23}(T_{22} - T_{23}) \\ g_3 &= b_{31}(T_{33} - T_{31}) + b_{32}(T_{33} - T_{32}) + b_{33}(T_{33} - T_{33}) \end{aligned} \quad (5.30)$$

All internal heat transfers from each component to each other are zero since they have the same temperature by definition, and $T_{11} - T_{11} = T_{22} - T_{22} = T_{33} - T_{33} = 0$. Therefore, Eq. 5.30 remains:

$$g_1 = b_{12}(T_{11} - T_{12}) + b_{13}(T_{11} - T_{13})$$

$$g_2 = b_{21}(T_{22} - T_{21}) + b_{23}(T_{22} - T_{23})$$
$$g_3 = b_{31}(T_{33} - T_{31}) + b_{32}(T_{33} - T_{32}) \tag{5.31}$$

Finally, by introducing in Eq. 5.23 different thermal properties of the three components, the multi-component equation is obtained:

$$\begin{aligned} c_1\rho_1\frac{\partial T_1}{\partial t} + c_2\rho_2\frac{\partial T_2}{\partial t} + c_3\rho_3\frac{\partial T_3}{\partial t} &= b_{12}(T_{12} - T_{11}) + b_{13}(T_{13} - T_{11}) \\ &+ b_{21}(T_{21} - T_{22}) + b_{23}(T_{23} - T_{22}) \\ &+ b_{31}(T_{31} - T_{33}) + b_{32}(T_{32} - T_{33}) + \lambda_{11}\frac{\partial^2 T_{11}}{\partial z^2} \\ &- v_{11}\rho_{11}c_{11}\frac{\partial T_{11}}{\partial z} + \lambda_{22}\frac{\partial^2 T_{22}}{\partial z^2} - v_{22}\rho_{22}c_{22}\frac{\partial T_{22}}{\partial z} \\ &+ \lambda_{33}\frac{\partial^2 T_{33}}{\partial z^2} \end{aligned} \tag{5.32}$$

Equation 5.32 subscripts can be updated to designate the components where the first component subscript ($\alpha = 1$) as the input pipe would be the subscript i (*input*), the second component representing the output pipe would have the subscript o (*output*) and the third component corresponding to the thermally enhanced grout would have the subscript g (*grout*). Note that the fluid properties in the convective components represent the cooling fluid, which is assigned the subscript $f_i \circ f_o$ depending on whether it is the inlet or outlet cooling fluid respectively, and the subscript also refers to the fluid velocity inside the pipe. All in all:

$$\begin{aligned} (c\rho)_i\frac{\partial T_i}{\partial t} + (c\rho)_o\frac{\partial T_o}{\partial t} + (c\rho)_g\frac{\partial T_g}{\partial t} &= b_{io}(T_o - T_i) + b_{ig}\left(T_g - T_i\right) + b_{oi}(T_i - T_o) \\ &+ b_{og}\left(T_g - T_o\right) + b_{gi}\left(T_i - T_g\right) + b_{go}\left(T_o - T_g\right) \\ &+ \lambda_i\frac{\partial^2 T_i}{\partial z^2} - (vc\rho)_{fi}\frac{\partial T_i}{\partial z} + \lambda_o\frac{\partial^2 T_o}{\partial z^2} \\ &- (vc\rho)_{fo}\frac{\partial T_o}{\partial z} + \lambda_g\frac{\partial^2 T_g}{\partial z^2} \end{aligned} \tag{5.33}$$

The domain of the partial volumes of the components, inlet pipeline (dV_i), the outlet pipe (dV_o) and the thermally enhanced grout (dV_g) allow the integration of the energy transferred between the control volume and the environment. It is necessary to specify the energy exchange surfaces between the internal components: between the inlet pipe and the outlet pipe (dS_{io}), between the outlet pipe and the thermally enhanced grout (dS_{og}), and between the inlet pipe and the thermally enhanced grout (dS_{ig}), etc. Taking these into consideration it follows that:

$$\int_{V_i} (c\rho)_i \frac{\partial T_i}{\partial t} dV_i + \int_{V_o} (c\rho)_o \frac{\partial T_o}{\partial t} dV_o + \int_{V_g} (c\rho)_g \frac{\partial T_g}{\partial t} dV_g$$
$$= \int_{S_{io}} b_{io}(T_o - T_i) dS_{io} + \int_{S_{ig}}^{1} b_{ig}(T_g - T_i) dS_{ig}$$
$$+ \int_{S_{oi}} b_{oi}(T_i - T_o) dS_{oi} + \int_{S_{og}} b_{og}(T_g - T_o) dS_{og}$$
$$+ \int_{S_{gi}} b_{gi}(T_i - T_g) dS_{gi} + \int_{S_{go}} b_{go}(T_o - T_g) dS_{go}$$
$$+ \int_{V_i} \left(\lambda_o \frac{\partial^2 T_i}{\partial z^2} - (vc\rho)_{f_o} \frac{\partial T_i}{\partial z} \right) dV_i$$
$$+ \int_{V_0} \left(\lambda_o \frac{\partial^2 T_o}{\partial z^2} - (vc\rho)_{f_o} \frac{\partial T_o}{\partial z} \right) dV_o$$
$$+ \int_{V_g} \lambda_g \frac{\partial^2 T_g}{\partial z^2} dV_g \tag{5.34}$$

5.3.3 Heat Transfer Equations for the Main Closed Geothermal Heat Exchanger Designs

5.3.3.1 Single U-Shaped Geothermal Heat Exchangers

Note that Eq. 5.34 is the sum of the partial energy balances of each of the components considered. Considering that in the single U-shaped geothermal heat exchanger design (Fig. 5.3a) there is no direct contact between the inlet and outlet pipes, the respective source terms are eliminated. For simplicity and ease of interpretation, Eq. 5.34 can be expressed by omitting the definition of the integration domain and dividing the equation into the different components by expressing three coupled partial differential equations:

$$(c\rho)_i \frac{\partial T_i}{\partial t} = b_{ig}(T_i - T_g) + \lambda_i \frac{\partial^2 T_i}{\partial z^2} - (vc\rho)_{f_i} \frac{\partial T_i}{\partial z} \tag{5.35}$$

Equation 5.35 refers to the energy balance of the partial component in the inlet pipe. The term to the left of the equality refers to the internal energy storage of the component. This term describes how the temperature of the internal pipe will vary as a function of its heat capacity. The first term to the right of the equality refers to the heat source/sink term for internal heat transfer between the inlet pipe and the thermally enhanced grout, as a function of the thermal connectivity (or thermal resistance) and the temperature difference between the pipe and its thermally enhanced grout environment. The second term on the right hand side of the equality refers to heat conduction in the vertical z direction, both through the pipe material and the cooling fluid inside the pipe. The component defined as the inlet pipe is actually a multi-component system of $n = 2$; the HDPE pipe material and the heat carrier fluid. Finally, the third summand of the right-hand term of the equality expresses the advection of heat along the pipe in the z direction.

$$(c\rho)_o \frac{\partial T_o}{\partial t} = +b_{og}\left(T_o - T_g\right) + \lambda_o \frac{\partial^2 T_o}{\partial z^2} - (vc\rho)_{f_o} \frac{\partial T_o}{\partial z} \tag{5.36}$$

Equation 5.36 refers to the energy balance of the partial component of the outlet pipe, and its interpretation is analogous to Eq. 5.35. It should be added to the above description, in the cooling scenario, that the inlet pipe will be larger than the thermally enhanced grout and, therefore, the storage will be negative, dissipating heat to the grout component. The source/sink term in this case will be negative since the inlet pipe temperature will be higher, acting as a heat sink. The conductive and advective term will be positive, since the inlet pipe of the system will tend to lower its temperature due to heat transfer to the thermally enhanced grout. This will generate heat input by conduction and advection. This heat input will tend to be stored in the internal pipe component, so that the heat transfer process to the thermally enhanced grout continues. It should be noted that the temperature of the coolant leaving the system ($z = dz$) will be lower than the inlet temperature. Therefore, if the outlet water from the inlet pipe flows directly into the outlet pipe, and the fluid temperature is still higher than the temperature of the thermally enhanced grout, it will continue to dissipate heat as it passes through the upstream flow in the system dz. Consequently, in the outlet pipe, the heat transfer process will be the same as described in the inlet pipe but of a smaller magnitude, since the temperature difference of the cooling fluid will become smaller with respect to the temperature of the thermally enhanced grout.

$$(c\rho)_g \frac{\partial T_g}{\partial t} = b_{gi}\left(T_g - T_i\right) + b_{go}\left(T_g - T_o\right) + \lambda_g \frac{\partial^2 T_g}{\partial z^2} \tag{5.37}$$

Finally, Eq. 5.37 refers to the heat balance in the thermally enhanced grout. Note that there is an additional source/sink term since this component is in direct contact with the inlet and outlet pipes. In a cooling scenario, both terms would be positive, since the inlet and outlet pipe temperatures are higher than those of the thermally enhanced grout, even though the inlet temperature is higher than the outlet temperature. There is only a conductive term in the direction of z, given that it is a material with very low hydraulic conductivity.

It should be noted that the thermally enhanced grout is the component that brings both pipes into contact with the ground (the environment of the control volume considered). Therefore, if there were a horizontal component in the control volume, the heat flow of the thermally enhanced grout with the ground by conduction and advection in the case of groundwater presence would be expressed.

5.3.3.2 Double U-Shaped Geothermal Heat Exchangers

The double U-shaped geothermal heat exchangers (Fig. 5.3b) have a total of five components: four pipes and the thermally enhanced grout. Applying Eq. 5.27 for $n = 4$ an equation is obtained as the sum of five partial derivative equations shown below, separately for clarity.

The partial energy balance of inlet pipes 1 and 2 are shown in Eqs. 5.38 and 5.39 respectively, where the subscripts $i1$ and $i2$ denote the two components.

$$(c\rho)_{i1}\frac{\partial T_{i1}}{\partial t} = b_{i1g}\left(T_{i1} - T_g\right) + \lambda_{i1}\frac{\partial^2 T_{i1}}{\partial z^2} - (vc\rho)_{f_{i1}}\frac{\partial T_{i1}}{\partial z} \tag{5.38}$$

$$(c\rho)_{i2}\frac{\partial T_{i2}}{\partial t} = b_{i2g}\left(T_{i2} - T_g\right) + \lambda_{i2}\frac{\partial^2 T_{i2}}{\partial z^2} - (vc\rho)_{f_{i2}}\frac{\partial T_{i2}}{\partial z} \tag{5.39}$$

Similarly, the partial energy balance of outlet pipes 1 and 2 are shown in Eqs. 5.40 and 5.41 respectively with their respective subscripts $o1$ and $o2$.

$$(c\rho)_{o1}\frac{\partial T_{o1}}{\partial t} = +b_{o1g}\left(T_{o1} - T_g\right) + \lambda_{o1}\frac{\partial^2 T_{o1}}{\partial z^2} - (vc\rho)_{f_{o1}}\frac{\partial T_{o1}}{\partial z} \tag{5.40}$$

$$(c\rho)_{o2}\frac{\partial T_{o2}}{\partial t} = +b_{o2g}\left(T_{o2} - T_g\right) + \lambda_{o2}\frac{\partial^2 T_{o2}}{\partial z^2} - (vc\rho)_{f_{o2}}\frac{\partial T_{o2}}{\partial z} \tag{5.41}$$

Note that the four pipes only interact internally with the thermally enhanced grout. This makes the source terms representing the internal interaction in the heat balance of the thermally enhanced grout more complicated by adding two additional terms.

$$\begin{aligned}(c\rho)_g\frac{\partial T_g}{\partial t} &= b_{gi1}\left(T_g - T_{i1}\right) + b_{go1}\left(T_g - T_{o1}\right) + b_{gi2}\left(T_g - T_{i2}\right)\\ &\quad + b_{gi2}\left(T_g - T_{i2}\right) + \lambda_g\frac{\partial^2 T_g}{\partial z^2}\end{aligned} \tag{5.42}$$

Annular Coaxial Geothermal Heat Exchangers

The coaxial geothermal heat exchangers (Fig. 5.3c) have a total of three components: two pipes and the thermally enhanced grout. Unlike in the U-shaped design, the outlet pipe is only in contact with the inlet pipe, without contact with the thermally enhanced grout. In addition, the inlet pipe is not only in contact with the thermally enhanced grout but also with the outlet pipe. Applying Eq. 5.27 and separating the equation in the coupled partial derivatives by components, the energy balance in the inlet pipe is given by:

$$(c\rho)_i \frac{\partial T_i}{\partial t} = b_{ig}\left(T_i - T_g\right) + b_{io}(T_i - T_o) + \lambda_i \frac{\partial^2 T_i}{\partial z^2} - (vc\rho)_{f_i} \frac{\partial T_i}{\partial z} \tag{5.43}$$

where a source/sink term is added for the direct interaction with the outlet pipe. The outlet pipe is only in contact with the inlet pipe, and its partial differential equation is:

$$(c\rho)_o \frac{\partial T_o}{\partial t} = +b_{oi}(T_o - T_i) + \lambda_o \frac{\partial^2 T_o}{\partial z^2} - (vc\rho)_{f_o} \frac{\partial T_o}{\partial z} \tag{5.44}$$

Finally, Eq. 5.45 is the differential equation coupled with the thermally enhanced grout component where it interacts only with the inlet pipe. Note that in annular coaxial systems the inlet pipe transfers energy with both the thermally enhanced grout and the central outlet pipe.

$$(c\rho)_g \frac{\partial T_g}{\partial t} = b_{gi}\left(T_g - T_i\right) + \lambda_g \frac{\partial^2 T_g}{\partial z^2} \tag{5.45}$$

5.3.3.3 Central Coaxial Geothermal Heat Exchangers

The difference with annular coaxial BHE is that the inlet pipe of the heat carrier fluid is now in the central pipe, and the fluid leaves the heat exchanger through the annular pipe. Therefore, the coupled equations of the inlet pipe, outlet pipe and thermally enhanced grout of this system will be respectively:

$$(c\rho)_i \frac{\partial T_i}{\partial t} = +b_{io}(T_i - T_o) + \lambda_i \frac{\partial^2 T_i}{\partial z^2} - (vc\rho)_{f_i} \frac{\partial T_i}{\partial z} \tag{5.46}$$

$$(c\rho)_o \frac{\partial T_o}{\partial t} = b_{og}\left(T_o - T_g\right) + b_{oi}(T_o - T_i) + \lambda_o \frac{\partial^2 T_o}{\partial z^2} - (vc\rho)_{f_o} \frac{\partial T_o}{\partial z} \tag{5.47}$$

$$(c\rho)_g \frac{\partial T_g}{\partial t} = b_{go}\left(T_g - T_o\right) + \lambda_g \frac{\partial^2 T_g}{\partial z^2} \tag{5.48}$$

5.3.4 Analytical Models of Heat Transfer in Closed Geothermal Heat Exchangers

Heat transfer between the heat carrier fluid and the underground environment depends on the heat carrier fluid circulation pipe assembly, the convective heat transfer occurring in the pipes and the thermal properties of the materials involved in the process. The heat transfer process is characterised by the fact that it is not instantaneous but the materials through which it propagates offer more or less resistance to the heat flow. All the thermal resistance offered by the materials involved in the process can be grouped together as a single thermal resistance of the BHE (R_s). In shallow geothermal energy problems, it is of special interest to know this resistance in the ratios between the heat fluxes per linear metre (q^o) [W m^{-1}] from the heat exchanger pipes, and the temperature differences between the fluid (T_f) in the pipes and the ground at the borehole wall (T_s).

The heat transfer process involved in the operation of a heat exchanger is complex, and it can take several years to reach steady state conditions. The complexity involved in studying the BHE and, at the same time, the induced response in the ground means that the heat transfer process is often analysed separately for different zones. The first region is the rock/sediment outside the BHE, where heat conduction and advection are considered in a transient regime. The study of the heat transfer processes in this zone makes it possible to obtain the temperature of the borehole wall, the surface in contact with the BHE, for each moment during the operation time. The second region typically differentiated in the overall analysis is the BHE itself, including all its components. In this analysis of the BHE, the aim is to determine the inlet and outlet temperatures of the heat carrier fluid circulating through the pipes as a function of the borehole wall temperature, the thermal resistance of the materials that make up the BHE, and the rate of heat dissipation/absorption imposed on the BHE during operation.

Due to the small size of a BHE, compared to the continuous geological environment (environment of the BHE), it is accepted that the temperature changes inside the heat exchanger are fast enough to be considered in equilibrium. If the rate of heat transferred to the geothermal heat exchanger from the heat pump is constant, a steady state is reached in a few hours, whereas in the geological environment it will take tens of hours to several days, even up to one or two years. This is an important simplification that facilitates the design of BHE. Although this simplification is generally considered adequate, more detailed analysis on the scale of a few hours may require consideration of transient models (Yavuzturk et al. 1999).

5.3.4.1 One Dimensional Model (Equivalent Diameter Model)

Taking into account that the BHE's length is significantly greater than its diameter, and that the stationary regime is achieved approximately instantaneously when compared to the surrounding terrain, it can be established that the temperature

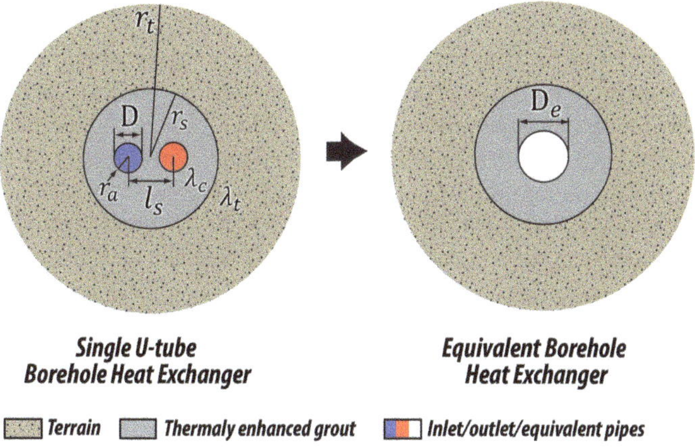

Fig. 5.10 Conceptual scheme of the equivalent diameter model according to Gu and O'Neal (1998)

gradient presents predominantly radial geometry. The heat flow has a strong horizontal component, and the vertical component is only significant in terms of heat advection through the pipes in which the heat carrier fluid circulates. For the shake of simplicity, it can be assumed that heat conduction is a two-dimensional problem. Equivalent pipe models reduce the symmetry even more by assuming BHE's heat transfer problem is one-dimensional (axisymmetric), where the temperature gradients only depend on the radius, not on the angle Fig. 5.10.

The equivalent diameter BHE method consists of finding an equivalent diameter whose generalised effect can be described by a single pipe which, as a function of its imaginary equivalent diameter, behaves similarly. The equivalent diameter D_e [m] with $D_e = aD$ where D [m] is the pipe diameter of a single U-shaped pipe and a [–] is a fit coefficient greater than unity (Fig. 5.10), in general close to $\sqrt{2}$. The temperature change induced by the heat carrier fluid of BHE considered as a one-dimensional steady state problem will be given by (Gu and O'Neal 1998):

$$\Delta T = \frac{q^o}{2\pi}\left[\frac{\ln\left(\frac{r_s}{\sqrt{r_a L_s}}\right)}{\lambda_c} + \frac{\ln\left(\frac{r_c}{r_s}\right)}{\lambda_t}\right] \tag{5.49}$$

Taking into account the definition of thermal resistivity R [K W^{-1}] expressed in Sect. 2.3.2 (Eq. 2.115), Eq. 5.49 which represents the ΔT and considering the equivalent radius as $r_e = \sqrt{r_a l_s}$ with $2r_a = D \leq l_s \leq r_s$, the equivalent thermal resistance R_e [K W^{-1}] of the BHE at steady state can be defined as:

$$R_e = \frac{\Delta T}{q^o} = \frac{1}{2\pi}\left[\frac{\ln\left(\frac{r_s}{r_e}\right)}{\lambda_g} + \frac{\ln\left(\frac{r_c}{r_s}\right)}{\lambda_t}\right] \tag{5.50}$$

Consequently, the equivalent radius heat exchanger r_e is independent of the thermal properties of the heat exchanger constituents, independent boundary conditions, or geometrical parameters other than the radius of the heat carrier fluid pipe (r_a), and the distance between the pipes (shank spacing) forming the single U-shaft (l_s). When $D = 2r_a = l_s$ then $D_e = D\sqrt{2}$ and $a = \sqrt{2}$ (Claesson and Dunand 1983).

Models of one-dimensional geothermal heat exchangers can be used for their constructive design by approximating the double U-shaped pipe as a single equivalent pipe. However, this approximation does not take into account the thermal interference between the pipes, known as thermal short-circuit and, therefore, this simplistic approximation will tend to underestimate the thermal resistance of the geothermal heat exchanger. The one-dimensional model neglects the loss of efficiency caused by the thermal short-circuit between the inlet and outlet pipes.

5.3.4.2 Two-Dimensional Model

Another possible approach to solve the heat transport problem in a BHE is to represent the heat carrier fluid circulation pipes as linear source/sink terms. Such a model was solved by Hellstrom (1991); he first considered the problem for a single pipe. In this case, the thermal resistance between the heat carrier fluid and the outer material would consist of three parts: (1) the thermal resistance to the heat convection R_c between the fluid and the inner surface of the pipe:

$$R_c = \frac{\left(T_{tb_{int}} - T_f\right)}{q^o} = \frac{1}{2\pi b r_{tb_{int}}} \tag{5.51}$$

where b [W m^{-2} K^{-1}] is the heat transfer coefficient per unit area, applying Newton's Law of Cooling (Eq. 5.25) and $r_{tb_{int}}$ is the internal radius of the pipe, $T_{tb_{int}}$ is the pipe internal wall temperature and T_f is the average temperature of the fluid flowing through the pipe. (2) The thermal resistance R_{tb} of the pipe through the pipe itself, which is given by the formula for steady-state heat conduction through an annular region:

$$R_{tb} = \frac{\left(T_{tb_{ext}} - T_{tb_{int}}\right)}{q^o} = \frac{1}{2\pi \lambda_{tb}} \ln\left(\frac{r_{tb_{ext}}}{r_{tb_{int}}}\right) \tag{5.52}$$

where λ_{tb} is the thermal conductivity of the pipe material, $r_{tb_{ext}}$ and $r_{tb_{int}}$ are the external and internal radio respectively. (3) The thermal resistance R_g of the thermally enhanced grout between the pipe and the ground, given by:

$$R_g = \frac{\left(T_s - T_{tb_{ext}}\right)}{q^o} = \frac{1}{2\pi \lambda_g} \ln\left(\frac{r_s}{r_{tb_{ext}}}\right) \tag{5.53}$$

where r_s is the radius of the borehole, λ_g is the thermal conductivity of the thermally enhanced grout, and T_s is the temperature at the borehole wall, i.e., the contact surface between the geothermal heat exchanger and the geological media. The thermal resistance of the heat exchanger R_I is given by:

$$R_I = R_c + R_{tb} + R_g \tag{5.54}$$

This gives the heat flux to be produced by the heat exchanger from a single pipe:

$$q^o = R_I\left(T_s - T_f\right) \tag{5.55}$$

For the general problem of n piping problem there will be n temperature increments (Eq. 5.55) with respect to the ground temperature T_t in the ground not affected by the heat exchanger at a sufficiently large distance r_t from the borehole.

$$\Delta T_{f_1}, \Delta T_{f_2}, \ldots, \Delta T_n = T_{f_1} - T_t, T_{f_2} - T_t, \ldots, T_{f_n} - T_t \tag{5.56}$$

Such temperature increases shall cause q_n^o [W m^{-1}] heat fluxes between the ground and the pipes, and the heat flux between the geothermal heat exchanger containing the pipes and the ground will be the sum of these $q_1^o, q_2^o, \ldots, q_n^o$ heat fluxes. The heat conduction problem is linear between the heat carrier fluid and the surface of the heat exchanger, where the thermally enhanced grout ends:

$$\sum_{i=1}^{n}\left(T_{f_i} - T_g\right) = \sum_{i=1}^{n} R_{g\,ij} \cdot q_i^o \tag{5.57}$$

With $j = 1, 2, \ldots, m$. The tensor $[R_{g\,ij}]$ represents the thermal resistances of the grout for the linear combination of heat fluxes taking place between the fluids and the grout edge. Since t the heat flux is given per linear metre of geothermal heat exchanger q_i^o [W m^{-1}] the thermal resistance of the grout has dimensions of [K m W^{-1}]. The tensor can be calculated from the solution of the two-dimensional steady-state heat transfer problem. The tensor describes the relationships between the temperature and heat flux conditions, between the heat exchanger pipes and the geological environment up to the edge of the heat exchanger (r_s), where the grout ends. The distance at which there is no thermal impact of the geothermal heat exchanger in the ground (r_t) is much larger than the radius of the geothermal heat exchanger (r_s). Initially r_t is unknown, so it $R_{g\,ij}$ depends on an unknown variable. This can be solved by taking the limit when the ratio r_s/r_t is large:

$$R_{g\,ij} = R_{g\,ij}^T + \frac{1}{2\pi\lambda_t}\ln\left(\frac{r_t}{r_s}\right) \tag{5.58}$$

Now $R_{g\,ij}^T$ does not depend on r_s. The temperature at the borehole wall T_s at the edge of the geothermal heat exchanger is defined as:

$$T_s - T_t = \sum_{i=1}^{n} q_i^o \cdot \frac{1}{2\pi \lambda_t} \ln\left(\frac{r_t}{r_s}\right) \tag{5.59}$$

Equation 5.59 describes how the sum of all heat fluxes in the BHE from the n pipes propagate through the geological media, from the BHE edge (r_s) to the terrain where the temperature change induced by BHE is zero (r_t). Combining Eqs. 5.57–5.59 gives:

$$\sum_{i=1}^{n} (T_{f_i} - T_s) = \sum_{i=1}^{n} R_{g\,ij}^{T} \cdot q_i^o \tag{5.60}$$

With $j = 1, 2, \ldots, m$. When $n = 2$ in the case of two pipes, i.e., a single U-shaped BHE, the model is referred to as a triangular circuit or circuit-Δ (Fig. 5.11).

For the case of two pipes ($n = 2$) Eq. 5.60 is:

$$\begin{cases} T_{f_1} - T_s = R_{g\,11}^{T} \cdot q_1^o + R_{g\,12}^{T} \cdot q_2^o \\ T_{f_2} - T_s = R_{g\,12}^{T} \cdot q_1^o + R_{g\,22}^{T} \cdot q_2^o \end{cases} \tag{5.61}$$

Solving the system of linear equations shown in Eq. 5.61 the heat fluxes as a function of temperatures are given:

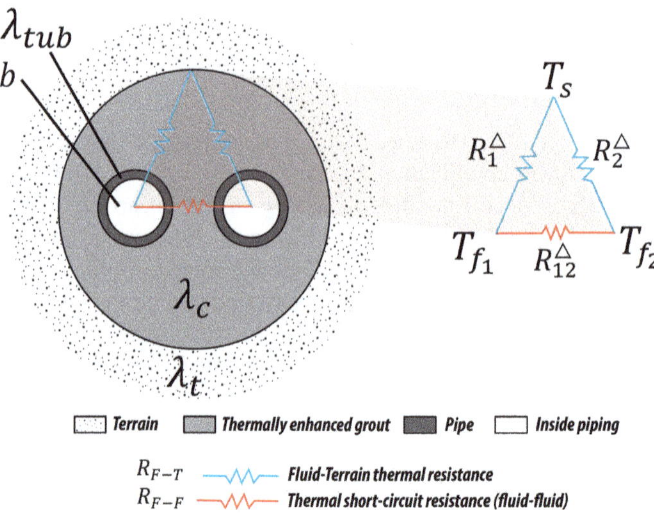

Fig. 5.11 Conceptual schematic of the circuit-Δ of a simple U-shaped borehole heat exchanger

$$q_1^o = \frac{R^T_{g\,22} - R^T_{g\,12}}{R^T_{g\,11}R^T_{g\,22} - \left(R^T_{g\,11}\right)^2}(T_{f_1} - T_s) + \frac{R^T_{g\,12}}{R^T_{g\,11}R^T_{g\,22} - \left(R^T_{g\,12}\right)^2}(T_{f_1} - T_{f_2}) \tag{5.62}$$

$$q_2^o = \frac{R^T_{g\,11} - R^T_{g\,12}}{R^T_{g\,11}R^T_{g\,22} - \left(R^T_{g\,12}\right)^2}(T_{f_2} - T_s) + \frac{R^T_{g\,12}}{R^T_{g\,11}R^T_{g\,22} - \left(R^T_{g\,12}\right)^2}(T_{f_2} - T_{f_1}) \tag{5.63}$$

Taking R_1^Δ, $R_2^\Delta \, y \, R_{12}^\Delta$ (Fig. 5.11) as:

$$\begin{aligned} R_1^\Delta &= \frac{R^T_{g\,11}R^T_{g\,22} - (R^T_{g\,11})^2}{R^T_{g\,22} - R^T_{g\,12}} \\ R_2^\Delta &= \frac{R^T_{g\,11}R^T_{g\,22} - (R^T_{g\,12})^2}{R^T_{g\,11} - R^T_{g\,12}} \\ R_{12}^\Delta &= \frac{R^T_{g\,11}R^T_{g\,22} - (R^T_{g\,12})^2}{R^T_{g\,12}} \end{aligned} \tag{5.64}$$

the two heat flows from the two pipes are obtained:

$$q_1^o = \frac{T_{f_1} - T_s}{R_1^\Delta} + \frac{T_{f_1} - T_{f_2}}{R_{12}^\Delta} \tag{5.65}$$

$$q_2^o = \frac{T_2 - T_s}{R_2^\Delta} + \frac{T_{f_2} - T_{f_1}}{R_{12}^\Delta} \tag{5.66}$$

The fluid-ground thermal resistance (R_{F-T}) (Fig. 5.11) is obtained assuming that $T_{f_1} = T_{f_2} = T_f$. Therefore, the thermal resistance between the fluid (T_f) and the edge of the BHE (T_s) consists of two coupled parallel resistances R_1^Δ and R_2^Δ :

$$R_{F-T} = \frac{R_1^\Delta R_2^\Delta}{R_1^\Delta + R_2^\Delta} \tag{5.67}$$

The thermal short-circuit resistance (R_{f-f}) (Fig. 5.11) consists of the parallel coupling of the thermal resistance R_{12}^Δ and the thermal resistance in series R_1^Δ and R_1^Δ :

$$\frac{1}{R_{f-f}} = \frac{1}{R_1^\Delta + R_2^\Delta} + \frac{1}{R_{12}^\Delta} \tag{5.68}$$

Returning to the system of linear equations of Eq. 5.61, subtracting both equations and considering the specific case in which all the heat from pipe one passes to pipe two ($q_{f-f} = q_1 = -q_2$):

$$T_{f_1} - T_{f_2} = q_{f-f} \cdot R_{f-f} = q_{f-f}\left(R^T_{g\,11} - 2R^T_{g\,12} + R^T_{g\,22}\right) \tag{5.69}$$

Clearing R_{f-f}:

$$R_{f-f} = R^T_{g\,11} - 2R^T_{g\,12} + R^T_{g\,22} \tag{5.70}$$

Considering the symmetry of the problem and a homogeneous and isotropic media, then $R^T_{g\,11} = R^T_{g\,22}$ and taking into account Eqs. 5.64, 5.67 and 5.70, it follows:

$$R^{\Delta}_1 = R^{\Delta}_2 = R^T_{g\,11} + R^T_{g\,12} \tag{5.71}$$

$$R^{\Delta}_{12} = \frac{\left(R^T_{g\,11}\right)^2 - \left(R^T_{g\,12}\right)^2}{R^o_{c\,12}} \tag{5.72}$$

$$R_{f-t} = \frac{1}{2}\left(R^T_{g\,11} + R^T_{g\,12}\right) \tag{5.73}$$

$$R_{f-f} = 2\left(R^T_{g\,11} - R^T_{g\,12}\right) \tag{5.74}$$

Thermal resistance $R^T_{g\,11}$ and $R^T_{g\,12}$ which control the heat flow from the pipes to the ground and between the pipes themselves, respectively, can be approximated by the linear heat source/sink model. This approximates each pipe to a line heat source/sink, in this case in steady state and applying the superposition principle. The solution to this problem is given by Hellstrom (1991):

$$R^T_{g\,11} = \frac{1}{2\pi\lambda_g}\left[\ln\left(\frac{r_s}{r_{tb}}\right) + \frac{\lambda_g - \lambda_t}{\lambda_g + \lambda_t}\cdot\ln\left(\frac{r_s^2}{r_s^2 - D^2}\right)\right] + R_{tb} \tag{5.75}$$

$$R^T_{g\,12} = \frac{1}{2\pi\lambda_g}\left[\ln\left(\frac{r_s}{2\mathrm{D}}\right) + \frac{\lambda_g - \lambda_t}{\lambda_g + \lambda_t}\cdot\ln\left(\frac{r_s^2}{r_s^2 + \mathrm{D}^2}\right)\right] \tag{5.76}$$

where D [m] is the distance from the pipes to the origin of coordinates (plane of symmetry). The distance separating the two pipes is $2\mathrm{D} = l_s$. R_{tub} is the thermal resistance of the pipe plus the convective resistance inside the pipe. The resistivities R^{Δ}_1 of the circuit-Δ according to Eq. 5.71 will be:

$$R^{\Delta}_1 = \frac{1}{2\pi\lambda_g}\left[\ln\left(\frac{r_s}{r_{tb}}\right) + \ln\left(\frac{r_s}{2\mathrm{D}}\right) + \sigma\ln\left(\frac{r_s^4}{r_s^4 - \mathrm{D}^4}\right)\right] + R_{tb} \tag{5.77}$$

Finally, the thermal resistance between the two pipes (Eq. 5.74) is given by:

$$R_{f-f} = \frac{1}{\pi\lambda_g}\left[\ln\left(\frac{2\mathrm{D}}{r_{tb}}\right) + \sigma\ln\left(\frac{r_s^2 + \mathrm{D}^2}{r_s^2 - \mathrm{D}^2}\right)\right] + 2R_{tb} \tag{5.78}$$

The effective thermal resistance between the fluid and the external wall of the BHE obtained with the two-dimensional model (R_{2D}) is defined as:

$$R_{2D} = \frac{T_f - T_s}{q_{2D}} = \frac{R^T_{g\,11} + R^T_{g\,12}}{2} \tag{5.79}$$

where $q_{2D} = 2q_1 = 2q_2$ and $T_f = T_{f_1} = T_{f_2}$.

This two-dimensional solution is a more complete approximation than the one-dimensional equivalent diameter model. It provides more information concerning the type of thermal resistance in the cross-sectional plane, thus serving as a basis for describing the thermal regime of a simple U-type geothermal heat exchanger. The main shortcoming of this two-dimensional model is that it does not take into account heat transfer in the vertical direction during the circulation of the heat carrier fluid. It also does not take into account the difference in fluid temperature between the inlet and outlet pipes and, therefore, the heat flows are different in each pipe. Furthermore, it does not take into account the temperature gain/loss of the heat carrier fluid along the U-shape at each depth resulting from absorbed/dissipated heat with the terrain.

5.3.4.3 Quasi-Three-Dimensional Model

Building on Hellstrom's (1991) solution for the two-dimensional problem, Zeng et al. (2003a) proposed a quasi-three-dimensional solution that takes into account the fluid temperature variation along the geothermal heat exchanger pipe. It is not considered fully three-dimensional because in order to maintain its analytical tractability, heat flow by conduction in the vertical direction is neglected. The energy balance equations for the heat carrier fluid rising at temperature T_{f_2} and falling at temperature T_{f_1} can be expressed as

$$-Q_f \rho_f c_f \frac{dT_{f_1}}{dz} = \frac{T_{f_1} - T_s}{R_2^{\Delta}} + \frac{T_{f_1} - T_{f_2}}{R_{12}^{\Delta}} \qquad 0 \le z \le H \tag{5.80}$$

$$Q_f \rho_f c_f \frac{dT_{f_2}}{dz} = \frac{T_{f_2} - T_s}{R_2^{\Delta}} + \frac{T_{f_2} - T_{f_1}}{R_{12}^{\Delta}} \qquad 0 \le z \le H \tag{5.81}$$

The boundary conditions imposed to solve these two equations are: at surface ($z = 0$) the fluid temperature of pipe one is the inlet temperature (T_{f_i}), i.e., $T_{f_1} = T_{f_i}$ and that at the bottom of the geothermal heat exchanger ($z = z_I$), the fluid temperature of pipe one becomes the temperature of pipe two ($T_{f_1} = T_{f_2}$). The solution is obtained by the Laplace transform (Eskilson and Claesson 1988; Zeng et al. 2003b) and is expressed in dimensionless terms. The dimensionless temperature Θ is defined as:

$$\Theta = \frac{T_f - T_s}{T_{f_i} - T_s} \tag{5.82}$$

The dimensionless depth Z [–] is determined as a function of the coordinate z descending in reference to the depth of the BHE (z_I) as:

$$\mathrm{Z} = \frac{z}{z_I} \tag{5.83}$$

For a symmetrical single U-shaped BHE, the dimensionless temperature distribution for pipe one (Θ_1) and pipe two (Θ_2) are determined by:

$$\Theta_1(Z) = \cosh(\beta Z) - \frac{R^T_{g11}}{\sqrt{\left(R^T_{g11}\right)^2 - \left(R^T_{g12}\right)^2}} \times \left[1 - \frac{R^T_{g12}}{R^T_{g11}} \cdot \frac{\cosh(\beta) - \sqrt{\frac{R^T_{g11} - R^T_{g12}}{R^T_{g11} + R^T_{g12}}} \cdot \sinh(\beta)}{\cosh(\beta) + \sqrt{\frac{R^T_{g11} - R^T_{g12}}{R^T_{g11} + R^T_{g12}}} \cdot \sinh(\beta)}\right] \cdot \sinh(\beta Z) \tag{5.84}$$

$$\Theta_2(Z) = \frac{\cosh(\beta) - \sqrt{\frac{R^T_{g11} - R^T_{g12}}{R^o_{c11} + R^o_{c12}}} \cdot \sinh(\beta)}{\cosh(\beta) + \sqrt{\frac{R^T_{g11} - R^T_{g12}}{R^o_{c11} + R^o_{c12}}} \cdot \sinh(\beta)} \cdot \cosh(\beta Z) + \frac{R^T_{g11}}{\sqrt{\left(R^T_{g11}\right)^2 - \left(R^T_{g12}\right)^2}} \left[\frac{\cosh(\beta) - \sqrt{\frac{R^T_{g11} - R^T_{g12}}{R^T_{g11} + R^T_{g12}}} \cdot \sinh(\beta)}{\cosh(\beta) + \sqrt{\frac{R^T_{g11} - R^T_{g12}}{R^T_{g11} + R^T_{g12}}} \cdot \sinh(\beta)} - \frac{R^T_{g12}}{R^T_{g11}}\right] \cdot \sinh(\beta Z) \tag{5.85}$$

where $\beta = \frac{H}{Q_f \rho_f c_f \sqrt{(R^T_{g11} + R^T_{g12})(R^T_{g11} - R^T_{g12})}}$.

Using Eqs. 5.84 and 5.85, the vertical temperature profile in a single-U BHE is obtained. Figure 5.12 shows two typical real-world profiles as a function of temperature and dimensionless depth. The figure shows a situation where a fluid is introduced from the surface at a given temperature, and as it descends it dissipates heat and its temperature is reduced to the depth of the heat exchanger (z_I). As the fluid rises, it may heat up again if there is no suitable thermal resistance.

Figure 5.12a shows a case where the thermal resistance between the pipes is half the thermal resistance between the pipes and the ground. Therefore, part of the heat dissipated by the descendent heat carrier fluid (pipe one) is transferred to the ascendant heat carrier fluid (pipe two) returning to the surface. The temperature of the ascending heat carrier fluid is not as low as at the bottom of the heat exchanger. This phenomenon of thermal interference between pipes is known as thermal short-circuit, and its occurrence significantly reduces the efficiency of the systems. Figure 5.12b shows a case where the thermal resistance between the pipes is five times higher than

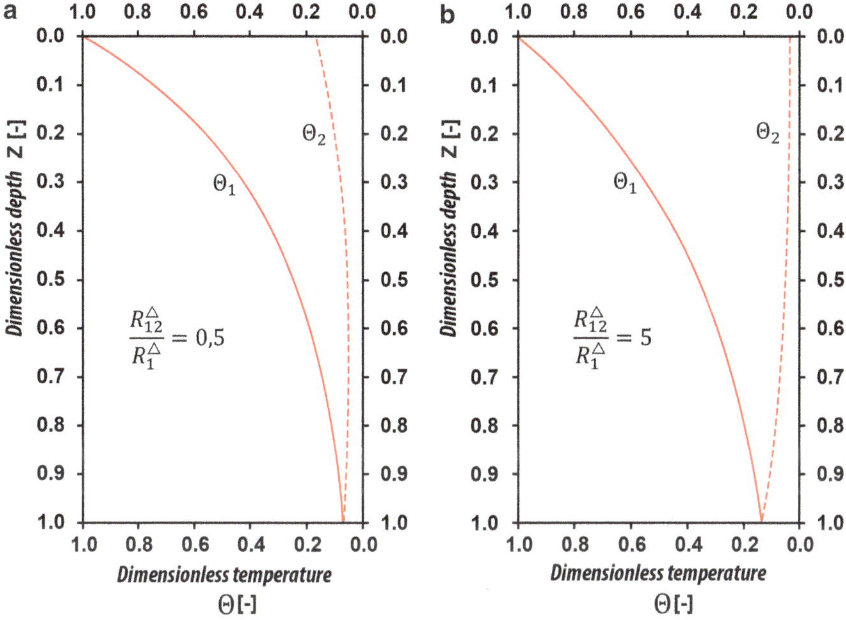

Fig. 5.12 Vertical temperature profile of the heat carrier fluid flowing through a single U-shaped borehole heat exchanger. **a** Case where the thermal resistance between the pipes is half the thermal resistance between the pipes and the ground. **b** The thermal resistance between pipes is five times the thermal resistance between pipes and ground. Data from Diao et al. (2004b)

the thermal resistance between the pipes and the ground. If the thermal connection between the pipes is prevented, the ascending fluid from the bottom of BHE is less affected by the inlet pipe and the ascendant fluid can still dissipate thermal energy.

The observed thermal short-circuit phenomenon can only be described by this quasi-three-dimensional model. In one-dimensional models it is completely masked or included behind the equivalent diameter/resistance concept. In the two-dimensional model, an average value for the whole heat exchanger can be obtained as R_{12}^{Δ}. In spite of the complexity of quasi-three-dimensional model, it is recommended for the design and analysis of geothermal heat exchangers (Diao et al. 2004b) when other analytical models of heat transfer in closed geothermal heat exchangers are considered.

For practical applications, the efficiency of a geothermal heat exchanger is defined as:

$$\eta = \frac{\Delta T_{F-F}}{\Delta T_{F-T}} = \frac{T_{f_i} - T_{f_o}}{T_{f_i} - T_s} = \frac{2\tanh(\beta)}{\sqrt{\frac{R_{g\,11}^T + R_{g\,12}^T}{R_{g\,11}^T - R_{g\,12}^T}}\tanh(\beta)} \tag{5.86}$$

Therefore, the efficiency of a geothermal heat exchanger will be determined by the temperature difference between the inlet and outlet fluid with respect to the potential temperature difference that could be achieved if the efficiency were 100%, i.e., where the outlet temperature was the temperature of the ground in contact with the heat exchanger. Finally, the effective thermal resistance of the heat exchanger in the quasi-three-dimensional model is given by:

$$\eta = \frac{z_I}{Q_f \rho_f c_f}\left(\frac{1}{\eta} - \frac{1}{2}\right) \tag{5.87}$$

5.4 Heat Transfer with the Ground

The temperature distribution in the ground in the vicinity of a geothermal heat exchanger can be described spatially and temporary by using the partial differential equation for heat diffusion (Eq. 2.77):

$$\nabla^2 T = \frac{1}{\alpha}\frac{\partial T}{\partial t} \tag{5.88}$$

Considering an infinite homogeneous and isotropic volume, where heat transfer occurs only through the conduction mechanism, heat will propagate radially from a linear heat source/sink (BHE). Under these conditions, it is reasonable to work in a radially symmetric coordinate system such as the cylindrical coordinate system. The Laplacian operator ∇^2 operator in this coordinate system is given by:

$$\nabla^2 = \frac{\partial^2}{\partial r^2} + \frac{1}{r}\frac{\partial}{\partial r} + \frac{1}{r^2}\frac{\partial^2}{\partial \theta^2} + \frac{\partial^2}{\partial z^2} \tag{5.89}$$

The heat diffusion equation in cylindrical coordinates $T(r, \theta, z, t)$:

$$\frac{1}{\alpha}\frac{\partial T}{\partial t} = \frac{\partial^2 T}{\partial r^2} + \frac{1}{r}\frac{\partial T}{\partial r} + \frac{1}{r^2}\frac{\partial^2 T}{\partial \theta^2} + \frac{\partial^2 T}{\partial z^2} \tag{5.90}$$

There are several solutions commonly used in shallow geothermal energy problems to solve Eq. 5.90 for the specific case of BHE.

The analytical models most commonly used in shallow geothermal energy to reproduce the thermal regime in the ground in response to a closed-loop vertical geothermal heat exchanger are presented below. The models presented are designed for geothermal exchangers operating for relatively long periods of time, from tens of hours to days. The short-term thermal response of the ground is ignored in order to make the models computationally and practically tractable in the field of shallow geothermal energy. These models are the basic tools used for the design of shallow geothermal installations (Stauffer et al. 2013). However, the short-term thermal response of the ground can be of great importance for performance of the heat pump and can influence optimisation of the electricity demand for the operation of shallow geothermal installations. For these cases, there are analytical and numerical models in the literature that cover the short-term thermal response for shallow geothermal installations (Al-Khoury and Bonnier 2006; Al-Khoury et al. 2010; Lamarche and Beauchamp 2007; Yavuzturk et al. 1999).

5.4.1 *Infinite Linear Source Model (ILS)*

The *infinite line source* (ILS) model is a solution to Eq. 5.90 in which a vertical infinite line source is considered to transfer heat to a continuous medium. The heat flow will occur radially in the horizontal plane with $\partial^2 T/\partial z^2 = 0$ (Carslaw and Jaeger 1986). The continuous medium is assumed to be homogeneous and isotropic where heat propagation is assumed to occur radially, resulting in fully circular isotherms. The temperature change produced by the infinite linear source, representing the BHE, is independent of the angle considered and therefore $\partial^2 T/\partial \theta^2 = 0$, there is no temperature change with a change of direction for a given radius. Finally, it is convenient to work with temperature changes with respect to the initial temperature T_0 [K] that allows working with an analytical model independent of the initial conditions. For this purpose, the variable τ [K] is defined as $\tau(r, t) = T(r, t) - T_0(r, 0)$. Taking these assumptions into account, Eq. 5.90 can be rewritten as:

$$\frac{1}{\alpha}\frac{\partial \tau}{\partial t} = \frac{\partial^2 \tau}{\partial r^2} + \frac{1}{r}\frac{\partial \tau}{\partial r} \tag{5.91}$$

The ILS model describes the heat flow in an infinite terrain where a constant heat flux is applied. In the case of shallow geothermal energy applications, this heat flow comes from a BHE, which is represented by an infinite line parallel to the vertical z axis. The initial conditions consider an undisturbed initial temperature T_0 [K] for any r:

$$\tau(r, t = 0) = 0 \tag{5.92}$$

The boundary conditions are given by:

$$T(r \to \infty, t) = 0 \tag{5.93}$$

$$-2\pi r \lambda \left.\frac{\partial \tau}{\partial r}\right|_{r \to 0} = q^o \tag{5.94}$$

where q^o [W m^{-1}] is the heat flux transferred with the domain by the BHE per linear metre of borehole. The solution of Eq. 5.91 with the boundary conditions of Eqs. 5.92, 5.93 and 5.94 are solved by applying the Laplace transform (Al-Khoury

2017), obtaining that the temperature change due to a BHE will be:

$$\tau(r,t) = T - T_0 = \frac{q^o}{4\pi\lambda}\int_{\frac{r^2}{4\alpha t}}^{\infty} -\frac{e^v}{v}dv \tag{5.95}$$

where the integral is the exponential integral function $E_1(x)$ defined as:

$$E_1(x) = \int_{x}^{\infty} -\frac{e^v}{v}dv \tag{5.96}$$

Equation 5.95 can be rewritten as:

$$\tau(r,t) = \frac{q^o}{4\pi\lambda}E_1\left(\frac{r^2}{4\alpha t}\right) \tag{5.97}$$

Note that Eqs. 5.96 and 5.97 are not defined for $r = 0$ and, therefore, the temperature increase at that point cannot be calculated. As time increases, the radius of influence will increase. Ingersoll and Plass (1948) showed that Eq. 5.97 can be used for heat injection from cylindrical source sources with an error of less than 2% for Fourier numbers $Fo > 20$ (Eq. 2.144):

$$t > \frac{20r_s^2}{\alpha} \tag{5.98}$$

Taking the characteristic distance r_s [m], the borehole radius of the BHE, for most exchangers Eq. 5.97 is valid for t greater than 10–20 h. For shorter times, the finite source/sink effect still has a significant effect on the temperature distribution around the heat exchanger.

Note that the exponential integral function (Eq. 5.96) tends to infinity with increasing time. This means that the ILS model does not reach a steady state and, therefore, has no stationary solution. However, for sufficiently large t the exponential integral function can be approximated to (Carslaw and Jaeger 1986):

$$T(r,t) \simeq T_0 + \frac{q^o}{4\pi\lambda}\left(\ln\left(\frac{4\alpha t}{r^2}\right) - \gamma\right). \tag{5.99}$$

where γ is the *Euler-Mascheroni* constant ($\gamma \approx 0.577215$). The errors generated are 2.5% for $Fo > 20$ and 10% for $Fo > 5$.

The ILS model (Eq. 5.97) predicts the thermal impact of a BHE in the terrain over time for a constant heat flux per linear metre, in a terrain where the convective heat transport mechanism is absent or negligible. This model is applied in soils where there is no groundwater flow or is negligible, i.e., saturated porous media with negligible groundwater velocity or unsaturated porous media.

The ILS model can be used to calculate the temperature distribution in the ground for time-varying heat fluxes and geothermal exchanger grids following the principle of temporal and spatial superposition, respectively. Figure 5.13 shows an example for a BHE dissipating heat at a constant rate of 40 W m^{-1} for 100 h in ground with thermal conductivity of 3.5 W m^{-1} K^{-1} and a volumetric heat capacity of 2.5 × 10^6 J m^{-3} K^{-1}.

The ILS model is often used because of its simplicity to make a first estimate of the thermal response of the ground to a BHE. When a field of geothermal heat exchangers exists, it is possible to estimate the ground temperature distribution by applying the spatial superposition principle of Eq. 5.97 with the borehole identification system schematized in Fig. 5.8.

$$\tau(P_j, t) = \frac{q_j^o}{4\pi\lambda} E_1\left(\frac{r_{s_j}^2}{4\alpha t}\right) + \sum_{\substack{i=1 \\ i \neq j}}^{n} \left[\frac{q_i^o}{4\pi\lambda} E_1\left(\frac{r_i^2}{4\alpha t}\right)\right] \tag{5.100}$$

Equation 5.100 indicates that the thermal impact at the coordinate point P_j, where the j-nth BHE is located, is the temperature increase produced by the j-nth BHE, calculated by the ILS model at the radius of the borehole, plus the sum of all temperature rises generated by the n exchangers in the geothermal field, calculated with the ILS model, taking as radius the distance separating each BHE from the grid of exchangers with the j-nth BHE. The ground thermal parameters λ and α are considered constant over the whole domain. The heat fluxes of each BHE can be different q_i^o for each i-nth for a time t since the BHE start their activity (Fig. 5.14)

Figure 5.15 shows an example of the application of the spatial superposition principle (Eq. 5.100) for a grid of nine 3 × 3 BHEs spaced seven metres apart. The nine geothermal exchangers dissipate heat at a constant rate of 40 W m^{-1} for ten years in terrain with a thermal conductivity of 3.5 W m^{-1} K^{-1} and a volumetric heat capacity of 2.5 × 10^6 J m^{-3} K^{-1}.

Up to this point, all equations considered heat fluxes q_0 as constant over time. However, in practice this is not generally the case, as buildings demand different heat loads throughout the year. It is possible to estimate the temperature distribution of ground with q^o variable over time, using a series of constant heat pulses and applying the time superposition principle (Bernier 2001; Lazzarotto 2015), which applied to the ILS model is:

$$\tau(r, t, q^o) = \frac{q_1^o}{4\pi\lambda} E_1\left(\frac{r^2}{4\alpha t_n}\right) + \sum_{i=1}^{n-1}\left[\frac{(q_{i+1}^o - q_i^o)}{4\pi\lambda} E_1\left(\frac{r^2}{4\alpha(t_n - t_{n-i})}\right)\right] \tag{5.101}$$

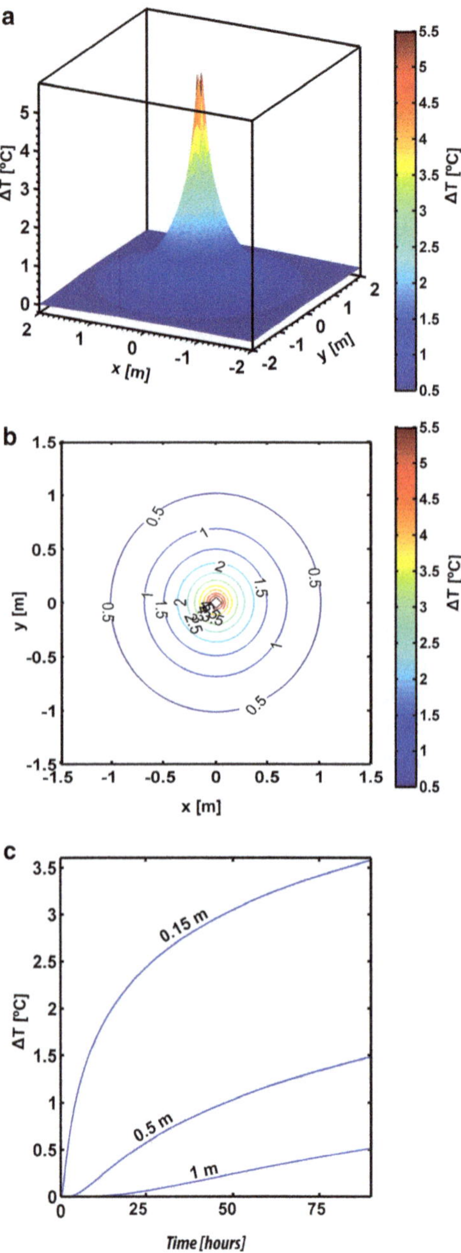

Fig. 5.13 Example of application of the infinite linear source (ILS) model for a BHE dissipating heat at a constant rate of 40 W m^{-1} for 100 h in a soil with thermal conductivity of 3.5 W m^{-1} K^{-1} and volumetric heat capacity of 2.5 × 10^{6} J m^{-3} K^{-1}. Three-dimensional (**a**) and plan (**b**) representation of the spatial distribution of the temperature increases produced. Evolution over time of the temperature increases at different distances *r* distance from the BHE (**c**)

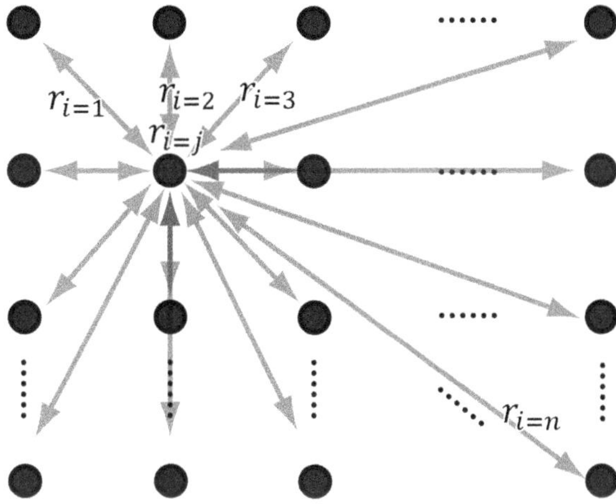

Fig. 5.14 Diagram of the principle of spatial superposition

where $q_1^o, q_2^o, \ldots, q_n^o$ are the heat fluxes of the BHE for the periods $t_1 - t_0, t_2 - t_1, \ldots, t_n - t_{n-1}$ respectively. Figure 5.16 shows an example of a geothermal heat exchanger operating over three periods ($n = 3$) of time ($t_1 - t_0$, $t_2 - t_1$ and $t_3 - t_2$) with three heat transfer rates q_1^o, q_2^o and q_3^o, respectively. Applying Eq. 5.101, it follows that:

$$\tau(r, t, q_0) = \frac{q_1^o}{4\pi\lambda} E_1\left(\frac{r^2}{4\alpha t_3}\right) + \frac{\left(q_2^o - q_1^o\right)}{4\pi\lambda} E_1\left(\frac{r^2}{4\alpha(t_3 - t_1)}\right) + \frac{\left(q_3^o - q_2^o\right)}{4\pi\lambda} E_1\left(\frac{r^2}{4\alpha(t_3 - t_2)}\right) \tag{5.102}$$

In this example, heat flows of 40, 63.33 and 25.71 W m^{-1} were taken, operating for 28 h, 28 h and 44 h respectively. The principle of superposition states that the temperature rise after 100 h of operation will be the sum of the temperature rise that would be produced by an exchanger operating for 100 h at 40 W m^{-1}, plus the temperature rise that would be produced by an exchanger operating at 23.33 W m^{-1} $(q_2^o - q_1^o)$ for 44 h $(t_3 - t_2)$, plus an exchanger operating at −37.62 W m^{-1} $\left(q_3^o - q_2^o\right)$ for 72 h (Fig. 5.16a). Solving Eq. 5.102 for a borehole radius of 0.15 m gives the results shown in Fig. 5.16b.

5.4.2 *Infinite Cylindrical Source (ICS) Model*

The ILS model is a simplification of the *infinite cylindrical source* (ICS) model. In the ICS model the heat source is the lateral surface of a cylinder of infinite length with the radius r_s [m] of the BHE. The analytical solution under these conditions is given by (Ingersoll et al. 1955):

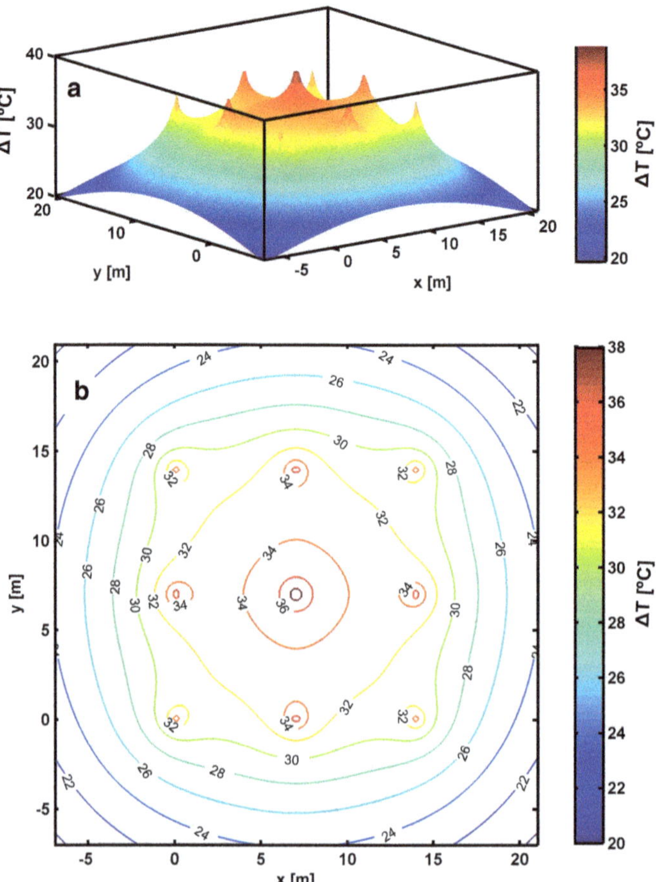

Fig. 5.15 Example of application of the infinite linear source (ILS) model for a grid of nine borehole heat exchangers dissipating heat at a constant rate of 40 W m^{-1} for ten years in terrain with thermal conductivity of 3.5 W m^{-1} K^{-1} and a volumetric heat capacity of 2.5 × 10^{6} J m^{-3} K^{-1}. Three-dimensional (**a**) and plan (**b**) representation of the spatial distribution of the temperature increases produced

$$\tau(r,t) = \frac{q^o}{\pi^2\lambda}\int_0^\infty \frac{e^{-\beta^2 Fo}-1}{J_1^2(\beta)+Y_1^2(\beta)} \times [J_0(\mathrm{R}\beta)Y_1(\beta) - J_1(\beta)Y_0(\mathrm{R}\beta)]\frac{d\beta}{\beta^2} \tag{5.103}$$

where R [–] is the dimensionless radius ($R = r/r_s$) in cylindrical coordinates; J_0 and J_1 are the first-species Bessel function for orders zero and one, respectively; and Y_0 and Y_1 are the second-species Bessel function for orders zero and one respectively. Simpler expressions of the ICS model can be obtained by expressing the ILS model in

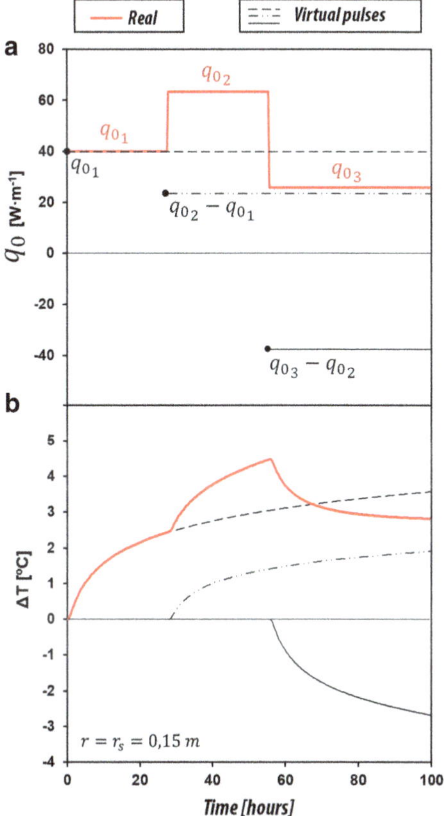

Fig. 5.16 Example of application of the time superposition principle for three heat dissipation pulses interpreted using the ILS model. **a** Heat flux change function of the borehole heat exchanger in three pulses with real values in red and virtual values in black. **b** Calculated temperature increase in the heat exchanger ($r_s = 0.15$ m) for each of the pulses and their summed value

radial coordinates for a linear source term at radius (r_s) and angle (θ_s) of the borehole. Integrating such a solution around a borehole radius circle yields the following more tractable expression (Man et al. 2010):

$$\tau(r,t) = -\frac{q^o}{4\pi\lambda}\int_0^{\pi}\frac{1}{\pi}E1\left(-\frac{r^2 + r_s^2 - 2rr_s\cos\theta_s}{4\alpha t}\right)\frac{1}{\pi}d\theta_s \tag{5.104}$$

Figure 5.17 shows the difference between the ILS and ICS models in terms of dimensionless temperature increase, expressed as $\Theta = 4\pi\lambda\tau/q^o$ as a function of Fourier Number interpreted as dimensionless time. It can be seen from Fig. 5.17 that the discrepancy between models is produced in the early calculation times and,

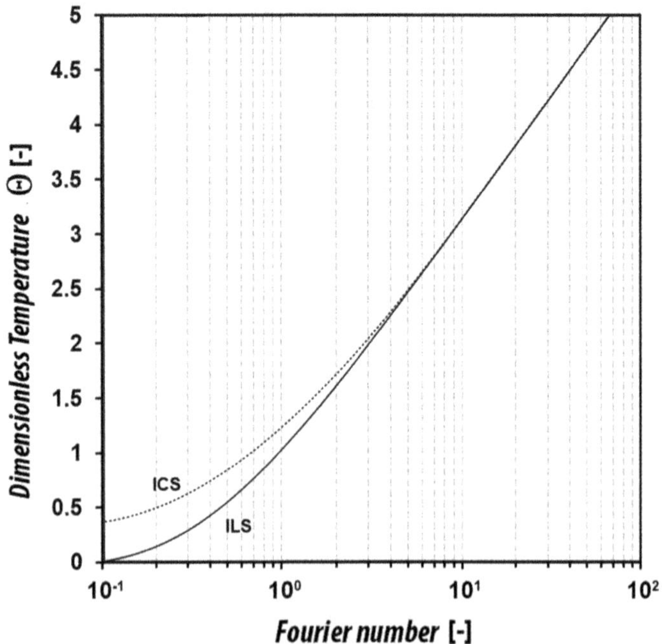

Fig. 5.17 Dimensionless temperature rise calculated with the ICS and ILS models with respect to the *Fourier Number* interpreted as dimensionless time. Data from Stauffer et al. (2013)

therefore, the ICS model will be preferable to the ILS model. However, for $Fo > 5$ the ILS model is valid (Diersch et al. 2011).

5.4.3 Finite Linear Source Model (FLS)

The *finite line source* (FLS) model represents a geothermal heat exchanger by a finite line of depth z_l [m] consisting of multiple point sources. The spatial and temporal distribution of ground temperature increases affected by a point source term with a constant heat flux is determined by (Al-Khoury 2017):

$$T(r,t) = \frac{q^o}{4\pi r\lambda}\text{erfc}\left(\frac{r}{2\sqrt{\alpha t}}\right) \tag{5.105}$$

where $erfc(x)$ is the error function. The temperature variation τ [K] in a layer of infinitesimal thickness $d\zeta$ [m] where heat exchanged $q_0 d\zeta$ will be:

$$\tau(r,t) = \frac{q^o}{4\pi r\lambda\zeta}\text{erfc}\left(\frac{r}{2\sqrt{\alpha t}}\right)d\zeta \tag{5.106}$$

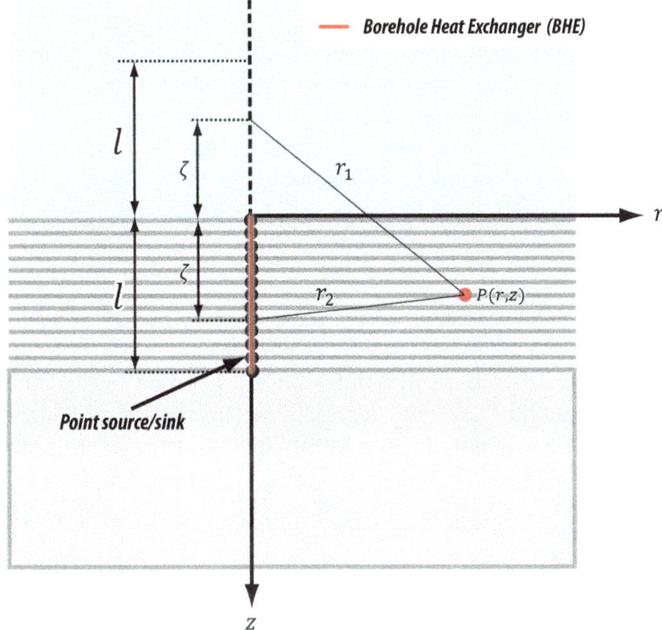

Fig. 5.18 Conceptual scheme of the finite line source (FLS) model

Considering a bundle of layers of infinitesimal thickness over an infinite domain (Fig. 5.18), the temperature in a point $P(r, z)$ can be described by integrating Eq. 5.67 along the source/sink points:

$$T(r, z, t) = \frac{q^o}{4\pi r\lambda}\int_0^l \frac{1}{r_1}\operatorname{erfc}\left(\frac{r_1}{2\sqrt{\alpha t}}\right)d\zeta \tag{5.107}$$

where $r_1 = \sqrt{r^2 - (z-\zeta)^2}$ with ζ corresponding to any source/sink point contained in the line representing the heat exchanger. To consider a semi-infinite domain in the z direction, a Dirichlet boundary condition of constant temperature at the ground surface is imposed. This is achieved by applying method of images (Carslaw and Jaeger 1986). The image has the same values as in the real domain, but with opposite sign. Using the image theory Eskilson (1987) introduced the FLS model as follows:

$$\tau(r, z, t) = \frac{q^o}{4\pi\lambda}\int_0^l \left[\frac{1}{r_1}\operatorname{erfc}\left(\frac{r_1}{2\sqrt{\alpha t}}\right) - \frac{1}{r_2}\operatorname{erfc}\left(\frac{r_2}{2\sqrt{\alpha t}}\right)\right]d\zeta \tag{5.108}$$

where $r_1 = \sqrt{r^2 - (z - \zeta)^2}$ The model is valid for times $5r_s^2/\alpha < t < t_s/10$ where t_s [s] is the time taken to reach the steady state, i.e., the time at which the system reaches equilibrium and the temperature field is constant over time. The time can range from a few hours to several years.

In contrast to the ILS model, the FLS model reaches the steady state (Fig. 5.19) and the analytical solution under these conditions can be described as (Philippe et al. 2009):

$$\tau(r,z) = \frac{q^o}{4\pi\lambda}\ln\left(\frac{\sqrt{r^2+(z-z_l)^2}-(z-z_l)}{\sqrt{r^2+(z+z_l)^2}+(z+z_l)}\cdot\frac{\sqrt{r^2+z^2}+z}{\sqrt{r^2+z^2}-z}\right) \tag{5.109}$$

Based on the model of Eq. 5.106 Claesson and Eskilson (1987) introduced the concept of a g-function G(x, y) to represent the borehole wall temperature of a geothermal heat exchanger as a dimensionless function. They described the borehole

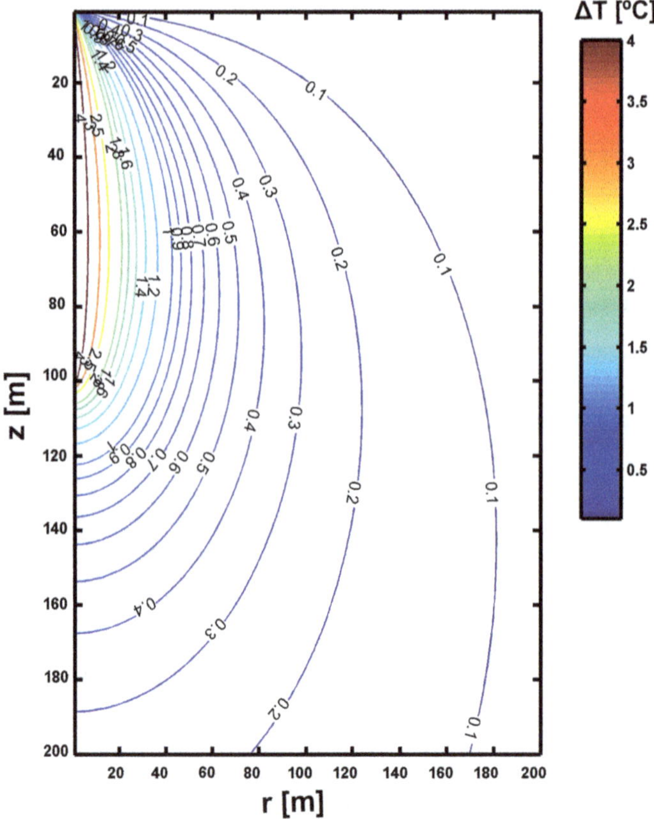

Fig. 5.19 Analytical steady-state solution of the FLS model (Eq. 5.72) for a 100 m deep borehole heat exchanger (l). $\lambda = 3.5$ W m^{-1} K^{-1} and $q^o = 40$ W m^{-1}

surface temperature T_s [ºC] as:

$$T_s - T_0 = \frac{q^o}{2\pi\lambda} G\left(\frac{t}{t_{ss}}, \frac{r_s}{l}\right) \tag{5.110}$$

where $t_{ss} = l^2/9\alpha$ is the time to reach steady state. Since the introduction of the g-function, this model has been widely used and has come to dominate research work related to the search for analytical and semi-analytical solutions of shallow closed-loop geothermal systems. Claesson and Eskilson approximated the g-function to:

$$G\left(\frac{t}{t_{ss}}, \frac{r_s}{l}\right) = \begin{cases} \ln\left(\frac{l}{2r_s}\right) + \frac{1}{2}\ln\left(\frac{t}{t_{ss}}\right) & 5r_s^2/\alpha < t \leq t_{ss} \\ \ln\left(\frac{l}{2r_s}\right) & t \geq l^2/9\alpha \end{cases} \tag{5.111}$$

5.4.4 *Moving Infinite Linear Source Model (MILS)*

The *moving infinite line source* (MILS) model applies the theory of moving heat sources (Carslaw and Jaeger 1986) to Eq. 5.94 of the ILS model in order to obtain the analytical solution to a constant infinite linear heat transfer moving through the medium at a constant velocity. Mathematically, the solution of this process is indistinguishable from that of a static heat source/sink where the medium is moving. The movement of a fluid inside the ground represents the movement of the medium surrounding the BHE, and the linear heat source/sink would be stationary. With this model, both the conductive and advective heat components can be taken into account, the latter due to groundwater flow. Therefore, for an infinite medium subjected to a groundwater flow of velocity v [m s^{-1}], in which there is a heat source/sink q^o the mathematical model that can be used is the transient solution of MILS model (Diao et al. 2004a):

$$\tau(x, y, t) = \frac{q^o}{4\pi\lambda} \exp\left[\frac{vx}{2\alpha}\right] \int_{\frac{r^2}{4\alpha t}}^{\infty} \exp\left[-\psi - \frac{v^2 r^2}{16\alpha^2 \psi}\right] \frac{d\psi}{\psi} \tag{5.112}$$

where α [m^2 s^{-1}] is the thermal diffusivity, $r = \sqrt{x^2 + y^2}$, $\psi = r^2/4\alpha \cdot (t - t')$ and $d\psi/\psi = dt'/(t - t')$. In this expression, the heat source/sink is located at coordinates $x_0 = y_0 = 0$. The MILS model has generally been used to calculate the thermal resistance of the ground, the subsurface temperature distribution in cases of heat extraction/dissipation in transient problems, as well as in situations where groundwater effects due to heat advection in shallow closed-loop geothermal installations are of interest (Diao et al. 2004a; Sutton et al. 2003). Figure 5.20 shows an example application of the MILS model for a BHE dissipating heat at a constant rate of 40 W m^{-1} for 417 days, in terrain with a thermal conductivity of 3.5 W m^{-1} K^{-1}, a volumetric heat capacity of 2.5 × 10^6 J m^{-3} K^{-1} and a groundwater velocity of 10^{-6} m s^{-1}. Under the same conditions, a 3 × 3 grid of BHEs is shown in Fig. 5.21, applying the principle of spatial superposition. In Eq. 5.112, when time tends to infinity, the steady-state MILS model is obtained:

$$\tau(x, y) = \frac{q^o}{2\pi\lambda} \exp\left[\frac{vx}{2\alpha}\right] K_0\left[\frac{v\sqrt{x^2 + y^2}}{2\alpha}\right] \quad (5.113)$$

where K_0 is the second-species, zero-order modified Bessel function.

5.4.5 Numerical Models

The use of the analytical models presented above is limited to simple geometries, initial and boundary conditions. These limitations can be overcome using numerical methods to solve heat conduction and advection problems in steady and/or transient regimes. The general heat transport modelling process is shown schematically in Fig. 5.22. Depending on the objectives of the work to be carried out, it will be necessary to adopt a physical model of heat transfer in the subsurface.

Where there is an interest in pure diffusive heat transport (heat conduction only), it is not usually considered exclusively since its numerical solution is formally identical to the solution of the diffusion equation. In general, solutions for pure heat conduction are obtained from solutions for advection–dispersion–diffusion problems by setting the groundwater flow velocity to zero. To solve the heat advection–dispersion–diffusion problem in a porous media, there are two different main approaches. The first approach consists of a fully coupled two-way solution dealing with non-linear systems, requiring an iterative approach. The temperature is influenced by the flow and, in turn, the flow is influenced by the temperature due to the fact that temperature modifies the hydraulic properties of density and hydraulic conductivity of the porous media. The second approach consists of an approximate solution in a linearised form, where the temperature is influenced by the flow, but the flow is not influenced by the temperature. Consequently, in this case, the hydraulic parameters are assumed constant. This unidirectional coupling of flow and heat transport is generally valid in shallow geothermal energy problems, where the dependence of the hydraulic parameters on temperature within the existing temperature ranges (approximately from 3 to 35 °C) does not represent a significant error in the obtained results (Lo Russo et al. 2018).

There has been a large body of modelling work on heat transport in porous media since the early 1970s (Stauffer et al. 2013). The most recent developments in the

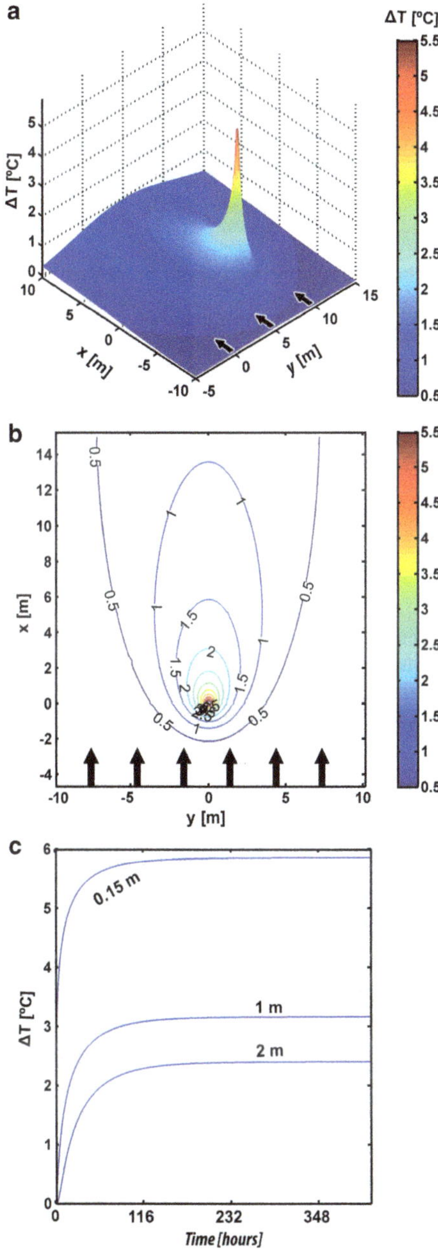

Fig. 5.20 Example of application of the moving infinite linear source (MILS) model for a borehole heat exchanger dissipating heat at a constant rate of 40 W m^{-1} for 417 days in soil with thermal conductivity of 3.5 W m^{-1} K^{-1}, volumetric heat capacity of 2.5×10^6 J m^{-3} K^{-1} and a groundwater velocity of 1×10^{-6} m s^{-1}. Three-dimensional (**a**) and plan (**b**) representation of the spatial distribution of the temperature increases produced is shown. Time evolution of the temperature increases at different *r* distances (0.15, 1 and 2 m) from the borehole heat exchanger (**c**)

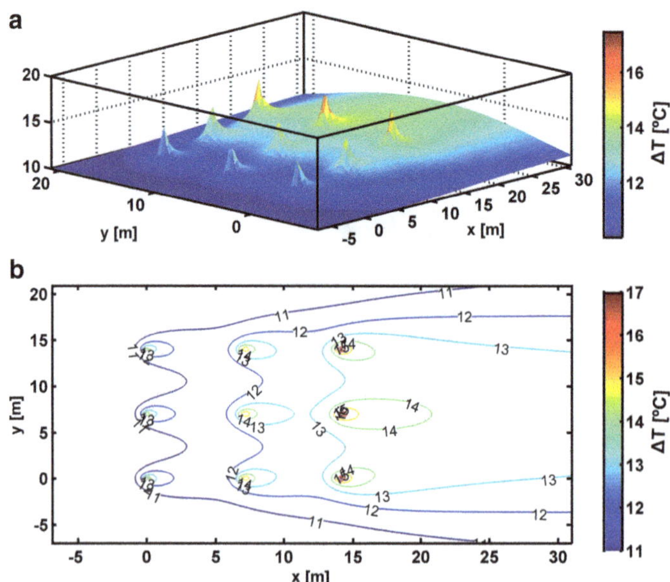

Fig. 5.21 Example of application of the moving infinite linear source (MILS) model for a grid of 3 × 3 borehole heat exchangers dissipating heat at a constant rate of 40 W m^{-1} each for ten years in terrain with a thermal conductivity of 3.5 W m^{-1} K^{-1}, volumetric heat capacity of 2.5 × 10^6 J m^{-3} K^{-1} and a groundwater velocity of 1 × 10^{-6} m s^{-1}. Three-dimensional (**a**) and plan (**b**) representation of the spatial distribution of the induced temperature increases

simulation of closed-loop shallow geothermal heat exchangers focus on reproducing the temperature fields in the ground. This is done by reproducing in detail the thermal processes within the geothermal heat exchanger pipe assembly, with inlet and outlet pipes for different configurations: single U-shaped, double U-shaped, coaxial, etc. (Biglarian et al. 2017; Lazzari et al. 2010; Lee and Lam 2008). Numerical models allow performance analysis of geothermal heat exchangers in order to determine the temperature of the heat carrier fluid according to operating conditions. This analysis is decisive for the design of a geothermal installation since it determines the temperature change both in the ground and in the heat exchanger during the lifetime of the system (Yang et al. 2010). Numerical models allow heat transfer analysis to predict the performance of a system based on the factors involved, such as ground thermal properties, groundwater flow, building heat load projection and boundary conditions of interest (ground surface effect, groundwater movement, possible ground freezing, soil moisture, ground heat storage, etc.) (Li and Lai 2015). The most relevant codes in scientific-commercial sectors for simulation of heat transfer in porous media, all coupled with groundwater flow include, among others: COMSOL (COMSOL 2020), FEFLOW (Diersch 2013), HST2D/3D (Kipp 1997), HYDROTHERM (Kipp et al. 2008) SEAWAT (Langevin et al. 2008) and SUTRA (Voss and Provost 2010).

Geothermal heat pumps with BHEs usually consist of one or more geothermal heat exchangers. In numerical methods, each geothermal heat exchanger and the

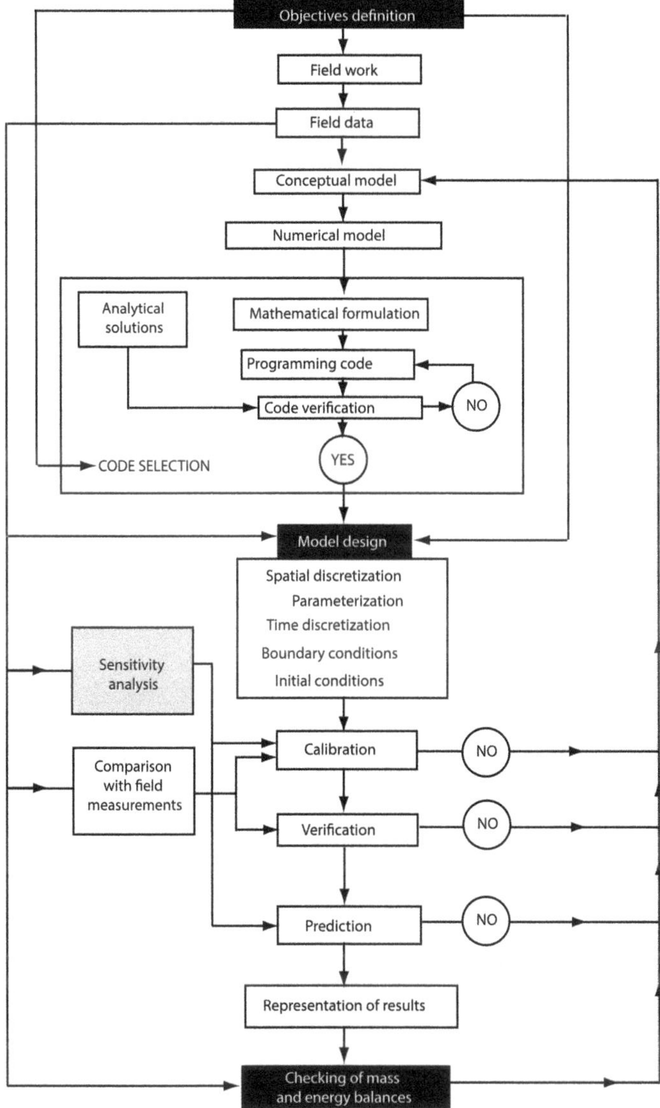

Fig. 5.22 Workflow in numerical modelling of heat transport. Modified from Anderson et al. (2015)

geothermal reservoir are described separately. The spatial coupling between them is usually done using the assembly technique of the finite element method (Al-Khoury et al. 2010). Due to the tightness of closed-loop geothermal heat exchangers and convection as the dominant mechanism inside them, this technique results in a large system of equations, which have to be stored and solved numerically. To obtain feasible computational requirements and CPU time, it is important to reduce the

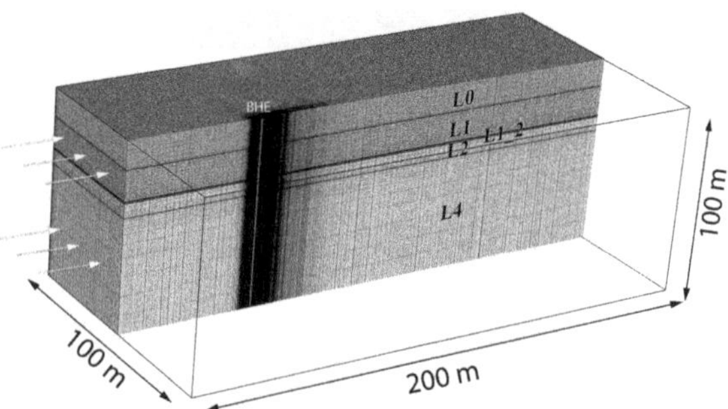

Fig. 5.23 Example of the application of a numerical model to reproduce the thermal response of ground (conduction and advection) with different layers (L0-4), and groundwater flow (white arrows) in the presence of a BHE represented by one-dimensional elements

number of elements. By taking advantage of the uniformity and cylindrical shape of the pipes, the equations can be discretised using one-dimensional elements to represent the heat exchangers (Fig. 5.23). This deviation from the standard finite element procedure avoids the need to discretise the individual components of closed-loop heat exchangers separately and, thus, saves an enormous amount of computational power.

5.5 Horizontal Closed-Loop Geothermal Heat Exchangers

Mechanical drilling of boreholes, although providing access to reservoirs with depths of more than 15 m where the temperature is stable all year round, is a major cost for small single-family house installations. Where surface space is not a constraint, it is possible to dig shallow trenches and install closed-loop heat exchangers in a horizontal arrangement. Although at these depths the reservoir is in a transitional thermal domain (Fig. 3.4) affected by surface atmospheric temperatures, it still offers better conditions than atmospheric conditions. Although the technology involved in horizontal heat exchangers is essentially very simple, there are some major challenges in modelling the physical processes that determine its efficiency. Compared to BHEs, less progress has been made in developing validated models for horizontal geothermal heat exchangers. Consequently, in practice, relatively simple design methods are often used with design tables and graphs with low accuracy.

5.5.1 Types of Horizontal Geothermal Heat Exchangers

Horizontal heat exchangers avoid the need for specialised drilling operations. Therefore, only excavation machinery commonly used in civil engineering and foundation construction is required for installation. Horizontal heat exchangers are usually light and flexible enough to be installed in trenches without other machinery. Although less common and requiring specialised machinery, directional drilling methods can be used for the installation of horizontal heat exchangers. This technique allows drilling with diameters of approximately 100 mm along a horizontal curved trajectory. The installation depth of horizontal heat exchangers using directional drilling methods can be controlled up to a depth similar to that of a conventional trench, i.e., one to two metres deep, and the typical lateral deviations produced are of up to ten metres over distances of 100 m.

Installation depths typically range from 1.2 to 2 m (Banks 2012) but can be found at greater depths. The simplest forms of horizontal heat exchangers consist of pairs of straight pipes with a loop or curved fitting at the end of the trench. More than one pair of pipes can be installed at different depths in the same trench (Fig. 5.24). To ensure good contact between the heat carrier fluid pipe and the ground, special care must be taken to backfill materials used and specific installation procedures must be adopted. The pipe assembly consists of 26–40 mm diameter polyethylene pipe (the same as in vertical heat exchangers) in a single U-shape or *slinky* coils (Fig. 5.25). The slinky type exchangers make it possible to increase the effective heat transfer surface area of the pipe per unit length of trench compared to the U-shaped exchangers. The

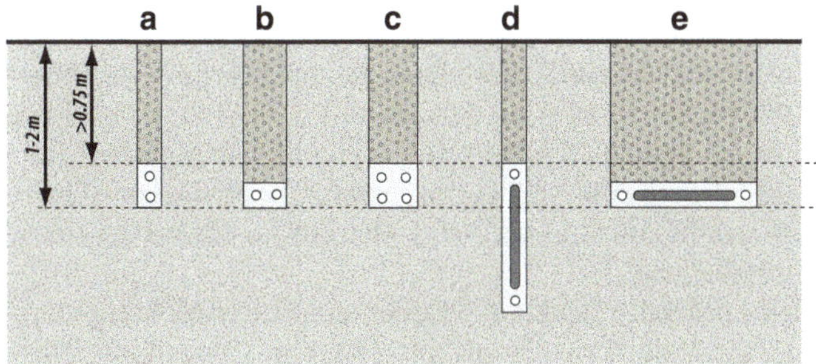

Fig. 5.24 Most common types of horizontal geothermal heat exchangers. A distinction is made between single U-shaped pipe (**a** and **b**) and double U-shaped pipe (**c**) horizontal heat exchangers, and vertical (**d**) and horizontal (**e**) slinky coils heat exchangers

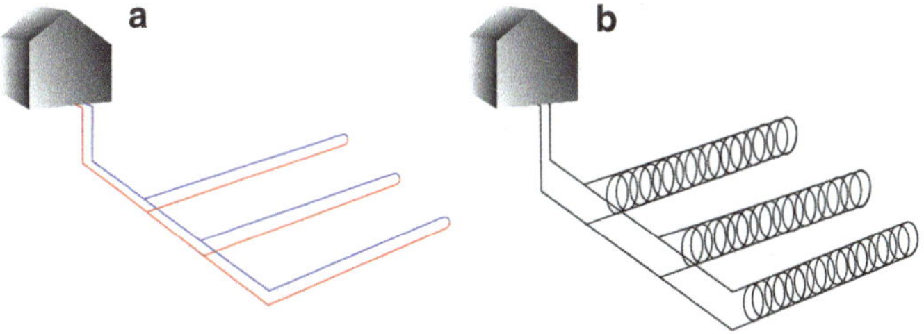

Fig. 5.25 Horizontal geothermal heat exchangers with single U-shaped pipe (**a**) and slinky coils (**b**)

disadvantage of this type of heat exchanger is the increased frictional pressure losses in the pipe and therefore represents an additional energy cost in regard to the drive pump of the heat carried fluid. Consequently, some care is required in optimising the design (Rees 2016).

5.6 Borehole Thermal Energy Storage (BTES)

Storing thermal energy is an ancient concept, from harvesting ice from frozen caves or rivers, to preserving food during the summer, to using *hot* water from thermal springs to warm up on *cold* days. The extraction, storage and trade of ice made up a whole industry in the nineteenth century. The concept of thermal storage is to store excess thermal energy for later use. The systems that carry out this storage are often referred to as accumulators. The technology involved makes it possible to regulate and balance heat demand on different time scales. Daily thermal storage systems between night and day, and seasonal storage systems between summer and winter are common. In addition, these systems are useful for regulating demand and smoothing demand peaks.

Most renewable energy sources are not continuous but are available intermittently. Heating and cooling thermal loads vary and the peaks of these loads do not necessarily coincide with the availability of renewable energy, making thermal storage necessary for systems to adapt to demand. This is done through *Thermal Energy Storage* (TES) systems. TES systems allow peak demand to be transferred to off-peak periods. These systems are particularly interesting when heat is generated from electrical energy, and the off-peak hours at night are used for the production of thermal energy. These devices can be found as accessory elements to controlling the operation of geothermal installations.

TES systems generally consist of a chiller (non-reversible heat pump) connected to a tank under adiabatic thermal insulation conditions. The early systems were based on the specific heat or sensible heat of the stored fluid. However, the most widely used TES systems today are those based on latent heat, using materials with a high enthalpy of phase change, which allows more energy to be stored per unit volume and whose temperature remains constant until the phase change is complete. There is a third type of system, less popular, that is based on chemical reactions capable of improving performance.

There are different strategies for operating TES systems. Short-term strategies store thermal energy in TES for daily, usually night-day, use. Long-term strategies are for seasonal use, usually winter-summer.

Seasonal Thermal Energy Storage (STES) systems store large amounts of thermal energy during one seasonal period for use in another season. In some cases, STES systems are coupled with heat pumps and solar thermal panels to improve the efficiency of the system. Parameters that influence the efficiency of TES systems are storage temperature, heat loss, the period between storage and use, and the storage medium. There are usually two storage media: solid storage media such as a sediment or rock, and liquid storage media, usually water (Dincer et al. 1997).

There are four main types of STES systems for indoor climate control: *hot water storage tanks* (HWST); *water-gravel pit storage* (WGPS) systems, also called artificial aquifers; and *underground thermal energy storage* (UTES) systems, which are divided into closed (BTES) and open (*ATES*) systems (Fleuchaus et al. 2018; Hesaraki et al. 2015). UTES systems are generally considered to have the highest capacity and long-term efficiency; however, their development is considered to be in its infancy, with great potential for future utilisation (Gao et al. 2017a).

HWST systems typically use reinforced concrete or thermally insulated steel tanks, filled with water as a storage medium. Although these systems have very low heat losses, their cost is very high. WGPS systems use gravel in addition to water as a thermal energy storage medium, with thermal insulation on the sides and top. They are cheaper than HWTS, but they are still expensive to build.

BTES systems use the natural terrain itself for thermal energy storage. In thermal installations with balanced annual heating and cooling loads, excess *heat* is stored from summer to winter and excess *cold* from winter to summer. BTES systems use closed-loop BHE arrays (Fig. 5.26). The main characteristics of BTES systems with balanced annual heat loads are given in VDI4640/3 (2001) and described below. Temperatures in the BHEs reach a steady state thermal regime relatively quickly

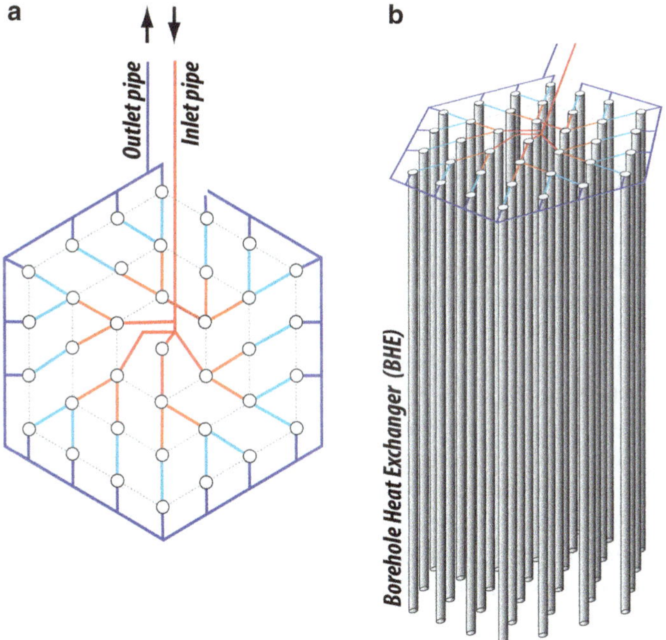

Fig. 5.26 Hexagonal grid in plan (**a**) and in perspective (**b**) of a borehole thermal energy storage (BTES) system with borehole heat exchangers

(2–3 years), with minor variations in daily and seasonal cycles throughout the year. Thermal interference between BHEs can occur during operation, but their effect on storage is minor compared to GSHP systems. Therefore, closer spacing between heat exchangers is common. To minimise the total surface area to volume ratio of the BHE array, a regular cylindrical or hexagonal grid is preferable (Fig. 5.26). A linear arrangement would be strongly discouraged for underground heat storage.

It is vital to understand the hydrogeology of the site in order to ensure efficient recovery of heat transferred to the ground. Thus, in an installation where there is heat advection by regional groundwater flow, the heat transferred with the ground would be displaced by advection down gradient, and it would not be recovered. Consequently, the higher the groundwater flow, the lower the percentage of heat recovery will be. Therefore, the hydraulic conductivity should be low to moderate, with a low hydraulic gradient. During heat storage operations, it is considered acceptable and even desirable for the heat exchanger temperatures to be as extreme as possible.

The total thermal storage E_{alm} [J] of a grid of vertical closed-loop geothermal heat exchangers with a morphology approximating to a cylinder (Fig. 5.26) of depth z [m] and a grid of equivalent radius r_e [m] can be approximated by (Banks 2012):

$$E_{alm} = C_e \pi r_e^2 z \Delta T \tag{5.114}$$

where C_e [J m^{-3} K^{-1}] is the equivalent volumetric heat capacity of the porous media described in Eq. 2.123, and ΔT [K] is the average temperature rise produced in the soil contained in the heat exchanger grid. The theoretical total thermal storage represents the maximum amount of heat that is transferred to the ground in summer and extracted in winter. Generally, the heat flux containing the thermal energy to be stored tends to enter in the subsurface through the central part of the heat exchanger grid (Fig. 5.26a) and is recovered from the outer edge (VDI4640/3, 2001). The inlet temperatures to the BHEs can be variable, depending on the climatic characteristics of the operating site. In northern countries, where there is a tendency to store summer heat for later extraction in winter, temperatures of the heat carrier fluid are generally considered acceptable in the range of 40–70 °C (VDI4640/3, 2001) and have been reported as high as 90 °C (Gabrielsson et al. 2000). At temperatures above 60 °C, it is necessary to consider the properties of the heat carrier fluid pipes and to take into account the temperature dependence of the thermal properties of the materials, as well as free convection processes of the groundwater.

BTES systems can be characterised by a recovery rate, calculated as the amount of energy recovered after one or more storage cycles divided by the amount of thermal energy initially transferred to the storage reservoir during those same cycles. A BTES installation should operate with recovery rates of at least 50%, rising to over 80% for larger installations (VDI4640/3, 2001).

5.7 Thermoactive Geostructures

In newly constructed buildings and underground structures, it is possible to incorporate closed-loop geothermal heat exchangers inside foundation elements (piles and walls) or in tunnel linings (Soga and Rui 2016). These are the so-called thermoactive (or thermally activated) piles, walls and tunnels (Fig. 5.27). These types of underground structures equipped with geothermal heat exchangers are collectively referred to as thermoactive geostructures. These types of geothermal heat exchangers take advantage of the high thermal conductivity of concrete and its capacity for heat transfer with the shallow subsurface. Among all geostructures, foundation piles are

Fig. 5.27 Conceptual scheme of thermoactive geostructures (1 and 2) as constituent elements of a shallow geothermal system (1, 2, 3 and 4) during shallow geothermal energy exploitation (5)

considered to be the most suitable medium for heat transfer with the surrounding ground.

In the study of thermoactive geostructures there are three coupling processes relevant to assessing their performance: (1) hydro-mechanical coupling within the heat exchanger pipes, (2) thermo-mechanical coupling in the substructure to describe the impact of temperature change on the mechanical stress state of concrete structures, and (3) thermo-hydro-mechanical (THM) coupling of the soil, for the study of interactions between temperature, displacements and groundwater pressure fields.

5.7.1 Thermoactive Piles

In a thermally active pile installation, a closed-loop pipe assembly is installed inside the structural piles of a building. The early thermoactive piles were mainly constructed as driven precast concrete piles in which the pipes, acting as geothermal heat exchangers in a closed loop, were integrated. Today, the technology has been extended to large diameter bored piles, which allow multiple U-shaped pipe loops to circulate the heat transfer fluid. In thermoactive bored piles, the pipes are made of high-density polyethylene and have a 20–25 mm diameter and a thickness of approximately two millimetres. The heat carrier fluid pipes are fixed to the steel reinforcement cage and then concreted. As with BHEs, the heat carrier fluid used in the primary circuit is water, salt solutions or, more commonly, water–glycol mixtures.

The heat carrier fluid mixture, flow rates and pipe diameters must be chosen to achieve turbulent flow conditions so that heat transfer can be effective. The overall thermal performance of a thermoactive pile depends on the position of the pipes within the pile, as well as the fraction volume represented by the concrete of the pile, relative to the size of the heat carrier fluid pipes (Gao et al. 2008).

5.7.2 Thermoactive Walls

The same principle followed in thermoactive piles can be applied to diaphragm walls and foundation walls of buildings, so they can provide thermal energy while supporting the structural load. Due to the large size of the thermoactive walls, more heat can be transferred with the ground compared to thermoactive pile systems. Inside the wall, the central part and the part facing the inside of the building are reserved for structural steel reinforcement cage, and the pipes acting as geothermal heat exchangers are positioned close to the ground in order to reduce the thermal resistance of the concrete and improve overall thermal performance of the energy wall. If the thermoactive wall is also used as a basement wall, then the inner side of the wall is covered with thermal insulation materials to ensure the heat from the heat exchangers is transferred to the ground and not to the basement. If the thermoactive wall belongs to an underground railway or metro station, there is the possibility of introducing another heat carrier fluid pipe system on the inside to extract heat in winter from both sides of the wall. In summer, there is no heat demand, and the thermoactive wall can be used to dissipate heat from the station to the ground, so that it is stored and can be extracted later in winter (Brandl 2006).

5.7.3 Thermoactive Tunnels

A metro train unit emits approximately 1.1 MW of heat through motors, air conditioning and braking systems (Nicholson et al. 2013). All of this heat generated on metro lines can be captured by thermoactive geostructures and reused in nearby buildings (Wilhelm and Rybach 2003). For this purpose, the heat exchanger tubes are inserted into the tunnel linings. For tunnels with prefabricated segments, a segmental lining with heat exchanger tubes can be prepared in advance. The work of connecting the systems to the heat pump is carried out after the construction of the tunnel so that the construction process is not affected.

The efficiency of a thermoactive tunnel can be influenced by both the ground temperature and the air temperature inside the tunnel. In short tunnels, the air temperature inside is usually controlled by the surface air temperature and they are thus called *cold tunnels*. In a long tunnel, the air temperature inside is usually higher than the air temperature at the surface, and these are called *hot tunnels*. Depending on the type of tunnel (*cold* or *hot*), there are different operating regimes of the geothermal installation. For *hot* tunnels, heat exchangers are used to extract heat from both sides of the tunnel wall (ground and internal air). Similar to thermoactive walls, in winter the heat generated by the tunnel can be reused for air conditioning of buildings while, in summer, heat is extracted from the air to cool it and improve the efficiency of the transport air conditioning system. The extracted heat can be also used as a DHW supply. If the land surrounding the tunnel is used for heat storage between the summer and winter seasons, the system becomes more efficient if the inner section of the tunnel is insulated.

References

Aifantis EC (1979) A new interpretation of diffusion in high-diffusivity paths—a continuum approach. Acta Metall 27(4):683–691. https://doi.org/10.1016/0001-6160(79)90019-1

Al-Khoury R (2017) Computational modeling of shallow geothermal systems. Multiphys Model 4

Al-Khoury R, Bonnier PG (2006) Efficient finite element formulation for geothermal heating systems. Part II: transient. Int J Numer Methods Eng 67(5):725–745. https://doi.org/10.1002/nme.1662

Al-Khoury R, Kölbel T, Schramedei R (2010) Efficient numerical modeling of borehole heat exchangers. Comput Geosci 36:1301–1315. https://doi.org/10.1016/j.cageo.2009.12.010

Anderson MP, Woessner WW, Hunt RJ (2015) Applied groundwater modeling: simulation of flow and advective transport. Elsevier Science

Banks D (2012) An introduction to thermogeology: ground source heating and cooling. Wiley, Chichester, West Sussex

Bernier MA (2001) Ground-coupled heat pump system simulation. ASHRAE Trans 107:605–616

Biglarian H, Abbaspour M, Saidi MH (2017) A numerical model for transient simulation of borehole heat exchangers. Renew Energy 104:224–237. https://doi.org/10.1016/j.renene.2016.12.010

Brandl H (2006) Energy foundations and other thermo-active ground structures. Géotechnique 56(2):81–122

Carslaw HS, Jaeger JC (1986) Conduction of heat in solids. Clarendon Press, Oxford, UK

Chiasson AD (2016) Geothermal heat pump and heat engine systems: theory and practice. Wiley, New York, USA

Claesson J, Dunand A (1983) Heat extraction from the ground by horizontal pipes: a mathematical analysis. Swedish Council for Building Research

Claesson J, Eskilson P (1987) Conductive heat extraction by a deep borehole, analytical study. Department of Mathematical Physics and Building Technology, Sweden

COMSOL (2020) COMSOL, version 5.5. Multiphysics reference manual. COMSOL Inc.

Diao NR, Zeng HY, Fang ZH (2004a) Improvement in modeling of heat transfer in vertical ground heat exchangers. HVAC&R Res 10(4):459–470. https://doi.org/10.1080/10789669.2004.10391114

Diao N, Li Q, Fang Z (2004b) Heat transfer in ground heat exchangers with groundwater advection. Int J Therm Sci 43(12):1203–1211. https://doi.org/10.1016/j.ijthermalsci.2004.04.009

Diersch HJ (2013) FEFLOW: finite element modeling of flow, mass and heat transport in porous and fractured media. Springer, Berlin, Heidelberg

Diersch HJG, Bauer D, Heidemann W, Rühaak W, Schätzl P (2011) Finite element modeling of borehole heat exchanger systems: Part 1. Fundamentals. Comput Geosci 37(8):1122–1135. https://doi.org/10.1016/j.cageo.2010.08.003

Dincer I, Dost S, Li X (1997) Performance analyses of sensible heat storage systems for thermal applications. Int J Energy Res 21(12):1157–1171. https://doi.org/10.1002/(sici)1099-114x(19971010)21:12%3c1157::aid-er317%3e3.0.co;2-n

Eskilson P (1987) Thermal analysis of heat extraction boreholes. University of Lund, Lund, Sweden

Eskilson P, Claesson J (1988) Simulation model for thermally interacting heat extraction boreholes. Numer Heat Transf 13(2):149–165. https://doi.org/10.1080/10407788808913609

Fleuchaus P, Godschalk B, Stober I, Blum P (2018) Worldwide application of aquifer thermal energy storage—a review. Renew Sustain Energy Rev 94:861–876. https://doi.org/10.1016/j.rser.2018.06.057

Gabrielsson A, Bergdahl U, Moritz L (2000) Thermal energy storage in soils at temperatures reaching 90 °C. J Sol Energy Eng 122(1):3–8. https://doi.org/10.1115/1.556272

Gao J, Zhang X, Liu J, Li K, Yang J (2008) Numerical and experimental assessment of thermal performance of vertical energy piles: an application. Appl Energy 85(10):901–910

Gao Y, Cheng Y, Nan S (2017a) Heat transfer performance of the underground CO_2 pipe in the direct expansion ground source heat pump. Energy Procedia 105:4955–4962. https://doi.org/10.1016/j.egypro.2017.03.989

Gao L, Zhao J, An Q, Wang J, Liu X (2017b) A review on system performance studies of aquifer thermal energy storage. Energy Procedia 142:3537–3545. https://doi.org/10.1016/j.egypro.2017.12.242

Gu Y, O'Neal D (1998) Development of an equivalent diameter expression for vertical U-Tubes used in ground-coupled heat pumps. ASHRAE Trans 104:347–355

Hellstrom G (1991) Ground heat storage. Thermal analysis of duct storage systems. University of Lund, Sweden

Hesaraki A, Holmberg S, Haghighat F (2015) Seasonal thermal energy storage with heat pumps and low temperatures in building projects—a comparative review. Renew Sustain Energy Rev 43:1199–1213. https://doi.org/10.1016/j.rser.2014.12.002

Hiller CC, Electric Power Research Institute, National Rural Electric Cooperative Association, International Ground Source Heat Pump Association (2000) Grouting for vertical geothermal heat pump systems: engineering design and field procedures manual. In: International Ground Source Heat Pump Association, Oklahoma State University, Stillwater, Okla

IGSHPA (2009) Residential and light commercial design and installation manual. International Ground Source Heat Pump Association

Ingersoll LR, Plass HJ (1948) Theory of the ground pipe heat source for the heat pump. ASHVE Trans 54:339–348

Ingersoll LR, Zobel OJ, Ingersoll AC (1955) Heat conduction with engineering, geological, and other applications. McGraw-Hill, New York, p 325

Kipp KLJ, Hsieh PA, Charlton SR (2008) Guide to the revised ground-water flow and heat transport simulator: HYDROTHERM—Version 3. U.S. Geological Survey Techniques and Methods 6-A25, p 160

Kipp KL (1997) Guide to the revised heat and solute transport simulator: HST3D—Version 2. USGS, Denver, CO

Lamarche L, Beauchamp B (2007) New solutions for the short-time analysis of geothermal vertical boreholes. Int J Heat Mass Transf 50:1408–1419. https://doi.org/10.1016/j.ijheatmasstransfer.2006.09.007

Langevin CD, Thorne DT Jr, Dausman AM, Sukop MC, Guo W (2008) SEAWAT Version 4: a computer program for simulation of multi-species solute and heat transport. U.S. Geological Survey Techniques and Methods (USGS), Reston, Virginia

Lazzari S, Priarone A, Zanchini E (2010) Long-term performance of BHE (borehole heat exchanger) fields with negligible groundwater movement. Energy 35(12):4966–4974. https://doi.org/10.1016/j.energy.2010.08.028

Lazzarotto A (2015) Developments in ground heat storage modeling. Doctoral thesis, comprehensive summary Thesis, KTH Royal Institute of Technology, Stockholm, xii, 73 pp

Lee CK, Lam HN (2008) Computer simulation of borehole ground heat exchangers for geothermal heat pump systems. Renew Energy 33(6):1286–1296. https://doi.org/10.1016/j.renene.2007.07.006

Li M, Lai ACK (2015) Review of analytical models for heat transfer by vertical ground heat exchangers (GHEs): a perspective of time and space scales. Appl Energy 151:178–191. https://doi.org/10.1016/j.apenergy.2015.04.070

Lo Russo S, Taddia G, Cerino Abdin E (2018) Modeling the effects of the variability of temperature-related dynamic viscosity on the thermal-affected zone of groundwater heat-pump systems. Hydrogeol J 26(4):1239–1247. https://doi.org/10.1007/s10040-017-1714-x

Man Y, Yang H, Diao N, Liu J, Fang Z (2010) A new model and analytical solutions for borehole and pile ground heat exchangers. Int J Heat Mass Transf 53(13):2593–2601. https://doi.org/10.1016/j.ijheatmasstransfer.2010.03.001

Nicholson DP, Chen Q, Pillai A, Chendorain M (2013) Developments in thermal pile and thermal tunnel linings for city scale GSHP systems. In: Proceedings of the 38th workshop on geothermal reservoir engineering

Nield DA, Bejan A (1999) Convection in porous media. Springer, New York, 546 pp. https://doi.org/10.1007/978-1-4757-3033-3

Philippe M, Bernier M, Marchio D (2009) Validity ranges of three analytical solutions to heat transfer in the vicinity of single boreholes. Geothermics 38(4):407–413. https://doi.org/10.1016/j.geothermics.2009.07.002

Rees S (2016) Horizontal and compact ground heat exchangers. In: Rees S (ed) Advances in ground-source heat pump systems. Woodhead Publishing, pp 117–156

Rybach L, Sanner B (2000) Ground-source heat pump systems—the European experience. GHC Bulletin, p 21

Soga K, Rui Y (2016) 7—Energy geostructures. In: Rees SJ (ed) Advances in ground-source heat pump systems. Woodhead Publishing, Oxford, UK, pp 185–221. https://doi.org/10.1016/B978-0-08-100311-4.00007-8

Soni SK, Pandey M, Bartaria VN (2016) Experimental analysis of a direct expansion ground coupled heat exchange system for space cooling requirements. Energy Build 119:85–92. https://doi.org/10.1016/j.enbuild.2016.03.026

Stauffer F, Bayer P, Blum P, Giraldo NM, Kinzelbach W (2013) Thermal use of shallow groundwater. Taylor & Francis, Abingdon, UK

Sutton MG, Nutter DW, Couvillion RJ (2003) A ground resistance for vertical bore heat exchangers with groundwater flow. J Energy Res Technol 125(3):183–189. https://doi.org/10.1115/1.1591203

UNE (2014) UNE100715-1:2014 guide for the design, implementation and monitoring of a geothermal system. Part 1: Vertical closed circuit systems. Asociacion Espanola de Normalizacion

VDI4640/3 (2001) Utilization of the subsurface for thermal purposes: underground thermal energy storage. In: V.D. Ingenieure, Düsseldorf, p 42

Voss C, Provost AM (2010) SUTRA, a model for saturated-unsaturated, variable-density groundwater flow with solute or energy transport. Version 2.2, USGS, Reston, USA

Wilhelm J, Rybach L (2003) The geothermal potential of Swiss Alpine tunnels. Geothermics 32(4):557–568. https://doi.org/10.1016/S0375-6505(03)00061-0

Yang H, Cui P, Fang Z (2010) Vertical-borehole ground-coupled heat pumps: a review of models and systems. Appl Energy 87(1):16–27. https://doi.org/10.1016/j.apenergy.2009.04.038

Yavuzturk C, Chiasson AD (2002) Performance analysis of U-tube, concentric tube, and standing column well ground heat exchangers using a system simulation approach. ASHRAE Trans 108:925–938

Yavuzturk C, Spitler J, Rees S (1999) A transient two-dimensional finite volume model for simulation of vertical U-tube ground heat exchanger. American Society of Heating, Refrigerating and Air-Conditioning Engineers Transactions, p 105

Zeng H, Diao N, Fang Z (2003a) Heat transfer analysis of boreholes in vertical ground heat exchangers. Int J Heat Mass Transf 46(23):4467–4481. https://doi.org/10.1016/S0017-9310(03)00270-9

Zeng H, Diao N, Fang Z (2003b) Heat transfer analysis of boreholes in vertical ground heat exchangers. Int J Heat Mass Transf 46:4467–4481. https://doi.org/10.1016/s0017-9310(03)00270-9

Chapter 6
Shallow Geothermal Systems with Open-Loop Geothermal Heat Exchangers

6.1 Shallow Geothermal Installations with Open-Loop Geothermal Heat Exchangers

Open-loop shallow geothermal installations are based on the use of groundwater (Fig. 6.1) as the heat carrier fluid for heat transfer between the geothermal heat pump and the underground thermal reservoir (ASHRAE 2016). Heat pumps connected to geothermal well systems are referred to as *groundwater heat pumps* (GWHP).

Groundwater is of great interest for shallow geothermal energy systems for one basic reason: its high energy density. Water has a high heat capacity, four times higher than other solid materials in the ground. In the case of groundwater bodies with good chemical quality, a significant amount of thermal energy can be obtained through one or two groundwater wells, much more than through closed-loop geothermal heat exchangers (Chiasson 2016). This is because heat transport by groundwater advection in aquifers is far superior to that by conduction. The advantages of using groundwater in open-loop heat exchangers compared to closed-loop heat exchangers are: (1) higher potential in energy density per unit area available; in thermal installations where the total area available for drilling of BHE is limited, the option of using geothermal wells is especially interesting; (2) lower construction cost (fewer boreholes), especially if the well is also built for drinking water supply; (3) the well drilling technology is highly developed; and (4) the collection temperatures are much more stable under a good exploitation design. The disadvantages are: (1) dependence on the chemical quality of the groundwater; (2) the need to evacuate the reject water from the heat pump; and (3) stricter regulation, which requires a groundwater concession and thermal discharge permits.

Groundwater can be fed directly into one of the heat exchangers (evaporator or condenser) of the heat pump, direct use installations. Normally, the pumped groundwater is not fed into a heat exchanger (evaporator or condenser) of the heat pump, but one or more plate heat exchangers are installed to create two secondary circuits, called indirect use installations. The first secondary circuit connects the geothermal production and injection wells with the plate heat exchanger. Pipes with specific

A. García Gil et al., *Shallow Geothermal Energy*, Springer Hydrogeology,
https://doi.org/10.1007/978-3-030-92258-0_6

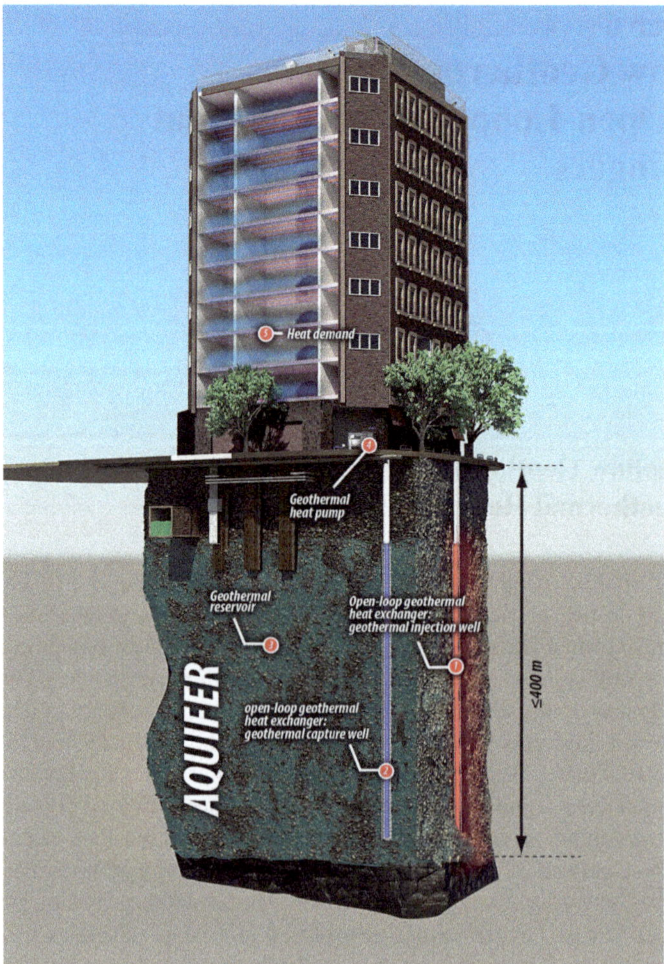

Fig. 6.1 Shallow geothermal system with open-loop geothermal heat exchanger

corrosion resistance characteristics (e.g., PVC) are usually considered in this circuit. The second secondary circuit connects the plate heat exchanger with the heat pump, with a controlled heat carrier fluid to ensure the durability of the heat pump. Although there is a loss of heat transfer efficiency, with the plate heat exchanger the ground water does not come into contact with the heat pump, extending its service life significantly. Direct use installations are only recommended for very small installations. The reason for this is that experience has shown that even with groundwater with favourable geochemical characteristics, problems of scaling due to mineral precipitation and corrosion often occur.

The installation shall be considered as a *unitary installation* if several water–air heat pumps are distributed in the building. It will be considered as a distribution *heating and cooling plant*, when a small number of large capacity water-to-water heat pumps are used, supplying *hot* and *cold* water to a distribution system of two or four pipe lines. Unitary installations are more common and efficient (Sarbu and Sebarchievici 2015).

6.2 Components of an Open-Loop Geothermal Heat Exchanger

The basic components of a shallow geothermal installation using groundwater are: geothermal wells for groundwater production (Fig. 6.2) and injection, and a groundwater drive pump and water-to-water plate heat exchanger adapted to the characteristics of the groundwater hydrochemistry. Although these components are in common use in many other areas outside of geothermal energy application, some special considerations are required, mostly related to the reactivity of groundwater.

The geothermal injection well is not strictly necessary if the water is not returned to the aquifer. The reject water from the heat pump can be discharged to a surface water body, such as a lake or river. Groundwater extraction pumps are usually submersible pumps, which not only allow the pumping of groundwater from relatively deep wells but also prevent oxygen from entering the system. In addition, it is advisable to introduce a sludge separator in the secondary groundwater circulation circuit to prevent the entry of solid particles from the entrainment (washing) of fines during the more or less turbulent flow of groundwater as it approaches the well screen.

Plate heat exchangers through which groundwater flows are susceptible to scaling, corrosion and clogging. To avoid damage to the exchanger, it is recommended that the groundwater should not have an electrical conductivity of more than 450 $\mu S\ cm^{-1}$. In addition, clogging processes can occur due to bacterial growth (iron bacteria).

6.3 Design, Construction and Operation

The design of shallow geothermal installations with open-loop geothermal heat exchangers depends on the thermal energy demand of the installation and the specific hydrogeological conditions of the intended site, including the geochemistry of the groundwater. The numberless possible demand scenarios and hydrogeological conditions make it impossible to make a general recommendation on the design of open-loop exchanger operated facilities.

For the design of this type of installation, hydrogeological knowledge of the area considered for its use is necessary. Deep piezometric levels may affect the efficiency of the installation, as greater electricity consumption will be necessary to pump the groundwater to the surface and, even if turbines are used during reinjection for energy recovery, the COP of the installation may be affected. Areas of poor transmissivity

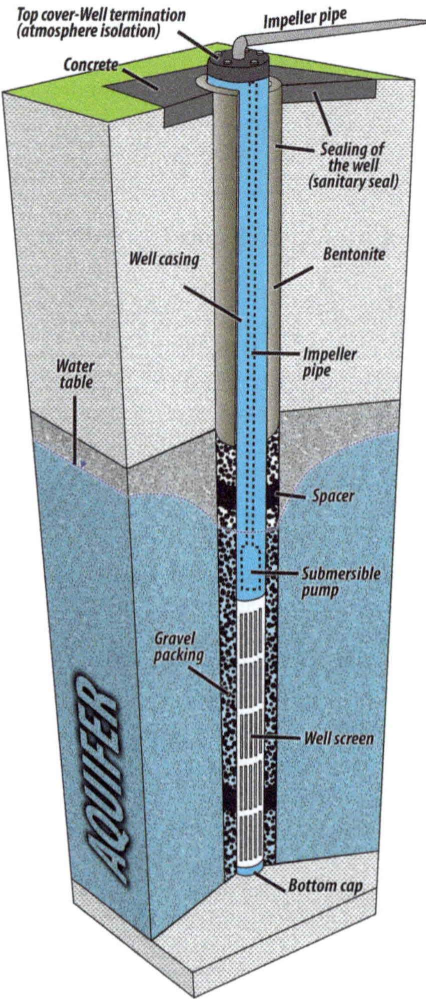

Fig. 6.2 Key elements of an open-loop geothermal heat exchanger: geothermal well

may require forced injection of the pumped water, which would increase the electricity consumption of the operation. Confined aquifers and shallow free aquifers can be favourable for geothermal exploitation, as the extraction-injection of groundwater has a low energy cost. An open-loop exchanger design without a hydrogeological study is a risk for installation and for the development of shallow geothermal energy in general. Consideration of hydrogeological properties in the design, construction and operation of a shallow open-loop geothermal installation is essential to ensure operational safety and a long lifetime.

The larger the geothermal installation, the more important is the correct design of the open-loop geothermal heat exchangers. In addition to size demand, it is necessary to take into account possible complex hydrogeological conditions and densely populated (installation density) areas where the design requires more attention.

In areas without hydrogeological complexity, low population density and low thermal power demands, the design of the open-loop geothermal heat exchanger (wells and their location) can be approximated by simplified conceptual models and analytical solutions associated with the assumed simplifications.

Thus, for example, a shallow geothermal installation for a house with a consumption of approximate 12 kWh thermal could be estimated as follows (Feuvre and Cox 2009): Operating 1800 h per year gives a heat demand of approximately 12 kWh year^{-1}. To transport this thermal energy to the heat pump by advection (Eq. 2.79) and assuming a temperature change of 5 °C, a flow rate of 13 L s^{-1} would be necessary, on the basis of which of the geothermal wells could be sized.

Regarding the design of the geothermal well layout, it is first necessary to estimate the width of the capture zone of the well and the width of the injection area (Fig. 6.3). The width l_{ca} [m] of the boundary of the pumping zone (capture or injection) is proportional to the pumping rate Q [m^3 s^{-1}] and is inversely proportional to the product of the Darcy velocity v_D [m s^{-1}] and the saturated thickness of the aquifer b [m] according to (DGG 2016):

$$l_{ca} = \frac{\mathrm{Q}}{v_D \mathrm{b}} \tag{6.1}$$

The temperature difference ΔT downstream of an injection point, with respect to the groundwater, decreases exponentially as a function of the groundwater residence time t_{re} [s] and the hydrogeological parameters according to:

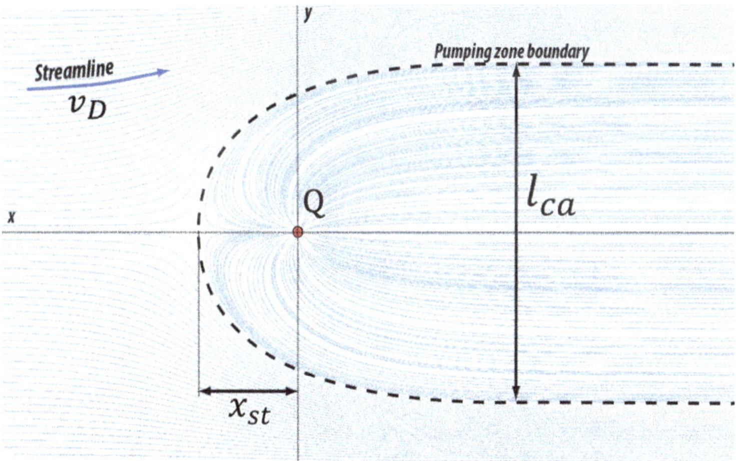

Fig. 6.3 Definition of the injection pumping zone boundary

$$\Delta T = \Delta T_{max} \cdot \exp\left[\frac{-\lambda t_{re}}{\phi_{ef}\rho_w c_w \mathrm{b}\,\mathrm{b}_{sup}}\right] \quad (6.2)$$

where ΔT_{max} [K] is the maximum initial temperature difference at $t_{re} = 0$; where the water injected at a given injection temperature and the groundwater at the background aquifer temperature are not yet mixed and their difference is maximum, λ [W m^{-1} K^{-1}] is the thermal conductivity, ϕ_{ef} [–] is the effective porosity, ρ_w [kg m^{-3}] is the density of water, c_w [J kg^{-3} K^{-1}] is the specific heat capacity of water, b [m] is the saturated aquifer thickness and b_{sup} [m] is the thickness of the covering layer above the aquifer. Note that it is possible to obtain the residence time by clearing Eq. 6.2.

Situations where the geothermal well doublet is aligned with the regional groundwater flow (Fig. 6.4) presenting a well spacing of a [m], it is possible to determine a_{cr} [m] as the critical separation distance between the captation and injection wells. Beyond this the extraction and injection cones interfere with each other and, consequently, a high risk of hydrothermal interference between wells (thermal self-interference) is generated.

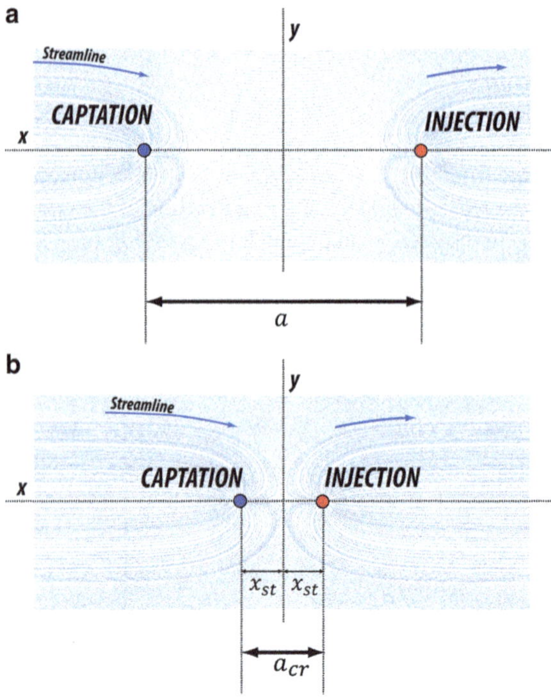

Fig. 6.4 General (**a**) and critical (**b**) separation distance between a geothermal captation well and an injection well. Well-doublet system configuring an open-loop geothermal heat exchanger for a shallow geothermal installation

For this purpose it is necessary to determine x_{st} [m] (Fig. 6.3) as the distance from the capture area in the direction of the adjoining well in both cases. Since the flow rate and the thermal parameters of the aquifer are the same, x_{st} will be of the same magnitude. Therefore $a_{cr} = 2x_{st}$. The critical separation distance a_{cr} between capture and injection well can be calculated from the following expression (Mehlhorn et al. 1981):

$$a_{cr} = \frac{2\mathrm{Q}}{\pi \, \mathrm{b} \, v_D} \tag{6.3}$$

In the absence of total alignment between wells and groundwater flow, the critical separation varies with the angle formed between the well axis (Fig. 6.5) and is minimal at angles of approximately 50° to the groundwater flow. Consequently, to minimise the potential for interference, it is recommended that shallow geothermal groundwater wells be sited at an angle of 50° to the groundwater flow. There are no analytical solutions for more complex situations such as multiple-captation/injection wells coexisting or realistic complex hydrogeological conditions. Therefore, design of geothermal well sites for groundwater thermal use will require numerical models.

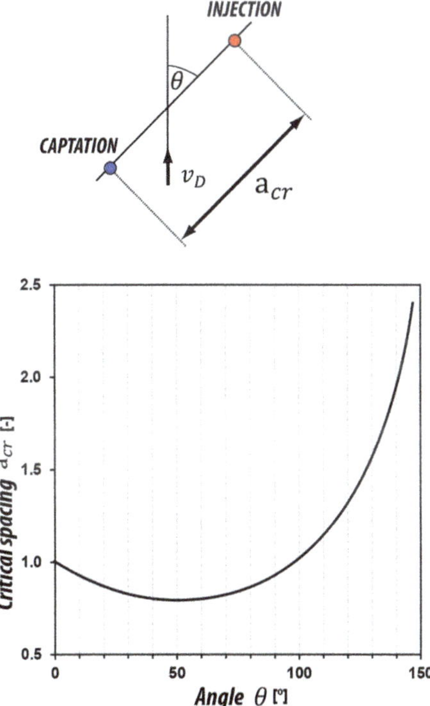

Fig. 6.5 Variation of the critical separation (a_{cr}) as a function of the angle (θ) formed between the groundwater flow (v_D) and the alignment of geothermal wells. Data obtained from DGG (2016)

6.4 Heat Transfer with the Ground

The groundwater flow and heat transport problems experienced in the thermal use of groundwater can only be solved analytically when idealised concrete cases are considered, in which various simplifications are assumed. The flow and transport parameters considered in the general equations, as well as the boundary conditions, are assumed to be homogeneous. In general, simple flow domains, initial and boundary conditions are considered.

Geothermal installations with open-loop geothermal heat exchangers usually present a single captation well and a single injection well (well-doublet system), although they can sometimes present multiple injection or caption wells in larger installations. A comprehensive review of available analytical solutions for open-loop heat exchangers can be found in Stauffer et al. (2013). The conclusion that can be drawn from this review is that the availability of analytical solutions to describe heat transfer in the ground using geothermal wells is very limited. Furthermore, a complete analytical solution for this problem does not exist to date. Only a hydraulic solution to the problem is available and can be used to assess recirculation rates between the wells and to describe the flow field during exploitation.

To analytically resolve the groundwater flow problem, it is necessary to assume an open-loop heat exchanger with multiple wells in a uniform and infinite two-dimensional aquifer. GWHP systems can be considered to be in a quasi-stationary regime during operation. Without considering surface recharge, and assuming an isotope medium with constant parameters of hydraulic permeability K [m s^{-1}] and saturated thickness of the aquifer b [m], it is found that for a single well, the velocity potential $\varphi = K\mathrm{b}$ will be determined by (Bear 1979):

$$\varphi(x, y) = \frac{\mathrm{Q}}{4\pi \mathrm{b}} \ln\left(\frac{(x - x_s)^2 + (y - y_s)^2}{\mathrm{R}^2}\right) \tag{6.4}$$

where Q [m^3 s^{-1}] is the pumping rate (Q > 0 injection), (x_s, y_s) are the coordinates of the borehole constituting the well, b [m] is the thickness of the aquifer, and R [m] is the radius of influence of the well.

In order to understand the possible hydraulic interactions between operating wells, it is interesting to identify the scalar function known as the stream function which, when derived, yields the components of the flow velocities in the field. Ψ which, when derived, gives the components of the flow velocities in the field. This makes it possible to obtain the flow lines that represent the groundwater trajectory in a stationary regime. The *stream function* $\Psi(x, y)$ for a well is expressed by:

$$\Psi(x, y) = \frac{\mathrm{Q}}{2\pi \mathrm{b}} \tan^{-1}\left(\frac{y - y_s}{x - x_s}\right) \tag{6.5}$$

A more realistic situation requires considering regional groundwater flow as a uniform flow field. Applying Darcy's Law (Eq. 2.148) for the two-dimensional problem in terms of velocity potentials:

$$\frac{Q}{A} = v_D = -K\nabla h(x, y) = -\nabla\varphi(x, y) \tag{6.6}$$

Considering that there is a regional uniform flow field v_D [m s^{-1}], the velocity potential will be determined by:

$$\varphi(x, y) = -v_D(x\cos\theta + y\sin\theta) \tag{6.7}$$

where θ is the angle conformed between the regional flow direction and the x axis (Fig. 6.5). The stream function for the uniform regional flow is:

$$\Psi(x, y) = -v_D(y\cos\theta - x\sin\theta) \tag{6.8}$$

By applying the superposition principle to Eqs. 6.5 and 6.8, it is possible to obtain the flow lines of a GWHP system operating in an environment subject to a uniform regional groundwater flow.

In the case of a GWHP system operating with a well doublet, the groundwater flow regime induced by the wells doublet where the captation rate is equal to the injection rate (a common situation) and where there is a uniform regional flow, the stream function will be determined by (DaCosta and Bennett 1960):

$$\Psi(x, y) = -v_D(y\ \cos\theta - x\ \sin\theta) + \frac{Q}{2\pi b}\tan^{-1}\left(\frac{ay}{\left(\frac{a}{2}\right)^2 - x^2 - y^2}\right) \tag{6.9}$$

where a [m] is the spacing between wells.

Figure 6.6 shows two examples of the application of Eq. 6.9 for a shallow geothermal well doublet in an aquifer with a hydraulic permeability of 30 m day^{-1}, a hydraulic gradient of 3×10^{-3}, a saturated thickness of ten meters and a capture-injection flow rate of 90 m^3 day^{-1}. Two alignment arrangements of the captation and injection wells are considered, one fully aligned with the regional flow (Fig. 6.6a) and the second one forming a 45° angle (Fig. 6.6b).

In these types of diagrams, there may be areas where the head contours indicate that the hydraulic gradients tend to zero in some points where the Darcy velocity is zero. These points are called *stagnation points*. Each well under operation will tend to generate a stagnation point. When there is enough spacing between wells conforming a wells-doublet, there are two stagnation points between the wells, depending on the pumping rate and hydraulic parameters. As long as the two stagnation points exist, there will be no recirculation between the wells. If all the parameters involved are kept constant and wells are moved closer to each other, the stagnation points will also move closer to each other. When the two stagnation points coexist, the well spacing is considered as the minimum critical recirculation distance.

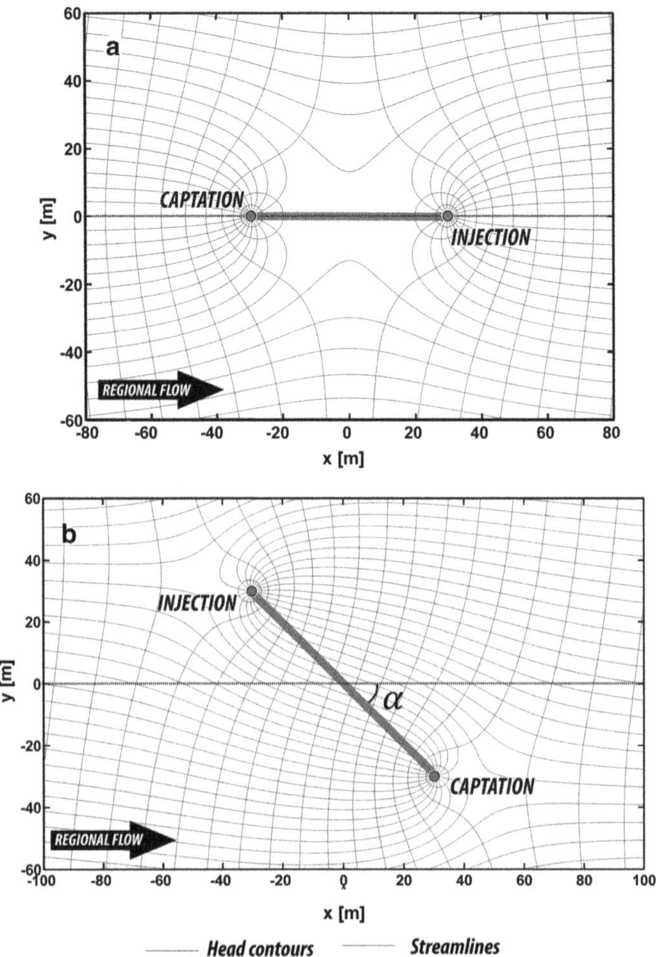

Fig. 6.6 Examples of groundwater flow fields related to the operation of a geothermal well-doublet where the wells are fully aligned with the regional flow (**a**) and forming a 45° angle (**b**). A regional groundwater flow as a uniform flow field is considered

Fig. 6.7 shows an example of the superposition principle for a GWHP system with three wells (two captation and one injection well) operating under a regionally uniform groundwater flow environment.

The presented analytical solutions are only useful for studying thermal–hydraulic interference phenomena between wells. However, this approach neglects heat conduction and hydrodynamic dispersion during heat transport, only considers heat advection mechanism. There is no exact analytical solution to simulate the thermal response of an aquifer, considering an injection well and regional flow (Stauffer et al. 2013). Therefore, the design of large/complex facilities requires the use of numerical models of groundwater flow and heat transport. A decade ago, numerical modelling

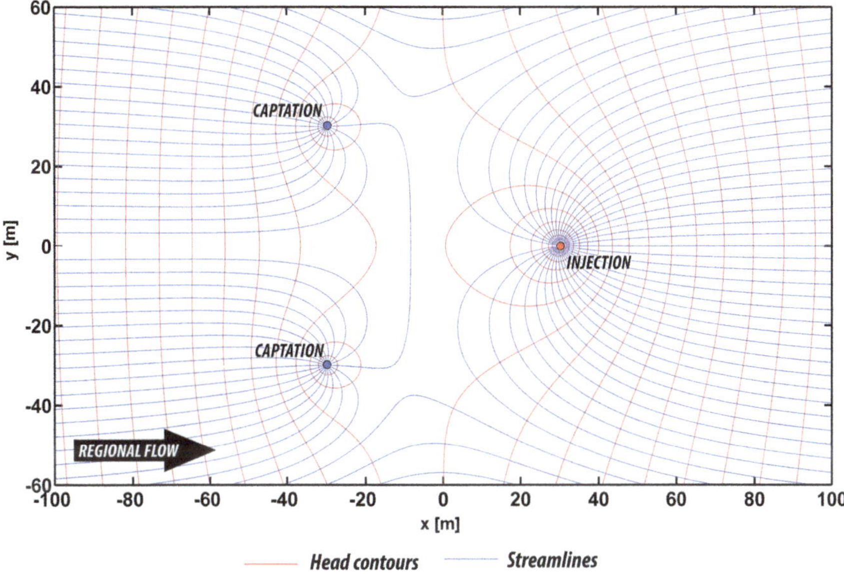

Fig. 6.7 Example of the application of the superposition principle for a groundwater heat pump system with three wells operating under a regionally uniform groundwater flow environment parallel to the x axis

was time-consuming and was considered unfeasible; today, computer power and software application have improved dramatically, reducing the time needed to perform such simulations.

6.5 Chemical Quality of Groundwater

The chemical quality of groundwater is a site-specific property and can vary significantly from place to place. According to Ellis (1998), when working with geothermal fluids it is necessary to take into account seven physicochemical parameters: pH, dissolved oxygen, carbon dioxide, hydrogen sulphide, ammonia, sulphate, chloride-TDS and suspended solid material. Depending on these parameters, groundwater is likely to cause different problems in the secondary circuit between the captation-injection wells and the plate heat exchanger in indirect use installations, or to the heat pump in direct use installations. These are mainly problems of corrosion and chemical and/or biological scaling.

In terms of corrosion, groundwater has a higher partial pressure of carbon dioxide, which means there is a tendency to have a low pH, and the lower the pH value, the more intense the corrosion process will be. With regard to the amount of *Total Dissolved Solids* (TDS) in groundwater, groundwater is considered fresh groundwater with TDS < 1000–2000 ppm, brackish groundwater for 1000–2000 ppm < TDS < 5000 ppm, saline groundwater with 5000 ppm < TDS < 40,000 ppm and brine up to 300,000 ppm. TDS is related to chloride ion as a conservative element in groundwater. Due to its high solubility it is very unlikely to precipitate and reduce its concentration along groundwater flow paths in aquifers. Chloride ion and salinity accelerate the corrosion of steel. Depending on the composition of the steel, corrosion resistance varies, e.g., 304 stainless steel tolerates chloride concentrations up to 140 ppm, while 316 stainless steel withstands chloride concentrations up to 400 ppm. Oxygen also accelerates the corrosion of steel. Groundwater has a lower oxygen partial pressure than atmospheric oxygen, so it is very important to remove air and other gases from the system circuits. There are other parameters that accelerate the corrosion of metals, e.g., if copper is attacked by hydrogen sulphide. In addition, it is not advisable to capture groundwater and inject it into different groundwater bodies. This is due to the fact that different hydrochemistry mixing could trigger reactions that are harmful to the well structure, as well as creating other environmental problems between water bodies of different chemical quality.

As far as clogging is concerned, high pH values favour chemical clogging processes (Rafferty 1999). There are many types of chemical precipitates that tend to clog pipes and plate heat exchangers, but calcium carbonate precipitates are the most common in GWHP systems. High pH values in groundwater will tend to have higher bicarbonate concentrations, and above 100 ppm the water will start to be considered fouling and therefore prone to cause calcium carbonate scales. The most reliable indicators for predicting potential scaling problems are the alkalinity and hardness of the groundwater. The potential for fouling and corrosion in GWHP systems is estimated from the *Ryznar stability index* (RSI) and the *Langelier saturation index* (LSI). Each of these indices is calculated from carbonate and calcium concentrations and pH.

6.5.1 Reducing the Lifetime of Open-Loop Geothermal Heat Exchangers

The construction and maintenance of geothermal wells is essentially the same as for groundwater wells for drinking water supply. In both cases, the components used are exposed to relatively hostile conditions of the environment that tend to degrade or make them unusable. The main degradation processes of well components are scaling and deposition of iron hydroxides, sandblasting, silt accumulation, corrosion, scaling and aluminium precipitates.

Depending on the chemistry of the groundwater and the nature of the aquifer being exploited, the problems that can occur in a well are of a different magnitude. This means that maintenance and prevention tasks will be different. Detrital alluvial aquifers are the most aggressive underground environment where it is common to find suspended fines problems, sand or silt slides, iron and manganese hydroxide scaling problems in the well screen and gravel, as well as corrosion problems.

These scenarios considers one to five-year maintenance and prevention strategies, including: desanding, use of stainless steel and oversizing of wells. Aquifers in hard fissured rocks have longer maintenance times, typically five to ten years, mainly due to fissure clogging. Corrosion processes will often occur when sandstones (hard rock) are involved. When soluble rocks such as limestone or gypsum configure the shallow geothermal reservoir, problems of encrustation of iron and manganese carbonates may occur. In endogenous rocks such as metamorphic rocks and basalts, the incrustations could also be of the ferro-magnesian type. Preventive measures also include the use of stainless steel piping and keeping groundwater flow velocities low, avoiding turbulent flow and cavitation in the submersible pumps. The physicochemical conditions of the aquifer in the secondary circuit must always be maintained, and any contact of the pumped water with atmospheric conditions must be avoided.

General strategies to prevent premature reduction of the well‘s lifespan are to design the gravel pack and well screen in such a way that a laminar flow is ensured at all times during operations. For this purpose, it is advisable to locate the submersible pump outside the gravel pack and screen section. In addition, it is desirable to design the well in such a way that, if there is a drop in the water table, the well screen and the submersible pump are not exposed to sub-aerial conditions. It is important that the gravel pack is made of inert material and is properly washed prior to installation. When using a metallic material for the screen, pump casing or sump casing, all of these shall be of materials that are not susceptible to inducing galvanic corrosion. Finally, the use of compounds with organic additives for well cleaning should be avoided. It is essential to regularly inspect geothermal wells, identifying incipient scaling or corrosion problems, among others. This is fundamental to preventing severe problems, which can be irreversible, and with significant costs to the construction of a well.

In addition to using theoretical hydrochemical models based on the thermodynamics of metal dissolution reactions (corrosion) and mineral and/or bacterial precipitation (scaling), it is possible to perform corrosion and scaling laboratory tests when deemed necessary, especially to prevent or solve problems with plate heat exchangers.

6.6 Numerical Modelling of Groundwater Flow and Heat Transport

Analytical and semi-analytical solutions to the heat transport equations in porous media require too many simplifications of the geometry, initial conditions and boundary conditions to deal with the different problems when real GWHP systems are considered. The intrinsic simplifications of the analytical models are a major limitation in terms of their application to real problems. In fact, as seen in the previous section, there is no analytical solution widely recognised as adequate for geothermal well doublets. The existing solutions partially solve the hydraulic problem, but there is no widely accepted coupled heat transport solution. Moreover, it is considered

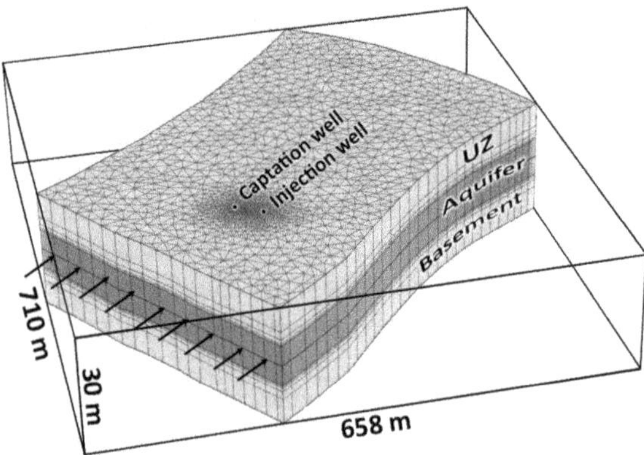

Fig. 6.8 Example of spatial discretisation using a finite element mesh for solving the heat conduction-advection-dispersion equation in the vicinity of a groundwater heat pump system operating a geothermal well doublet

that in practice there is a high complexity associated with the hydrogeology of the problem and the engineering of a GWHP system. This complexity includes aspects such as multi-layered aquifers, multiple-well installations with transient operating regimes for each geothermal well, transient groundwater flow regime and third party installations. These limitations can be overcome by the use of numerical models, as they are effective in describing complicated geometry and complex initial and boundary conditions.

To numerically solve the heat transfer equations in porous media, and to simulate the operation of GWHP systems, the finite difference and finite element methods are mainly used (Fig. 6.8). The finite difference method has historically been the most widely used (Clauser 2012; Eskilson and Claesson 1988; Lee and Lam 2008; Śliwa and Gonet 2004). The finite element method is a robust and versatile method for generating complex computational meshes, but it requires a higher degree of specialized personnel for its implementation and handling. However, currently the finite element method has gained prominence (Al-Khoury and Bonnier 2006; Al-Khoury et al. 2005, 2010; Diersch 2013; Diersch et al. 2011; Muraya 1994; van der Meer et al. 2009).

Currently, models reproducing the thermal regime of the subsurface exposed to the operation of shallow geothermal installations with open-loop geothermal heat exchangers pose significant challenges to finite element code developers. The mesh sizes required to solve three-dimensional hydrogeological problems are in the order of millions of elements. Even with the computational power available today, this problem is quite demanding. The use of numerical models is currently focused on the study of the performance of open-loop geothermal heat exchangers, in particular on the study of thermal interference between geothermal wells in heterogeneous media and complex transient boundary conditions.

The study of terrain hydraulic and thermal parameters using numerical models shows that there are certain parameters controlling heat plume development from geothermal wells. Lo Russo et al. (2012) observed that $\pm$ 20% variability of groundwater flow-related hydraulic parameters influenced the development of thermal plumes after one year of exploitation, in particular those related to the advective component of heat flow (hydraulic conductivity and hydraulic gradient, porosity, etc.). In a similar study, Piga et al. (2017) used parametric variations (hydraulic and thermal) of $\pm$ 50% for 55-year simulations to identify the Darcy velocity as the most influential parameter in the development of the generated thermal plumes, both in its length and width. This study also identified that thermal conductivity and dispersivity play an important role in the long term, while variations in heat capacity have minor effects on the size of the thermal plume. Numerical models have also been used to study actual injection flow rates and temperatures. These operating parameters vary considerably over time, depending on the energy demand required by the thermal installation. Using numerical models, and to improve the prediction of thermal plumes generated by geothermal installations, there are different works using real, equivalent daily (Muela Maya et al. 2018), monthly and seasonal injection temperature and flow rate data. These concluded that the calculated plumes were consistent with experimental data in a downstream piezometer when using injection temperature data on an hourly, daily or monthly resolution, while seasonal averages are not adequate (Lo Russo et al. 2014). Thermal feedback and thermal recirculation phenomena have been studied using numerical models by Casasso and Sethi (2015) to develop a tool for the design of geothermal installations with geothermal wells, taking this phenomenon into account.

Another aspect of the use of city-scale numerical models is the management of shallow geothermal resources in urban environments. They allow us to understand the thermal regime under cities with a large number of geothermal systems and to study the thermal impacts generated, depending on any given boundary condition, as well as to study the effects of shallow geothermal resource management measures. Models of cities such as London (Herbert et al. 2013), Basel (Mueller et al. 2018), Zaragoza (García-Gil et al. 2014b) (Fig. 6.9) or Turin (Sciacovelli et al. 2014) are of interest.

6.7 Aquifer Thermal Energy Storage (ATES)

The concept of underground thermal storage developed in Sect. 5.6 is not exclusive to *borehole thermal energy storage* (BTES) installations. When in so-called UTES systems, geothermal wells are used for seasonal heat storage in low-permeability soils, and they are referred to as ATES (*aquifer thermal energy storage*) (Fleuchaus et al. 2018; Hesaraki et al. 2015).

Currently, shallow geothermal ATES installations are well developed in the Netherlands, Sweden, Canada, France, Germany and the USA (Bloemendal et al. 2014; Possemiers et al. 2014; Sommer et al. 2015). ATES installations are used for

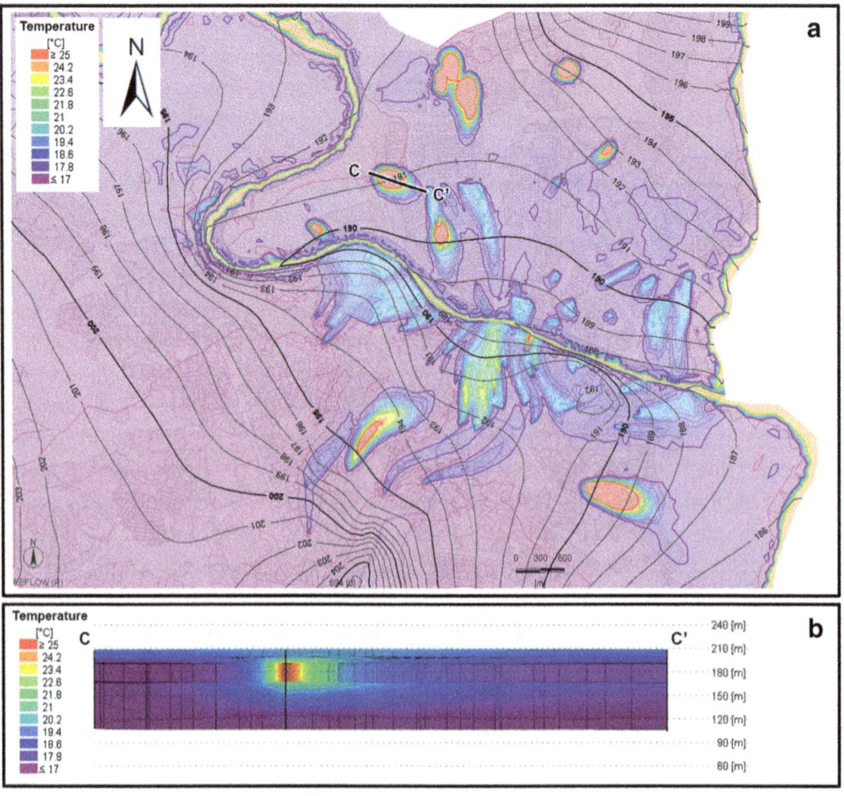

Fig. 6.9 Example of a 3D numerical model of groundwater flow and heat transport in finite elements at a city scale of the Zaragoza metropolitan area. Calculated temperature distribution in the aquifer in plan view (**a**) and calculated temperatures in cross section (**b**)

thermal storage of heat from solar power plants, industrial processes and cogeneration plants (Pinel et al. 2011; Xu et al. 2014). Research on these systems has focused on estimating their potential, taking into account hydrogeological and climatic factors, as well as on improving the recovery rates of stored energy (Gao et al. 2017a).

6.7.1 Thermal Performance in ATES Systems

The performance of an ATES installation will depend on its ability to transfer thermal energy to the ground and reverse the heat transfer process for heat recovery. The irreversibility of the process is clear; however, part of the heat transfer process will be reversible. The main indicators of thermal performance are described below.

The *intensity of thermal interference* between geothermal wells during the heat transfer process is the first performance indicator in ATES installations. It refers to temperature change in one operating well due to temperature change in another well. The parameter that best represents the intensity of the thermal interference is the arrival time of the heat front t_b [s] between two wells forming a doublet (Banks 2009; Gringarten and Sauty 1975):

$$t_b = \frac{\pi \mathrm{b} a^2 C_e}{3\mathrm{Q} C_w} \tag{6.10}$$

where b [m] is the thickness of the low permeability body used for thermal storage, a [m] is the separation distance between wells, Q [m^3 s^{-1}] is the captation/injection flow rate in the doublet of wells, C_e [J m^{-3} K^{-1}] and C_w [J m^{-3} K^{-1}] are the equivalent volumetric heat capacity of the aquifer and groundwater, respectively. In general, the shorter the arrival time of the heat front, the higher the thermal interference and the less stored thermal energy can be recovered (Gao et al. 2013; Lee et al. 2010). However, some authors consider that a small and controlled thermal interference can be favourable in increasing heat recovery from the geothermal reservoir by 30–40% (Bakr et al. 2013; Sommer et al. 2013, 2015).

A second indicator is the *heat recovery ratio* κ [–], represents the ratio of recovered thermal energy to the total thermal energy E_{rq} [J] initially transferred and stored E_{alm} [J] in the shallow geothermal reservoir (Bakr et al. 2013; Rosen 1999; Sommer et al. 2015).

$$\kappa = \frac{E_{rq}}{E_{alm}} \tag{6.11}$$

where the energy recovered E_{rq} [J] is calculated as the sum of the abstracted water flow rates considering the temperature difference between the abstracted groundwater T_o [K] and the background temperature of the aquifer T_B [K], the volumetric heat capacity of the groundwater C_w[J m^{-3} K^{-1}] for an extraction time t [s].

$$E_{rq} = \int C_w \mathrm{Q} |T_o - T_\mathrm{B}| dt \tag{6.12}$$

The stored energy E_{alm} [J] is calculated as the sum of the injected water flow rates considering the temperature difference between the injected water T_i [K] and the bottom temperature of the aquifer T_B [K], the volumetric heat capacity of the water injected C_w [J m^{-3} K^{-1}] at an injection time of t [s].

$$E_{alm} = \int C_w \mathrm{Q} \cdot |T_i - T_\mathrm{B}| dt \tag{6.13}$$

The heat recovery ratio can generate confusion since it only takes energy as the conceptual basis for the indicator. It is more accurate to use the concept of exergy, as it takes into account the environment of the geothermal reservoir and the energy transferable until thermal equilibrium with the environment is reached (Dincer et al.

1997; Rosen and Dincer 2003). The *exergy efficiency* Λ [–] is defined as the ratio of recovered to stored exergy expressed as (Alzahrani and Dincer 2015; Rosen 1999):

$$\Lambda = \frac{Ex_{rq}}{Ex_{alm}} \tag{6.14}$$

where Ex_{rq} [J] and Ex_{alm} [J] are the recovered and stored exergies, respectively. Finally, the smaller the energy imbalance between the season of energy dissipation in the reservoir (cooling) and the season of heat absorption (heating) in the shallow geothermal reservoir, the higher the storage efficiency. Therefore, if an ATES installation is used for both cooling and heating, it is convenient to define the indicator *energy balance ratio* ψ [–] expressed mathematically as (Sommer et al. 2013, 2014):

$$\psi = \frac{E_d^{rq} - E_{ab}^{rq}}{E_d^{rq} + E_{ab}^{rq}} \tag{6.15}$$

where *E* [J] is the energy recovered (rq) during dissipation (d) of heat in the cooling season and absorbed (ab) during the heating season.

6.8 Thermal Use of Mine Water

A variant of the ATES installation, known as a *Cavern Thermal Energy Storage* (CTES) installation, consists of the thermal utilisation of cavernous cavities (natural or anthropogenic) in the subsoil flooded with groundwater. The most promising case of CTES, due to large controlled volume of available stored water, is the use of abandoned mine galleries close to urban environments (Loredo et al. 2016). Geothermal energy exploitation using mine water can involve a heat storage activity and/or directly exploit the geothermal energy associated with mine water without storage.

Extensive areas of flooded mines underlie many populated regions, providing an economic asset to extract shallow geothermal energy. Mine water can be used as a stable thermal reservoir in both operating and abandoned mines.

Open-loop geothermal heat exchangers could be used to extract the mine water. Since mine water is usually problematic, either corrosive or susceptible to scaling, an additional secondary circuit is often used to isolate the mine water from the geothermal heat pump. For this purpose, plate heat exchangers are used to connect the two secondary circuits (indirect use system). The reject water from the plate heat exchanger can be discharged into a surface water body (river, lake, etc.) or injected back into the flooded mine shafts. The volume of mine drifts usually involves such a large flooded volume that it behaves as an ideal heat source/sink. Consequently, it is possible to capture and inject large volumes of water without affecting the heat capacity of the reservoir. However, it is recognised that there is a finite amount of shallow geothermal energy stored in the mine for long-term management.

In some case studies it is estimated that about 30% of mine water could be extracted and returned annually to the mines without concern for long-term heat transfer effects (Watzlaf and Ackman 2006). However, to determine the potential of mine water volume for heating and cooling, it is necessary to take into account the number of buildings and infrastructure that can benefit from shallow geothermal energy, considering other factors, e.g., natural recharge rate of mine water, on-site heat exchange, etc. In addition, it is necessary to know the rate of mine water renewal by regional groundwater. Another important factor is the 3D interconnectivity of mine galleries as a network of reservoirs. It is necessary to take into account that 3D mapping of the existing galleries does not always exist, and it is difficult to know the exact geometry of the installations in order to calculate the existing geothermal potential.

From a sociological point of view, broad social acceptance is needed for the development of such projects. The renewable character of geothermal energy and the high economic profitability that offers to the users needs to promoted in such type of activities. The quality characteristics of coal mine water usually have high loads of dissolved solids and low pH. For this reason, mine water that is pumped to the surface must either be treated under sealed conditions or treated for environmental quality improvement in mine water treatment plants. It is important to note that if mine water is pumped and returned to the mine under isolated conditions, scaling and corrosion can be minimised. It is important to avoid aerial exposure of mine water and to maintain the original conditions of the geothermal reservoir throughout the process (as in any open-loop geothermal installation). The water in the mine is likely to be in chemical equilibrium with the underground environment of the mine. Changes in the physicochemical conditions during pumping will trigger corrosion or scaling processes. In addition, corrosion resistant materials (e.g., titanium) can be incorporated into the secondary system (pipes and plate heat exchangers) to further reduce the risk of corrosion.

The efficient use of mine water can be optimised by applying numerical heat transport models (Renz et al. 2009). Depending on the case study, it may be necessary to account for variable density coupled processes and to consider the thermal dependence of physical properties of the geological materials that make up the shallow geothermal reservoir. Therefore, simultaneous modelling of groundwater flow and heat transport is necessary. Some examples of modelling of geothermal mine water exploitation have been reviewed by Loredo et al. (2016). The authors highlight the need for the provision of in-depth knowledge of the hydrogeological behaviour of the system in order to predict its long-term behaviour during geothermal exploitation tasks. The difficulties associated with this modelling exercise stem from the geometric complexity of the galleries and underground mine spaces. The associated complexity will produce different types of groundwater flow. Inside the galleries it may be laminar and/or turbulent, while inside geological formations present in the mining environment this flow will be laminar through fractured media, mine voids and porous media. Consequently, the code selected to solve the relevant equations must be able to cope with this complexity of scale and flow conditions inherent to this type of system.

References

Al-Khoury R, Bonnier PG (2006) Efficient finite element formulation for geothermal heating systems. Part II: transient. Int J Numer Methods Eng 67(5):725–745. https://doi.org/10.1002/nme.1662

Al-Khoury R, Bonnier PG, Brinkgreve RBJ (2005) Efficient finite element formulation for geothermal heating systems. Part I: steady state. Int J Numer Methods Eng 63(7):988–1013. https://doi.org/10.1002/nme.1313

Al-Khoury R, Kölbel T, Schramedei R (2010) Efficient numerical modeling of borehole heat exchangers. Comput Geosci 36:1301–1315. https://doi.org/10.1016/j.cageo.2009.12.010

Alzahrani A, Dincer I (2015) Performance assessment of an aquifer thermal energy storage system for heating and cooling applications. J Energy Res Technol 138:011901–011901. https://doi.org/10.1115/1.4031581

ASHRAE (2016) ASHRAE handbook 2016: HVAC systems and equipment: SI edition. ASHRAE

Bakr M, Oostrom N, Sommer W (2013) Efficiency of and interference among multiple aquifer thermal energy storage systems: a Dutch case study. Renew Energy 60:53–62. https://doi.org/10.1016/j.renene.2013.04.004

Banks D (2009) Thermogeological assessment of open-loop well-doublet schemes: a review and synthesis of analytical approaches. Hydrogeol J 17(5):1149–1155. https://doi.org/10.1007/s10040-008-0427-6

Bear J (1979) Hydraulics of groundwater. McGraw-Hill International Book Co., New York

Bloemendal M, Olsthoorn T Boons F (2014) How to achieve optimal and sustainable use of the subsurface for aquifer thermal energy storage. Energy Policy 66:104–114

Casasso A, Sethi R (2015) Modelling thermal recycling occurring in groundwater heat pumps (GWHPs). Renew Energy 77:86–93. https://doi.org/10.1016/j.renene.2014.12.003

Clauser C (2012) Numerical simulation of reactive flow in hot aquifers: SHEMAT and processing SHEMAT. Springer, Berlin, Heidelberg

Chiasson AD (2016) Geothermal heat pump and heat engine systems: theory and practice. Wiley, New York

DaCosta JA, Bennett RR (1960) The pattern of flow in the vicinity of a recharging and discharging pair of wells in an aquifer having areal parallel flow. IAHS Publ 52:524–536

DGG (2016) Shallow geothermal systems: recommendations on design, construction, operation and monitoring. In: Deutsche Gesellschaft für Geotechnik Deutsche Geologische Gesellschaft GV (ed) Shallow geothermal systems: recommendations on design, construction, operation and monitoring. Wiley, Berlin, p 312. https://doi.org/10.1002/9783433606674.ch8

Diersch HJ (2013) FEFLOW: finite element modeling of flow, mass and heat transport in porous and fractured media. Springer, Berlin, Heidelberg

Diersch HJG, Bauer D, Heidemann W, Rühaak W, Schätzl P (2011) Finite element modeling of borehole heat exchanger systems: Part 1. Fundamentals. Comput Geosci 37(8):1122–1135. https://doi.org/10.1016/j.cageo.2010.08.003

Dincer I, Dost S, Li X (1997) Performance analyses of sensible heat storage systems for thermal applications. Int J Energy Res 21(12):1157–1171. https://doi.org/10.1002/(sici)1099-114x(19971010)21:12%3c1157::aid-er317%3e3.0.co;2-n

Ellis P (1998) Materials selection guidelines. In: Lund JW, Lienau PJ, Lunis BC (eds) Geothermal direct use engineering and design guidebook. Klamath Falls, pp 191–209

Eskilson P, Claesson J (1988) Simulation model for thermally interacting heat extraction boreholes. Numer Heat Transf 13(2):149–165. https://doi.org/10.1080/10407788808913609

Feuvre PL, Cox CSJ (2009) In: Directorate E (ed) Ground source heating and cooling pumps in England and Wales: state of play and future trends (environment agency report). Environment Agency, United Kingdom, p 81

Fleuchaus P, Godschalk B, Stober I, Blum P (2018) Worldwide application of aquifer thermal energy storage—a review. Renew Sustain Energy Rev 94:861–876. https://doi.org/10.1016/j.rser.2018.06.057

Gao L, Zhao J, An Q, Wang J, Liu X (2017) A review on system performance studies of aquifer thermal energy storage. Energy Procedia 142:3537–3545. https://doi.org/10.1016/j.egypro.2017.12.242

Gao Q, Zhou X, Jiang Y, Chen X-L, Yan Y-Y (2013) Numerical simulation of the thermal interaction between pumping and injecting well groups. Appl Therm Eng 51:10–19. https://doi.org/10.1016/j.applthermaleng.2012.09.017

García-Gil A, Vázquez-Suñe E, Schneider EG, Sánchez-Navarro JÁ, Mateo-Lázaro J (2014) The thermal consequences of river-level variations in an urban groundwater body highly affected by groundwater heat pumps. Sci Total Environ 485–486:575–587. https://doi.org/10.1016/j.scitotenv.2014.03.123

Gringarten AC, Sauty JP (1975) A theoretical study of heat extraction from aquifers with uniform regional flow. J Geophys Res (1896–1977) 80(35):4956–4962. https://doi.org/10.1029/JB080i035p04956

Herbert A, Arthur S, Chillingworth G (2013) Thermal modelling of large scale exploitation of ground source energy in urban aquifers as a resource management tool. Appl Energy 109:94–103. https://doi.org/10.1016/j.apenergy.2013.03.005

Hesaraki A, Holmberg S, Haghighat F (2015) Seasonal thermal energy storage with heat pumps and low temperatures in building projects—a comparative review. Renew Sustain Energy Rev 43:1199–1213. https://doi.org/10.1016/j.rser.2014.12.002

Lee CK, Lam HN (2008) Computer simulation of borehole ground heat exchangers for geothermal heat pump systems. Renew Energy 33(6):1286–1296. https://doi.org/10.1016/j.renene.2007.07.006

Lee Y, Yoon W, Jeon J, Koo M-H, Keehm Y (2010) Numerical modeling of aquifer thermal energy storage system. Energy 35:4955–4965. https://doi.org/10.1016/j.energy.2010.08.029

Lo Russo S, Gnavi L, Roccia E, Taddia G, Verda V (2014) Groundwater Heat Pump (GWHP) system modeling and Thermal Affected Zone (TAZ) prediction reliability: influence of temporal variations in flow discharge and injection temperature. Geothermics 51:103–112. https://doi.org/10.1016/j.geothermics.2013.10.008

Lo Russo S, Taddia G, Verda V (2012) Development of the thermally affected zone (TAZ) around a groundwater heat pump (GWHP) system: a sensitivity analysis. Geothermics 43:66–74. https://doi.org/10.1016/j.geothermics.2012.02.001

Loredo C, Roqueñí N, Ordóñez A (2016) Modelling flow and heat transfer in flooded mines for geothermal energy use: a review. Int J Coal Geol 164:115–122. https://doi.org/10.1016/j.coal.2016.04.013

Mehlhorn H, Spitz K-H, Kobus H (1981) Kurzschlussströmung zwischen Schluck- und Entnahmebrunnen – Kritischer Abstand und Rückströmrate. Wasser Und Boden 33(4):170–174

Muela Maya S et al (2018) An upscaling procedure for the optimal implementation of open-loop geothermal energy systems into hydrogeological models. J Hydrol 563:155–166. https://doi.org/10.1016/j.jhydrol.2018.05.057

Mueller MH, Huggenberger P, Epting J (2018) Combining monitoring and modelling tools as a basis for city-scale concepts for a sustainable thermal management of urban groundwater resources. Sci Total Environ 627:1121–1136. https://doi.org/10.1016/j.scitotenv.2018.01.250

Muraya NK (1994) Numerical modelling of the transient thermal interface of vertical U-tube heat exchangers. TX, USA

Piga B, Casasso A, Pace F, Godio A, Sethi R (2017) Thermal impact assessment of groundwater heat pumps (GWHPs): rigorous vs. simplified models. Energies 10(9):1385

Pinel P, Cruickshank C, Beausoleil-Morrison I, Wills A (2011) A review of available methods for seasonal storage of solar thermal energy in residential applications. Renew Sustain Energy Rev 15:3341–3359. https://doi.org/10.1016/j.rser.2011.04.013

Possemiers M, Huysmans M, Batelaan O (2014) Influence of aquifer thermal energy storage on groundwater quality: a review illustrated by seven case studies from Belgium. J Hydrol: Reg Stud 2:20–34

Rafferty K (1999) Scaling in geothermal heat pump systems. U.S. Department of Energy, Idaho, p 60

Renz A, Rühaak W, Schätzl P, Diersch H-JG (2009) Numerical modeling of geothermal use of mine water: challenges and examples. Mine Water Environ 28(1):2–14. https://doi.org/10.1007/s10230-008-0063-3

Rosen MA (1999) Second-law analysis of aquifer thermal energy storage systems. Energy 24(2):167–182

Rosen MA, Dincer I (2003) Exergy methods for assessing and comparing thermal storage systems. Int J Energy Res 27(4):415–430. https://doi.org/10.1002/er.885

Sarbu I, Sebarchievici C (2015) Ground-source heat pumps: fundamentals, experiments and applications.https://doi.org/10.1016/b978-0-12-804220-5.0000

Sciacovelli A, Guelpa E, Verda V (2014) Multi-scale modeling of the environmental impact and energy performance of open-loop groundwater heat pumps in urban areas. Appl Therm Eng 71(2):780–789. https://doi.org/10.1016/j.applthermaleng.2013.11.028

Śliwa T, Gonet A (2004) Theoretical model of borehole heat exchanger. J Energy Res Technol 127(2):142–148. https://doi.org/10.1115/1.1877515

Sommer W, Valstar J, Leusbrock I, Grotenhuis T, Rijnaarts H (2015) Optimization and spatial pattern of large-scale aquifer thermal energy storage. Appl Energy 137:322–337. https://doi.org/10.1016/j.apenergy.2014.10.019

Sommer W, Valstar J, van Gaans P, Grotenhuis T, Rijnaarts H (2013) The impact of aquifer heterogeneity on the performance of aquifer thermal energy storage. Water Resour Res 49(12):8128–8138. https://doi.org/10.1002/2013wr013677

Sommer WT et al (2014) Thermal performance and heat transport in aquifer thermal energy storage. Hydrogeol J 22(1):263–279. https://doi.org/10.1007/s10040-013-1066-0

Stauffer F, Bayer P, Blum P, Giraldo NM, Kinzelbach W (2013) Thermal use of shallow groundwater. Taylor & Francis, Abingdon, UK

van der Meer FP, Al-Khoury R, Sluys LJ (2009) Time-dependent shape functions for modeling highly transient geothermal systems. Int J Numer Meth Eng 77(2):240–260. https://doi.org/10.1002/nme.2414

Watzlaf GR, Ackman TE (2006) Underground mine water for heating and cooling using geothermal heat pump systems. Mine Water Environ 25(1):1–14. https://doi.org/10.1007/s10230-006-0103-9

Xu J, Wang RZ, Li Y (2014) A review of available technologies for seasonal thermal energy storage. Sol Energy 103:610–638

Chapter 7
Obtaining Terrain Thermal Parameters

The study of technical and economic factors that influence the design and performance of shallow geothermal installations confirms the importance of geological, hydrogeological and thermal conditions. It also reveals a lack of correlation between subsurface characteristics and the design of shallow geothermal installations (Blum et al. 2011). This lack of correlation suggests that subsurface characteristics are still not adequately considered during the planning and design of shallow geothermal installations, especially in small installations. Failure to take ground characteristics into account leads to under- or over-dimensioning installations and, thus, to long-term impacts on maintenance costs and payback time of such installations. Borehole logging and sampling for laboratory tests are more representative than theoretical values found in the literature but may not be representative of the terrain, especially when subsoil heterogeneity is significant or when unconsolidated materials lose texture/structure properties while sampling.

Prior to the design of shallow geothermal installations, it is necessary to have an initial estimation of the hydraulic and thermal parameters of the terrain, as well as economic, technical and design parameters. The thermal parameters necessary for the design of shallow geothermal installations can be obtained from different sources, with different representativeness. For a first iteration, thermal data can be obtained from theoretical values in bibliographic tables (Sect. 2.3) and thematic maps (Bertermann et al. 2013; Sáez Blázquez et al. 2017). In Spain, the Geological Survey of Spain (IGME) and some regional geological surveys have publicly available geo-referenced databases with values of thermal and hydraulic properties of the most relevant geological formations. Another way to obtain thermal and hydraulic parameters is from laboratory tests. In particular, thermal conductivity values can be obtained from borehole cores or field samples. A third way to obtain the thermal and hydraulic parameters of the terrain is from different hydraulic and thermal response tests in situ, which are much more representative (and also less economical), such as thermal response tests (TRT) or thermal tracer tests (TTT). Broadly speaking, it can be assumed that the parameters obtained from theoretical tables and laboratory techniques are sufficiently representative for small installations, and the parameters

A. García Gil et al., *Shallow Geothermal Energy*, Springer Hydrogeology,
https://doi.org/10.1007/978-3-030-92258-0_7

obtained in the field at a decametric scale are aimed at large installations, where an error in the estimation of the parameters has greater consequences for a given project.

The equations governing heat transport in the terrain (Sect. 2.2) depend primarily on thermal conductivity and hydraulic permeability, as well as heat capacity and storativity of the ground. These are the most important parameters for heat transport and, consequently, for design of open and closed-loop geothermal heat exchangers in shallow geothermal systems.

7.1 Estimation of Laboratory Thermal Parameters

Laboratory estimations of thermal conductivity and heat capacity of rocks and sediments is standard practice in many laboratories (Alrtimi et al. 2014; Jackson and Taylor 1986). Two main types of methodologies are distinguished, depending on the thermal regime to which laboratory samples are subjected during estimation of thermal parameters (Clarke et al. 2008). The first includes methods based on the study of samples subjected to a constant heat flux, until the heat transport acquires a steady state regime in thermal equilibrium, and only then experimental measurements are taken that allow estimation of the thermal properties. The second type includes those methods that measure the thermal response of the samples under a transient regime, from the time when heat begins to be transferred to the sample until it reaches the steady state regime.

Transient procedures are simpler, but stationary methods are considered more accurate. Of all the stationary methods, the *guarded hot plate* (GHP) test stands out as the most accurate for measuring thermal conductivity (Xamán et al. 2009). As far as transient procedures are concerned, the single thermal needle method is the only standardised transient method for measuring thermal conductivity in sediments and rocks (ASTM 2000). It should be noted that the use of stationary and transient procedures is not always coincident, with discrepancies in soil thermal conductivity values of up to 50% (10–20% on average), depending on whether the method used was stationary or transient (Alrtimi et al. 2014). This discrepancy is attributed to the heat transfer mechanism through the porous sample and structure anisotropy, with a different grain arrangement depending on the sample direction, typical of fine-grained sediments. The same techniques performed in different laboratories in different countries on reference or standard samples may produce differences in results (Salmon 2001). Therefore, to ensure good performance of any instrument used to estimate thermal conductivity of the soil, it is necessary to accurately control the environmental conditions during the measurement process. The precision and accuracy of estimation of thermal conductivity in these tests will depend, to a large extent, on the precision and accuracy of the measurement and control of the process conditions. For this reason, it is common to find differences between experimentally obtained values and values tabulated in the literature when using steady-state measurement procedures.

In all methods it is necessary to measure the thermal gradient ($\nabla T = dT/dx$) downward between the heat source and the heat sink source (see Fig. 2.5). Usually, temperature measurements are taken along the sample to estimate the temperature gradient. Knowing the gradient ∇T [–] and the heat flow q_x^* [W m^{-2}], thermal conductivity is obtained (Eq. 2.50). The latter is simply the constant of proportionality between heat flow and temperature gradient.

The existing methods were originally developed for building materials used for thermal insulation of building structures, where temperature measurements could be made on large homogeneous samples. Their direct use on rocks and sediment is not possible in all cases, since soil samples are difficult to obtain and the nature of the samples is uneven. Consequently, several equipment configurations have been established to measure thermal conductivity of different geological materials under different conditions (Abuel-Naga et al. 2006). The precision and accuracy of each measuring instrument depends entirely on precise and accurate estimation of the heat flux passing through the sample, and this estimation is compromised by the amount of radial heat losses along the length of the sample. Environmental thermal interference is the main factor capable of significantly influencing laboratory tests aimed at studying one-dimensional heat flow (Prunty and Horton 1994). To solve this problem, Zhou et al. (2006) proposed the use of a buffer made of the same geological material as the sample to construct a ground capsule with limited radial heat losses.

7.1.1 Tests for the Estimation of Thermal Conductivity in the Laboratory

Major tests to determine the thermal conductivity of sediment and rock samples in the laboratory are the following (Clarke et al. 2008):

- ***ASTM E1225 Standard test method for thermal conductivity of solids by means of the guarded-comparative-longitudinal heat flow technique***. Test method consisting of inserting a sample between two discs of a reference-known material. The heat flow per unit area transferred through the sample material is not known exactly, but the measured temperature gradient is matched to the temperature gradient generated in a standard material of known conductivity, allowing accurate quantification of the heat flow through the sample. This is a comparative method, since it compares the conductivity of the sample material being measured with that of a known material acting as a laboratory standard.
- ***ASTM C518 Standard test method for steady-state thermal transmission properties by means of a heat flow meter apparatus***. This test method uses a heat flow meter instrument to stablish a steady state heat flux through flat slab specimens. The heat flow meter instrument used is simple, rapid, and applicable to a wide range of test specimens. The method requires heat flow meter calibration within the range of heat flows expected, similar types of materials, similar thermal

conductances, similar thicknesses and temperature gradients as expected for the test specimens. Adequate calibration provides the instrument precision and bias.

- ***Protected hot plate test (ASTM C177).*** Test used to measure the thermal conductivity of large samples of thermal insulation materials for building construction. This method is inappropriate for geological samples due to the large sample size required.
- ***Hot wire method (ASTM C1113).*** Method suitable for use with rock and sediment samples. A heating element is placed in a pre-drilled axial hole in a sample, and the radial temperature difference across the sample is measured. In this case, the outside temperature may be the ambient temperature.
- ***Thermal needle probe procedure (ASTM D5334).*** It uses a thermal probe consisting of a needle-shaped heating filament that is introduced into the sample of unconsolidated material (sediment) to transfer a controlled current of heat into the test sample. The probe itself includes both the heating element and the thermocouple temperature sensor. It records and monitors the temperature variation over time, until steady-state conditions are reached; subsequently, the thermal conductivity is determined. This method is used on sediment samples in the laboratory but can also be incorporated into geophysical borehole rigs and in-situ soil testing. The probe is calibrated with material standards of known properties and is therefore also a comparative method. The method is not very accurate according to ASTM (American Society for Testing and Materials). The interpretation of the test is based on the infinite linear heat source (ILS) model assuming the sample as an infinite homogeneous medium (Eq. 5.97).
- ***Dual needle heat pulse method (DNHP).*** Similar to the thermal needle probe procedure but, unlike the single needle method, it includes a second filament with an additional thermocouple temperature sensor to obtain the thermal response at a distance from the thermal probe and its time evolution. The test consists of performing a heat pulse with the heated thermal probe and measuring the thermal response over time (Nope and de Santiago Buey 2014). It is a transient procedure and, therefore, it is possible to obtain the thermal diffusivity, differentiating it from the heat capacity (Bristow et al. 1994). This method does not have a standardised ASTM procedure. In this case, as in the previous case, interpretation of this test is based on the ILS model.

7.2 Thermal Response Test (TRT)

The *Thermal Response Test* (TRT) is an in situ test for indirect determination of the effective thermal conductivity of the ground and the thermal resistance of a BHE (Morgensen 1983). The thermal resistance of the BHE refers to the heat transfer inside the closed-loop heat exchanger with the heat carrier fluid pipe assembly (see Fig. 5.11). TRTs are usually performed prior to the construction of a shallow geothermal installation with BHEs, in order to design the number, arrangement and depth of the BHEs to be built in the future installation. In general, a TRT consists of monitoring heat transfer between a vertical closed-loop geothermal heat exchanger and the terrain (Fig. 7.1). Depending on the thermal conductivity of the ground and the thermal resistance of the geothermal heat exchanger, heat transfer will be more or less efficient. For this purpose, a heat carrier fluid is circulated through a closed-loop geothermal heat exchanger in a terrain, initially under natural conditions, to which a constant amount of heat is transferred by means of an electric resistance (or other type of device). This heat transfer to the heat carrier fluid is carried out by a TRT unit at the surface (Fig. 7.2). Constant heat transfer to the heat carrier fluid is maintained for a variable time period of 10–70 h, and the inlet and outlet temperatures of the heat carrier fluid to the geothermal heat exchanger are recorded at specific time intervals. By keeping the flow rate of the heat carrier fluid in the circuit constant, the temperature difference between the outlet and inlet is proportional to the heat transfer transferred with the ground by the geothermal heat exchanger. TRT units equipped with a reversible heat pump allow, in addition to standard heat dissipation tests, TRTs of heat absorption by cooling the ground (Witte 2001).

In addition to conventional TRT units, different types of *Enhanced Geothermal Response Test* (EGRT) have been developed to measure temperatures along the geothermal heat exchanger at different depths, in order to estimate the vertical distribution of thermal properties (Raymond et al. 2010; Wagner and Rohner 2008). In early EGRTs, temperature measurements were made at different depths using thermocouple-type sensors but, later, distributed temperature sensors (DTS) were introduced along the entire length of the exchanger (between the pipe assembly and the borehole wall) using fibre optics and a thermoactive cable capable of heating the ground (Luo et al. 2015).

7.2.1 Performance of TRTs

A TRT requires a borehole equipped with a BHE, usually in single or double U-shape, and filled with a thermally enhanced grout. The grout is left to stand for several days until it hardens and the exothermic reaction associated with solidification of the grout is complete. When the heat exchanger is in thermal equilibrium with the underground environment, before performing the TRT, the average temperature in the heat exchanger is determined by recirculating the heat transfer fluid, without transferring heat to the heat carrier fluid in the TRT unit. Simply by activating the water pump, the temperature in the closed circuit is homogenised. Before remobilisation, it may also be advisable to measure the initial temperature profile inside the BHE. When designing a TRT, obtaining the initial temperature is the second most important factor, after thermal conductivity.

The standard TRT is initiated by the constant transfer of heat, within the range of 20–80 W m^{-1}, to the circulating heat carrier fluid at a constant flow rate in the closed-loop BHE. Both the heat transferred to the heat carrier fluid and the flow rate must be kept constant throughout the test to facilitate interpretation of the results obtained. Although, from a theoretical point of view, no more heat should be transferred to the

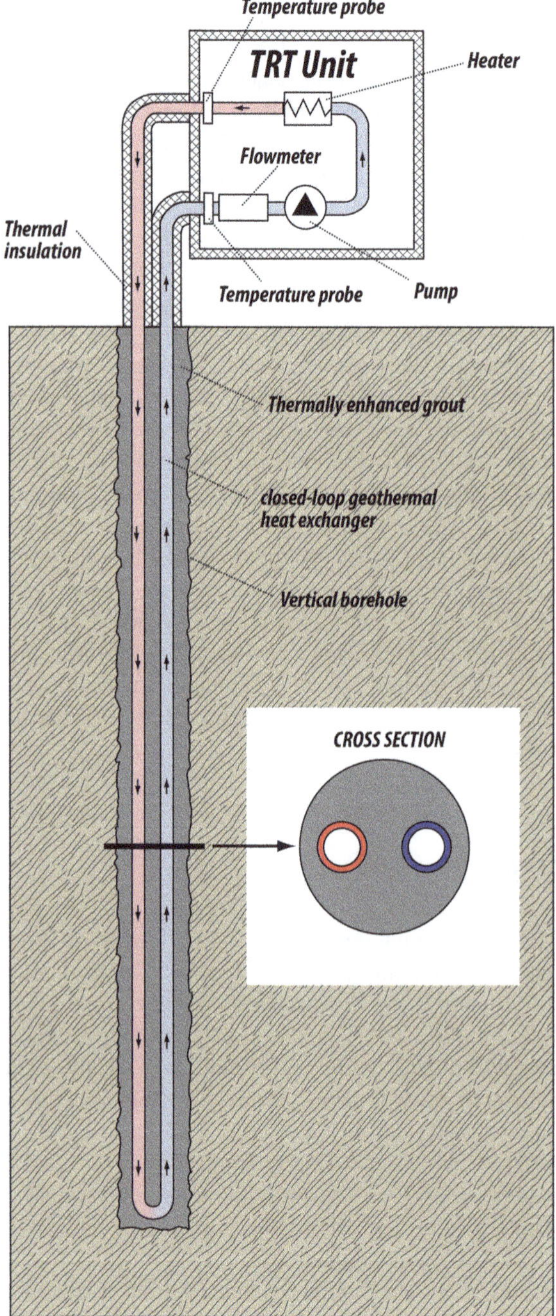

Fig. 7.1 Conceptual scheme of a TRT with a single U-shaped borehole heat exchanger

Fig. 7.2 Photograph of a TRT unit during a thermal response test with a double U-shaped borehole heat exchanger. Photograph courtesy of Gregor Götzl of the Geological Survey of Austria (*Geologische Bundesanstalt*)

heat carrier fluid at the surface than is transferred by the TRT unit, this is not always possible due to different surface environmental factors such as weather conditions, direct sunlight or rainwater infiltration. Therefore, the external piping and the TRT unit itself must be thermally insulated (Fig. 7.1). In addition, in order to assess the effectiveness of the thermal insulation, it is recommended the ambient air temperature is measured throughout the test.

During realisation of a standard TRT, the following parameters are continuously monitored and recorded: flow rate of the heat carrier fluid in the closed-loop, temperature of the heat carrier fluid at the inlet and outlet of the BHE, temperature difference between the circulation pump and the heating/cooling system, reference temperature inside the TRT unit and ambient air temperature. The representation of the obtained data consists of curves showing the ambient temperature and the evolution of the heat carrier fluid temperature, both at the inlet and outlet over time (Figs. 7.3a and 7.4a). The ambient air temperature is usually measured in TRTs to be plotted, together with the exchanger inlet and outlet temperatures, in order to check whether the test data is influenced by the ambient surface temperatures. Figure 7.3b shows an example of a TRT significantly affected by surface atmospheric conditions. Different methods exist to assess and correct for the effects of ambient conditions on TRT data (Bandos et al. 2011; Kavanaugh et al. 2000).

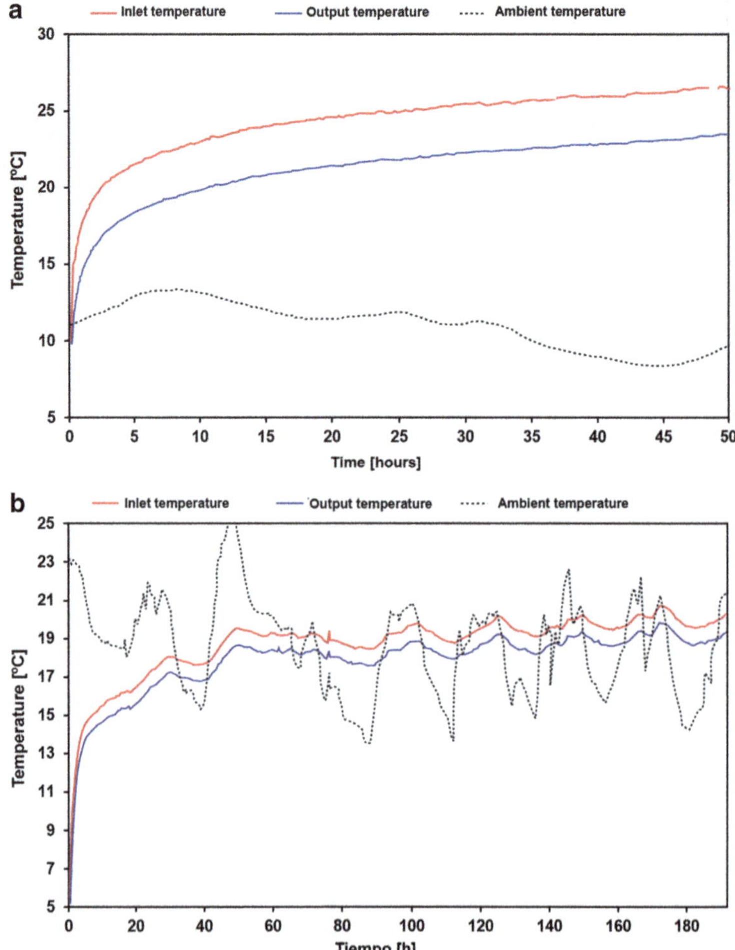

Fig. 7.3 **a** Results obtained from a standard TRT using a 100 m deep closed simple U-shaped geothermal heat exchanger in Denmark. Data from Poulsen and Alberdi-Pagola (2015). **b** Example of results of a TRT significantly affected by changing atmospheric conditions during a test of approximately one week. Data from Vieira et al. (2017)

The total duration of a TRT can become a matter of controversy, as different periods can be considered as appropriate depending on the authors, ranging from 12 to 20 h (Smith and Perry 1999), 30 h (Gehlin and Hellström 2003) or more than 50 h (Austin et al. 2000). Longer test periods, although more expensive, are desirable to eliminate diurnal variations. Beier and Smith (2003) provided a graphical method to determine the minimum TRT duration as a function of the expected thermal properties of the soil and heat exchanger resistivity. Based on a numerical model, Signorelli et al. (2007) concluded that, under ideal conditions during a TRT, a duration of at least 50 h is required.

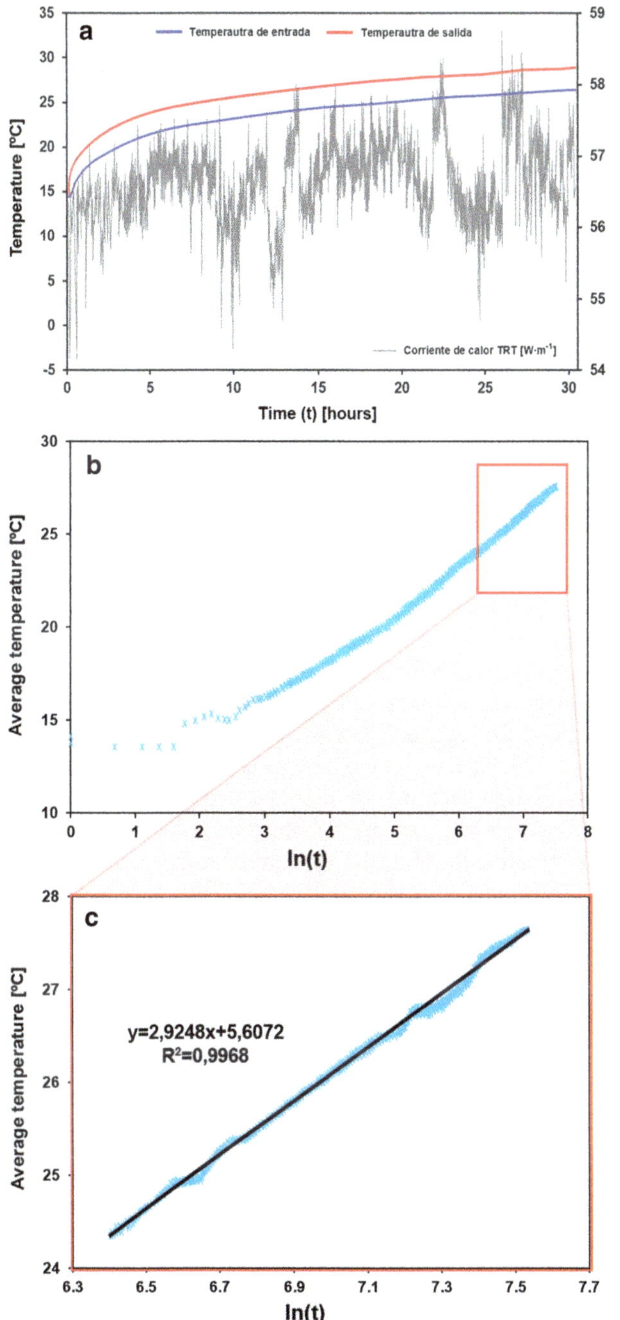

Fig. 7.4 Procedure for obtaining thermal conductivity from conventional TRT. Data Kurevija et al. (2017)

7.2.2 Interpretation of Results Obtained from TRT Testing

It is not possible to determine the thermal parameters of the terrain directly, so it is necessary to apply a mathematical model. In this case, it must be capable of reproducing the results of a TRT and deducing, by trial and error, the parameters that best fit the predictions of the mathematical model to the experimental data obtained in the field. This process is known as calibration and involves testing parameters in a controlled manner until a so-called objective function is reduced. The objective function describes the difference between the measured data and the results obtained from an analytical or numerical model describing the heat transport in the system. The iterative algorithm to reduce the objective function during the calibration process involves solving the direct or inverse problem.

Solving the direct problem consists of solving a mathematical expression that describes the physical system and is able to predict the response in terms of the most important state variables (e.g., temperature) from given values of the model parameters. TRTs can be interpreted through the use of numerical and analytical models. The inverse problem consists of a mathematical optimisation process capable of obtaining the parameter values of the direct model and also capable of reproducing the experimental values measured in the field. Therefore, solving the inverse problem involves choosing a physical model and iteratively applying it to the direct problem until the desired parameters are obtained.

The accuracy of the thermal parameters obtained from the inverse problem solving methods will depend on different factors, including (1) the quality of the data measured during the TRT, (2) the suitability of the heat transport model (and groundwater flow model in case heat advection is produced) to solve the direct problem in each iteration, and (3) the accuracy, efficiency and robustness of the optimisation algorithm used to minimise the objective function. The data measured in the field may be affected by ambient noise (generated by other heat transport processes of lesser magnitude in the environment) or instrumental noise due to electrical equipment.

The choice of the physical model to solve the direct problem is an important factor in the calibration procedure. There are two determining factors in choosing the most suitable physical model: accuracy and computational efficiency. Simplified models are easy to handle but, in many cases, they are far from adequate to describe the thermal processes involved in the system under study, e.g., the analytical ILS model. More complex models, such as finite element or finite difference numerical models, reproduce better thermal processes but are relatively computationally inefficient for use in iterative algorithms.

The optimisation technique selected to minimise the objective function is another important factor in the calibration process. The different mathematical optimisation techniques are specially designed to solve non-linear equations, such as the objective function. The more unknown parameters to calibrate at the same time, the more difficult it is to reduce the objective function and the less likely it is to obtain a unique solution (uniqueness). Therefore, it is advisable to minimise the number of unknown parameters to be calibrated.

7.2.2.1 Analytical Models

As has been shown, the inverse problem can be complex and costly, especially if there is heat advection from regional groundwater flow. Both conventional TRTs and EGRTs are usually interpreted using the ILS model, which assumes an isotropic, homogeneous and infinite medium, an infinite and constant heat source and conduction as the only heat transport mechanism (see Sect. 5.4.1). Alternative models can also be applied, e.g., the infinite cylindrical source (ICS) model (see Sect. 5.4.2), the finite linear source (FLS) model (see Sect. 5.4.3) and the moving infinite linear source (MILS) model, the latter in case heat advection by groundwater flow is required (see Sect. 5.4.4).

When studying low permeability or unsaturated terrains, it may be appropriate to use the ILS model to solve the inverse problem and interpret the results of a TRT. The ILS model developed in Sect. 5.5.1 is applicable when the test lasts sufficient time and can be approximated by (Carslaw and Jaeger 1986):

$$T(r,t) \simeq T_0 + \frac{q^o}{4\pi\lambda}\left(\ln\left(\frac{4\alpha t}{r^2}\right) - \gamma\right) \tag{7.1}$$

where T_0 [K] is the initial temperature, q^o [W m^{-1}] is the heat flux transferred by the TRT, λ [W m^{-1} K^{-1}] is the thermal conductivity of the terrain, α [m^2 s^{-1}] is the thermal diffusivity of the terrain, r [m] is the radial distance to the BHE, t [s] is the time and γ [–] is the Euler-Mascheroni constant ($\gamma \approx 0.577215$). At the borehole wall with $r = r_s$ the average temperature $\overline{T}$ [K] can be described as (Wagner and Clauser 2005):

$$\begin{aligned}\overline{T} &= T(r_s,t) + q^o R_I \\ &= \frac{q^o}{4\pi\lambda}\ln(t) + q^o R_I + \frac{q^o}{4\pi\lambda}\left(\ln\left(\frac{4\alpha}{r_s^2}\right) - \gamma\right) + T_0\end{aligned} \tag{7.2}$$

where R_I [K W^{-1}] is the thermal resistance between the geothermal heat exchanger and the borehole wall. The average temperature between the inlet pipe T_i [K] and outlet pipe T_o [K] is $\overline{T} = (T_i - T_o)/2$. Taking the following auxiliary variables from Eq. 7.2:

$$\mathrm{a} = \frac{q^o}{4\pi\lambda} \tag{7.3}$$

$$\mathrm{c} = q^o R_I + \frac{q^o}{4\pi\lambda}\left(\ln\left(\frac{4\alpha}{r_s^2}\right) - \gamma\right) + T_0 \tag{7.4}$$

Equation 7.2 can be rewritten in combination with Eqs. 7.3 and 7.4 as

$$\overline{T} = \mathrm{a}\ln(t) + c \tag{7.5}$$

Equation 7.5 indicates that the average temperature at the borehole wall is proportional to time, on a natural logarithmic scale, plus a constant c. The proportionality constant a will be higher as the heat flow from the geothermal heat exchanger positively increases, but the lower the constant a, the higher the thermal conductivity. In other words, the better the medium conducts heat, the lower the temperature at the wall of the heat exchanger will be for the same heating rate (for a heating TRT). Equation 7.5 is the equation of a straight line and, therefore, the representation of the mean temperature in ordinates versus the neperian logarithm of time will tend to a straight line. The slope of the straight line a $= \Delta\overline{T}/\Delta\ln(t)$ can be obtained graphically (Fig. 7.4b, c) or by a linear regression method. Taking Eq. 7.3 into account, the thermal conductivity of the ground can be obtained as:

$$\lambda = \frac{q^o \Delta\ln(t)}{4\pi\Delta\overline{T}} \tag{7.6}$$

Note that in the early times of the test (Fig. 7.4c), as explained in Sect. 5.4.1, Eq. 7.1 is not valid. In the first time steps of the test, the heat transfer process only affects the immediate surroundings of the BHE, which has nothing to do with a homogeneous isotropic medium, especially if the regime is strongly transient. As time increases and the heat spreads through the terrain, the local effect of the BHE becomes less relevant, until the ILS model is an effective model for sufficiently long times, where the ILS model is a reasonable approximation. Equation 7.6 can be used to estimate the minimum test time needed to use the ILS model with an error of less than 2%. To use Eq. 7.6, it is necessary to make a prior estimate of the effective thermal conductivity and heat capacity of the terrain.

$$t > \frac{20r_s^2}{\alpha} \tag{7.7}$$

Knowing the thermal conductivity, the thermal resistance of the geothermal heat exchanger can be obtained from Eqs. 7.2 and 7.4 as:

$$R_I = +\frac{\mathrm{c} - T_0}{q^o} - \left(\frac{1}{4\pi\lambda}\left[\ln\left(\frac{4\alpha}{r_s^2}\right) - \gamma\right]\right) \tag{7.8}$$

Using the time superposition principle (Fig. 5.16), it is possible to develop a method of interpreting time-varying heat flow TRTs (Raymond et al. 2011b). The main advantage of being able to interpret time-varying heat flow TRTs is to be able to analyse the temperature recovery of the terrain after a TRT is conducted.

The main limitations, due to the assumptions made in the ILS model applied to the assessment of conventional TRT, are the following (Stauffer et al. 2013):

1. It is impossible to evaluate the first hours of a TRT for the thermal characterisation of the terrain.
2. Assuming constant heat transfer by the TRT unit is questionable, since it is usually difficult to ensure in practice. Although it complicates the analysis,

the time superposition principle can be applied to overcome this problem and variable heat flows can be considered.

3. The ILS model assumes that there is no vertical upward heat flow due to geothermal gradient. The influence of the geothermal gradient was studied by Wagner et al. (2012) using numerical methods, showing that if the ILS model is used, typical geothermal gradients (0–52 K km^{-1}) lead to an underestimation of the effective thermal conductivity of the ground and the thermal resistivity of the heat exchanger. The estimated error could even exceed 10% for gradients of 52 K km^{-1} (Wagner et al. 2013).
4. The assumption in the ILS model that the subsurface medium is homogeneous, isotropic and infinite is always questionable, as geothermal heat exchangers typically penetrate different geological layers with different hydraulic and thermal properties. It has been shown that heterogeneity in geology, including stratification, can lead to overestimation of the effective thermal conductivity of the ground (Raymond et al. 2011b). In cases of significant heterogeneity in the geology, EGRT or numerical models could be used to study the vertical distribution and heterogeneity of thermal properties (Raymond and Lamarche 2013).
5. Regional horizontal groundwater flow is not considered in the ILS model, although it has been shown in numerous studies that the influence of groundwater flow tends to overestimate the effective thermal conductivity (Witte 2001). Horizontal groundwater velocities greater than 0.1 m day^{-1} have a significant influence on the TRT results (Signorelli et al. 2007). Therefore, it is advisable to take into account the hydrogeological characteristics of the ground where a TRT is to be performed. It is important to assess the confidence of the thermal parameters obtained, and especially when the terrain is to be used as a thermal energy storage site, where a moderate groundwater flow would cause significant loss of energy stored in the ground.
6. Vertical groundwater flow can also influence the results of a TRT. Heat convection processes can occur within the BHE itself, when the pipe assembly is directly inserted in the borehole without grouting (BHE in hard rock), in inadequately constructed BHEs, or in BHEs filled with sand instead of thermally enhanced grouting (Sanner 2005).

Comprehensive analysis of the error committed in TRTs has shown that the measurement and theoretical errors (model and parameter errors) are around 5% for thermal conductivity and between 10 and 15% for BHE resistance (Witte 2013).

7.2.2.2 Numerical Models

Numerical models allow the incorporation of temporal and spatial aspects mostly ignored by analytical models, such as regional and local groundwater flow between wells, specific geometry of the geothermal exchanger used in the TRT, or heterogeneities of the thermal and hydraulic properties in the ground (Fig. 7.5). Numerical

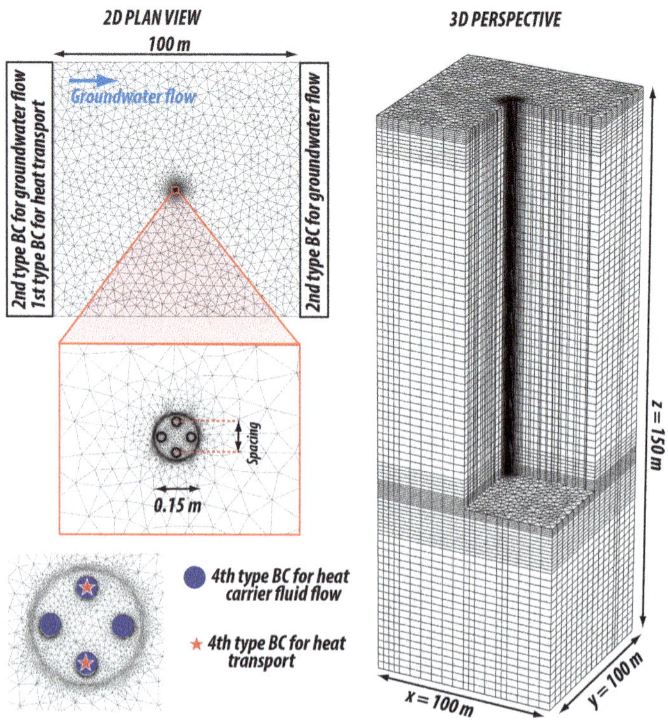

Fig. 7.5 Example of domain discretisation for a finite element numerical model for the estimation of thermal parameters from a thermal response test with a double U-shaped borehole heat exchanger. BC stands for boundary condition

models allow working with a large amount of data and information and, although their use is generally considered more time-consuming than the use of analytical models, graphical interfaces are becoming more user-friendly and less time-consuming to use. An extensive review of the numerous numerical models used for the interpretation of TRTs coupled to mathematical algorithms designed for the automatic calibration of thermal and hydraulic parameters is given by Stauffer et al. (2013). One of the early codes was SBM (*Superposition Borehole Model*), capable of simulating the three-dimensional temperature field of one or more BHEs (Eskilson 1986). In 1996, the SBM code was integrated into the commercial software (*TRNSBM*) package for transient ground heat transfer simulation, which included a parameter estimation procedure using the generic optimisation package *GenOpt* (Witte and van Gelder 2006). In 1999, a one-dimensional finite-difference numerical model of a closed-loop geothermal heat exchanger was developed (Shonder and Beck 1999). This model was coupled, similarly to the *TRNSBM*, to an automatic calibration code for parameter estimation from TRT data. The novelty of this model was that the temperatures and flow rates in the inlet and outlet pipes of the exchanger are simulated using the cylindrical source model. The authors of the code showed that it was even more accurate at the initial times of the TRT compared to the analytical infinite cylindrical source (ICS) model, therefore allowing the use of thermal response data from the beginning of the TRT. In the same year, a two-dimensional transient finite volume numerical model was developed (Yavuzturk and Spitler 1999), also with a parameter estimation module. This 2D model also used a parameter estimation algorithm, including the thermal conductivity of the ground and the thermal resistivity of the heat exchanger. In 2005, a new inversion technique for thermal parameter estimation from TRTs was developed, using the three-dimensional finite difference code *SHEMAT* (Wagner and Clauser 2005) which, in addition to effective thermal conductivity of the soil and the resistivity of the heat exchanger, made it possible to estimate the heat capacity of the soil. However, although heat capacity is generally variable within ±20% for the same type of geological material in their analysis, this uncertainty only resulted in a ±2% variation of the heat exchanger outlet temperature. In 2007, other authors developed the code called *FRACTure*, as a three-dimensional finite element model for TRT analysis (Signorelli et al. 2007). Other numerical codes emerged from this model, such as *HydroGeoSphere* (Raymond et al. 2011a) and *FEFLOW* (Diersch et al. 2011), in which groundwater flow and heat transport were simulated in a coupled manner, allowing interaction with BHEs. The *FEFLOW* numerical strategy was developed by Al-Khoury et al. (2005). Currently, *FEFLOW* code is considered to be the most robust and reliable modelling software for solving the groundwater flow and heat transport problem, both scientifically and in use by public administrations worldwide. Using *FEFLOW*, the effects of pipe layout within the vertical borehole (shank spacing), different non-uniform geothermal gradients and thermal dispersivity have been investigated (Wagner et al. 2012). The results showed that the shank spacing and non-uniform geothermal gradients exert only minor effects of less than 10% on the estimated conductivities and resistivities. However, the authors observed that, for a constant groundwater flow rate, different values of thermal dispersivity can have a significant impact on the evaluation of the thermal resistivity of the closed-loop geothermal heat exchanger.

7.3 Thermal Tracer Test (TTT)

For decades, heat has been considered as a tracer in groundwater (Anderson 2005; Saar 2011). When hydraulic tests are performed considering heat as a tracer, they are called *thermal tracer tests* (TTT). Initially, its application as a tracer was to characterise the river-aquifer relationship (Vogt et al. 2010), to investigate the effects of climate change (Brouyère et al. 2004), to identify preferential flow zones or fracture zones in boreholes (Pehme et al. 2013) or to study the dynamics of past anthropogenic environmental impacts (Engelhardt et al. 2013). The first applications of the use of heat as a tracer to characterise ground thermal properties in shallow geothermal energy were made to study the thermal storage of heat in low-permeability geological formations by means of open-circuit geothermal heat exchangers, i.e., in ATES systems (Sauty et al. 1982).

In shallow geothermal energy problems, two types of TTT are distinguished: rapid response tests and long-term tests (Wagner et al. 2014). Long-term TTT are carried out to evaluate the performance of ATES systems by injecting large volumes of *hot* water (thousands of cubic meters) and monitoring long-term (months to years) changes in groundwater temperature (Molz et al. 1978; Sauty et al. 1982; Wu et al. 2008; Xue et al. 1990). The ultimate objective of long-term TTT is to assess the thermal energy storage capacity in the subsurface and/or to estimate the recovery rate of thermal energy relative to that introduced. Rapid response TTTs are those tests carried out to obtain the thermal and hydraulic parameters of the ground (Ma et al. 2012; Shook 2001; Wagner et al. 2014) by injecting water at a temperature as different as possible from the ground for short periods of time and then monitoring the temperature changes in piezometers or wells in the vicinity.

7.4 Field Estimation of Hydraulic Parameters

The variability of thermal conductivity is usually within the same order of magnitude and could be considered as relatively constant compared to hydraulic conductivity. Therefore, whenever heat convection is present, the determination of hydraulic conductivity is the most important parameter to determine heat transport in the ground. Furthermore, the effect of porosity on the parameterisation, both hydraulic and thermal, to estimate the equivalent behaviour of the porous medium is of great importance, as discussed in Sect. 2.3. Thus, hydraulic characterisation of the ground is crucial for the design of shallow geothermal installations, especially those with open-loop heat exchangers (Banks 2012). Far from an exhaustive review of hydrogeological methods to parameterise aquifers, reference is made here to basic manuals in this discipline (e.g., Custodio et al. 1996; Fetter 2018). Of all the hydraulic characterisation methods available in hydrogeology, pumping tests configure a standard procedure in the design of shallow geothermal installations with open-loop heat exchangers. They allow, among other things, the estimation of hydraulic conductivity and the storage coefficient of a free aquifer by well pumping and measuring the hydraulic response in a nearby piezometer(s). In the case of ATES systems where low-permeability dominates, more detailed techniques may be required, such as hydraulic tomography (Bohling and Butler 2010; Mejías et al. 2009; Renard et al. 2009) or *direct-push* methods (Butler et al. 2002). These methodologies allow a detailed spatial distribution of hydraulic permeability to be obtained.

References

Abuel-Naga HM, Bergado DT, Chaiprakaikeow S (2006) Innovative thermal technique for enhancing the performance of prefabricated vertical drain during the preloading process. Geotext Geomembr 24(6):359–370. https://doi.org/10.1016/j.geotexmem.2006.04.003

Al-Khoury R, Bonnier PG, Brinkgreve RBJ (2005) Efficient finite element formulation for geothermal heating systems. Part I: steady state. Int J Numer Methods Eng 63(7):988–1013. http://doi.org/10.1002/nme.1313

Alrtimi A, Rouainia M, Manning DAC (2014) An improved steady-state apparatus for measuring thermal conductivity of soils. Int J Heat Mass Transf 72:630–636. https://doi.org/10.1016/j.ijheatmasstransfer.2014.01.034

Anderson MP (2005) Heat as a ground water tracer. Ground Water 43(6):951–968. https://doi.org/10.1111/j.1745-6584.2005.00052.x

ASTM (2000) ASTM D 5334 standard test method for determination of thermal conductivity of soil and soft rock by thermal needle probe procedure. ASTM International, West Conshohocken, PA

Austin AW, Yavuzturk C, Spitler J (2000) Toward optimum sizing of heat exchangers for ground-source heat pump systems-development of an in-situ system and analysis procedure for measuring ground thermal. ASHRAE Trans 106(1):831–842

Bandos TV, Montero Á, Fernández de Córdoba P, Urchueguía JF (2011) Improving parameter estimates obtained from thermal response tests: effect of ambient air temperature variations. Geothermics 40(2):136–143. https://doi.org/10.1016/j.geothermics.2011.02.003

Banks D (2012) An introduction to thermogeology: ground source heating and cooling. Wiley, Hoboken

Beier RA, Smith MD (2003) Minimum duration of in-situ tests on vertical boreholes. ASHRAE Trans 109(3):475–486

Bertermann D et al (2013) ThermoMap—an open-source web mapping application for illustrating the very shallow geothermal potential in Europe and selected case study areas

Blum P, Campillo G, Kölbel T (2011) Techno-economic and spatial analysis of vertical ground source heat pump systems in Germany. Energy 36(5):3002–3011. https://doi.org/10.1016/j.energy.2011.02.044

Bohling GC, Butler JJ Jr (2010) Inherent limitations of hydraulic tomography. Groundwater 48(6):809–824. https://doi.org/10.1111/j.1745-6584.2010.00757.x

Bristow KL, Kluitenberg GJ, Horton R (1994) Measurement of soil thermal properties with a dual-probe heat-pulse technique. Soil Sci Soc Am J 58(5):1288–1294. https://doi.org/10.2136/sssaj1994.03615995005800050002x

Brouyère S, Carabin G, Dassargues A (2004) Climate change impacts on groundwater resources: modelled deficits in a chalky aquifer, Geer basin, Belgium. Hydrol J 12(2):123–134. https://doi.org/10.1007/s10040-003-0293-1

Butler J Jr, Healey J, McCall W, Garnett E, Loheide S (2002) Hydraulic tests with direct-push equipment. Ground Water 40:25–36. https://doi.org/10.1111/j.1745-6584.2002.tb02488.x

Carslaw HS, Jaeger JC (1986) Conduction of heat in solids. Clarendon Press, Oxford

Clarke BG, Agab A, Nicholson D (2008) Model specification to determine thermal conductivity of soils. Proc Inst Civ Eng Geotech Eng 161(3):161–168. https://doi.org/10.1680/geng.2008.161.3.161

Custodio E, Llamas MR (1996) Hidrología subterránea. Ed. Omega, Barcelona

Diersch HJG, Bauer D, Heidemann W, Rühaak W, Schätzl P (2011) Finite element modeling of borehole heat exchanger systems: Part 1. Fundamentals. Comput Geosci 37(8):1122–1135. https://doi.org/10.1016/j.cageo.2010.08.003

Engelhardt I et al (2013) Suitability of temperature, hydraulic heads, and acesulfame to quantify wastewater-related fluxes in the hyporheic and riparian zone. Water Resour Res 49(1):426–440. https://doi.org/10.1029/2012wr012604

Eskilson P (1986) Superposition borehole model, manual for computer code. Department of Mathematical Physics, University of Lund, Lund, Sweden

Fetter CW (2018) Applied hydrogeology, 4th edn. Waveland Press, USA

Gehlin SEA, Hellström G (2003) Influence on thermal response test by groundwater flow in vertical fractures in hard rock. Renew Energy 28(14):2221–2238. https://doi.org/10.1016/S0960-1481(03)00128-9

Jackson RD, Taylor SA (1986) Thermal conductivity and diffusivity, methods of soil analysis, pp 945–956. http://doi.org/10.2136/sssabookser5.1.2ed.c39

Kavanaugh SP, Xie L, Martin C (2000) TRP-1118—investigation of methods for determining soil and rock formation thermal properties from short term field test. American Society of Heating, Refrigerating and Air-Conditioning Engineers (ASHRAE), Atlanta, GA, USA

Kurevija T, Macenić M, Borović S (2017) Impact of grout thermal conductivity on the long-term efficiency of the ground-source heat pump system. Sustain Cities Soc 31:1–11. https://doi.org/10.1016/j.scs.2017.02.009

Luo J et al (2015) Experimental investigation of a borehole field by enhanced geothermal response test and numerical analysis of performance of the borehole heat exchangers. Energy 84:473–484. https://doi.org/10.1016/j.energy.2015.03.013

Ma R, Zheng C, Zachara JM, Tonkin M (2012) Utility of bromide and heat tracers for aquifer characterization affected by highly transient flow conditions. Water Resour Res 48(8). http://doi.org/10.1029/2011WR011281

Mejías M, Renard P, Glenz D (2009) Hydraulic testing of low-permeability formations: a case study in the granite of Cadalso de los Vidrios, Spain. Eng Geol 107(3):88–97. https://doi.org/10.1016/j.enggeo.2009.05.010

Molz FJ, Warman JC, Jones TE (1978) Aquifer storage of heated water: Part I—a field experiment. Groundwater 16(4):234–241. https://doi.org/10.1111/j.1745-6584.1978.tb03230.x

Morgensen P (1983) Fluid to duct wall heat transfer in duct system heat storage. In: Proceedings of the international conference on surface heat transfer in theory and practice, Stockholm, pp 652–657

Nope F, de Santiago Buey C (2014) Medida de la Conductividad Térmica del suelo en laboratorio. Fundamentos físicos, aplicaciones geotérmicas y relaciones con otros parámetros del suelo. Ingeniería Civil 175:97–104

Pehme PE, Parker BL, Cherry JA, Molson JW, Greenhouse JP (2013) Enhanced detection of hydraulically active fractures by temperature profiling in lined heated bedrock boreholes. J Hydrol 484:1–15. https://doi.org/10.1016/j.jhydrol.2012.12.048

Poulsen SE, Alberdi-Pagola M (2015) Interpretation of ongoing thermal response tests of vertical (BHE) borehole heat exchangers with predictive uncertainty based stopping criterion. Energy 88:157–167. https://doi.org/10.1016/j.energy.2015.03.133

Prunty L, Horton R (1994) Steady-state temperature distribution in nonisothermal, unsaturated closed soil cells. Soil Sci Soc Am J 58(5):1358–1363. https://doi.org/10.2136/sssaj1994.03615995005800050011x

Raymond J, Lamarche L (2013) Simulation of thermal response tests in a layered subsurface. Appl Energy 109:293–301. https://doi.org/10.1016/j.apenergy.2013.01.033

Raymond J, Robert G, Therrien R, Gosselin L (2010) A novel thermal response test using heating cables. In: Proceedings of the World Geothermal Congress, Bali, Indonesia, pp 1–8

Raymond J, Therrien R, Gosselin L, Lefebvre R (2011a) Numerical analysis of thermal response tests with a groundwater flow and heat transfer model. Renew Energy 36(1):315–324. https://doi.org/10.1016/j.renene.2010.06.044

Raymond J, Therrien R, Gosselin L, Lefebvre R (2011b) A review of thermal response test analysis using pumping test concepts. Groundwater 49(6):932–945. https://doi.org/10.1111/j.1745-6584.2010.00791.x

Renard P, Glenz D, Mejias M (2009) Understanding diagnostic plots for well-test interpretation. Hydrogeol J 17(3):589–600. https://doi.org/10.1007/s10040-008-0392-0

Saar MO (2011) Review: geothermal heat as a tracer of large-scale groundwater flow and as a means to determine permeability fields. Hydrogeol J 19(1):31–52. https://doi.org/10.1007/s10040-010-0657-2

Sáez Blázquez C et al (2017) Thermal conductivity map of the Avila region (Spain) based on thermal conductivity measurements of different rock and soil samples. Geothermics 65:60–71. https://doi.org/10.1016/j.geothermics.2016.09.001

Salmon D (2001) Thermal conductivity of insulations using guarded hot plates, including recent developments and sources of reference materials. Meas Sci Technol 12(12):R89–R98. https://doi.org/10.1088/0957-0233/12/12/201

Sanner B, Hellström G, Spitler J, Gehlin S (2005) Thermal response test—current status and world wide application. World Geothermal Congress, Antalya, Turkey, pp 1436–1445

Sauty JP, Gringarten AC, Menjoz A, Landel PA (1982) Sensible energy storage in aquifers: 1. Theoretical study. Water Resour Res 18(2):245–252. https://doi.org/10.1029/WR018i002p00245

Shonder JA, Beck JV (1999) Determining effective soil formation properties from field data using a parameter estimation technique. ASHRAE Trans 105:458–466

Shook GM (2001) Predicting thermal breakthrough in heterogeneous media from tracer tests. Geothermics 30(6):573–589. https://doi.org/10.1016/S0375-6505(01)00015-3

Signorelli S, Bassetti S, Pahud D, Kohl T (2007) Numerical evaluation of thermal response tests. Geothermics 36(2):141–166. https://doi.org/10.1016/j.geothermics.2006.10.006

Smith M, Perry R (1999) In situ testing and thermal conductivity testing. In: Proceedings of the 1999 GeoExchange technical conference and expo, Oklahoma

Stauffer F, Bayer P, Blum P, Giraldo NM, Kinzelbach W (2013) Thermal use of shallow groundwater. Taylor & Francis, London

Vieira A et al (2017) Characterisation of ground thermal and thermo-mechanical behaviour for shallow geothermal energy applications. Energies 10(12):2044

Vogt T, Schneider P, Hahn-Woernle L, Cirpka OA (2010) Estimation of seepage rates in a losing stream by means of fiber-optic high-resolution vertical temperature profiling. J Hydrol 380(1):154–164. https://doi.org/10.1016/j.jhydrol.2009.10.033

Wagner R, Clauser C (2005) Evaluating thermal response tests using parameter estimation for thermal conductivity and thermal capacity. J Geophys Eng 2(4):349–356. https://doi.org/10.1088/1742-2132/2/4/s08

Wagner R, Rohner E (2008) Improvements of thermal response tests for geothermal heat pumps. In: IEA heat pump conference, Zürich, Switzerland

Wagner V, Bayer P, Kübert M, Blum P (2012) Numerical sensitivity study of thermal response tests. Renew Energy 41:245–253. https://doi.org/10.1016/j.renene.2011.11.001

Wagner V, Blum P, Kübert M, Bayer P (2013) Analytical approach to groundwater-influenced thermal response tests of grouted borehole heat exchangers. Geothermics 46:22–31. https://doi.org/10.1016/j.geothermics.2012.10.005

Wagner V et al (2014) Thermal tracer testing in a sedimentary aquifer: field experiment (Lauswiesen, Germany) and numerical simulation. Hydrogeol J 22(1):175–187. https://doi.org/10.1007/s10040-013-1059-z

Witte HJL (2001) Geothermal response test with heat extraction and heat injection: examples of application in research and design of geothermal ground heat exchangers. European workshop on geothermal response tests, Lausanne, Switzerland

Witte HJL (2013) Error analysis of thermal response tests. Appl Energy 109:302–311. https://doi.org/10.1016/j.apenergy.2012.11.060

Witte HJL, van Gelder GJ (2006) Geothermal response tests using controlled multipower level heating and cooling pulses (MPL-HCP): quantifying ground water effects on heat transport around a borehole heat exchanger. In: Proceedings of the tenth international conference on thermal energy storage, New Jersey, USA

Wu X, Pope GA, Shook GM, Srinivasan S (2008) Prediction of enthalpy production from fractured geothermal reservoirs using partitioning tracers. Int J Heat Mass Transf 51(5):1453–1466. https://doi.org/10.1016/j.ijheatmasstransfer.2007.06.023

Xamán J, Lira L, Arce J (2009) Analysis of the temperature distribution in a guarded hot plate apparatus for measuring thermal conductivity. Appl Therm Eng 29(4):617–623. https://doi.org/10.1016/j.applthermaleng.2008.03.033

Xue Y, Xie C, Li Q (1990) Aquifer thermal energy storage: a numerical simulation of field experiments in China. Water Resour Res 26(10):2365–2375. https://doi.org/10.1029/WR026i010p02365

Yavuzturk C, Spitler JD (1999) Short time step response factor model for vertical ground loop heat exchangers. ASHRAE Trans 105(2):475–485

Zhou J et al (2006) Method for maintaining one-dimensional temperature gradients in unsaturated, closed soil cells. Soil Sci Soc Am J 70(4):1303–1309. https://doi.org/10.2136/sssaj2005.0336N

Chapter 8
Environmental Impacts

The first statement that should be highlighted is the fact that shallow geothermal technology does not pose a generalized risk to the environment. Only specific situations must be considered, related to specific hydrogeological settings and the construction of geothermal heat exchangers (open and closed) to maintain pristine conditions of aquifers and underground environments. The following subsections discuss the potential impacts related to negligent use and/or use by untrained personnel of this renewable technology.

The exploitation of shallow geothermal resources can be associated with different impacts on the environment, which can be direct and indirect. Direct environmental impacts refer to the environmental changes produced by the systems in the ground, physical, chemical or biological changes produced in the environment itself where the exploitation of geothermal resources takes place, and within the shallow geothermal reservoir. Indirect environmental impacts refer to changes in the environment elsewhere, where an element (electricity, borehole drilling, coolant, etc.) necessary for shallow geothermal exploitation is produced.

8.1 Thermal Impacts

The ground temperature in the vicinity of a geothermal heat exchanger may be affected by thermal anomalies generated by another geothermal heat exchanger. In the case of closed-loop heat exchangers, an exchanger affected thermally by another nearby exchanger will observe that the heat carrier fluid circulating inside the exchanger will behave differently. If both systems are performing the same activity (e.g., heat absorption from the shallow geothermal reservoir), the interference will tend to be negative and the system affected by the presence of other heat exchangers will observe that the temperature of the surrounding heat carrier fluid is less *hot* than expected, thus losing thermal efficiency. In other words, the adjoining exchanger will

A. García Gil et al., *Shallow Geothermal Energy*, Springer Hydrogeology,
https://doi.org/10.1007/978-3-030-92258-0_8

have already additionally cooled the ground, and the capacity to absorb heat will be lower the larger the anomaly generated by the neighbouring heat exchanger.

The risk of potential interference has been identified in different settings (Ferguson 2009): between geothermal installations with closed-loop heat exchangers located in high-density areas of shallow geothermal installations (usually in urban areas) by numerical modelling (Hecht-Mendez et al. 2010; Rivera et al. 2016). There are very few documented cases of widespread interference between installations with closed-loop heat exchangers. The monitoring of ground temperature changes in areas with a high density of installations is necessary to establish the space requirements between individual systems and to determine the feasible densities of these systems. However, the definition of these minimum distances between systems is still under discussion. Fasci et al. (2018) proposed a method to assess the size of the area of influence of a BHE by considering the absolute distance between thermal interferences in a central borehole surrounded by two consecutive BHE rings. The study of thermal interference between closed-loop installations has called into question the long-term sustainability of the exploitation of the shallow geothermal energy available in urban areas with a high density of installations. This is because a long period of time is needed to reach the steady state of the thermal regime under BHEs operations. Several decades are needed to ensure that the thermal impacts are stable over time and that the operation regimes of the installations are sustainable in the long term. Fascì et al. (2019) showed that BHEs in areas of high BHE density should be 2, 3 or 6 times deeper than in fully insulated systems for lifetimes of 15, 25 and 50 years, operating with the same thermal loads and exploitation regimes throughout the year. This trend to make exchangers deeper (≥200 m) to avoid interference problems is already observed in Sweden (SGU 2016). In addition, the inverse relationship between depth and distance between closed-loop exchangers has been investigated (Kurevija et al. 2012). The interference in this case has been evaluated numerically in terms of heat transfer rate per metre of exchanger to estimate the associated efficiency loss (Gultekin et al. 2016). The results obtained indicate that a distance of 4.5 m between heat exchangers is sufficient to avoid total efficiency losses below 10% with an operating regime of 2400 continuous hours and a high thermal conductivity value of 4 W m^{-1} K^{-1} in worst-case scenarios. However, Law and Dworkin (2016) showed that a distance of six metres between BHEs, as recommended by the *American Society of Heating, Refrigerating and Air-Conditioning Engineers* (ASHRAE 2016) may be insufficient to avoid negative thermal interference, especially when installations have large differences in *cooling and heating* loads throughout the year. In any case, none of these studies have taken into account heat advection by groundwater flow, even though it is an important factor in determining the spacing between closed-loop shallow geothermal heat exchangers (Alcaraz et al. 2016; Fujii et al. 2005). Other authors have proposed the interconnection of facilities through a fifth generation ultra-low temperature *district heating and cooling network* (5G DHC) as an alternative option to minimise and even eliminate thermal interferences (Verda et al. 2012).

The thermal interference between open-loop heat exchangers (geothermal wells) occurs when the production (pumping) well(s) extract groundwater affected by other geothermal injection well(s) of the installation itself or another installation in the vicinity. In the first case, the interference generated is known as thermal self-interference or *intra-system thermal interference*, and in the second case, the interference is known as *inter-system thermal interference*. In both cases, the efficiency of the installations is reduced and, in extreme cases, their economic and energy efficiency can be compromised. Thermal interferences may even make shallow geothermal heat pump systems lose their renewable character if the increase in energy consumption is associated with an increase in GHG emissions. This is possible if electricity consumption increases with a high fossil fuel mix. Evidence of thermal self-interference has been documented in case studies in the Po-Venice plains region (Italy), related to a low hydraulic gradient and the lack of available space in densely urbanised historic urban centres, where an optimal separation between geothermal production and injection wells is not possible (Galgaro and Cultrera 2013). Two types of self-interference are differentiated (Milnes and Perrochet 2013): (1) *thermal feedback*, when the injection temperature is fixed and the captation temperature varies but it does not condition the injection temperature (Gringarten and Sauty 1975a; Lippmann and Tsang 1980; Stauffer et al. 2013) and (2) *thermal recycling*, when the temperature change between the captation and injection temperature is superimposed, thus conditioning the injection temperature to the captation temperature. To describe the thermal recycling phenomenon mathematically, an algorithm based on a semi-analytical solution or an approximation by numerical methods is necessary (Luo and Kitanidis 2004; Strack 1989). The thermal feedback phenomenon has been studied by particle tracking methods in open-loop heat exchanger installations with complex configurations of several collection-injection wells (Ferguson 2006).

Inter-system thermal interference, on the other hand, occurs when the captation temperature of a geothermal well from one facility is affected by a geothermal injection well from a different facility, i.e., by an external thermal plume (Andrews 1978). There are documented cases of thermal interference between open systems, most notably the cases of Manitoba in a carbonate aquifer in Canada (Ferguson and Woodbury 2006) and the urban alluvial aquifer of Zaragoza (Spain) (García-Gil et al. 2014). Urich et al. (2010) analysed the impact of urban housing distribution on the massive use of domestic geothermal wells, taking into account the distances of influence between systems due to intersystem thermal interference and hydraulic impacts associated with pumping and subsoil water injections. The authors estimated, according to the proposed exploitation scenarios, a reduction of between 10 and 70% of the demanded thermal energy.

The subsurface thermal regime can be complex (Chap. 2), especially in urban areas where the existing subsurface energy transfer processes of natural and anthropogenic origin have the potential to affect (positively or negatively) shallow geothermal system performance. Some of the most important energy transfer processes in urban environments include the river-aquifer relationship (García-Gil et al. 2014), surface-soil energy transfer due to atmospheric temperature variations (Kupfersberger et al. 2017), the SUHI effect (Bayer et al. 2019; Benz et al. 2015; Zhu et al. 2010b) involving heat transfer between building structures and city infrastructure and the ground (Attard et al. 2016b; Epting et al. 2017b), among others. In this context, numerical models of groundwater flow and heat transport, in combination with high-resolution monitoring of ground/groundwater temperature and shallow geothermal installations, have proven to be the most appropriate tools capable of reproducing thermal interferences in an environment with complex transient boundary conditions. Numerical models configure a design tool, e.g., for predicting temperature changes in the subsurface and for assessing the efficiency of a system projected to operate for a lifetime of 20 years (Rybach and Mongillo 2006). However, the models are more commonly used to assess thermal impacts and long-term sustainability studies of large and/or complex facilities, e.g., the London Underground (Arthur et al. 2010). A numerical model used to reproduce the subsurface thermal regime in central London (Herbert et al. 2013) demonstrated for the first time that it was feasible to create and manage numerical models at the urban district scale, capable of providing an estimate of the probability of thermal interference, as well as the corresponding efficiency and feasibility in that environment. More complex city-scale models have introduced more boundary conditions, such as those for the cities of Zaragoza (García-Gil et al. 2015b), Basel (Mueller et al. 2018), or Torino (Sciacovelli et al. 2014).

Although thermal interference between ATES systems can also lead to a reduction in the recovery efficiency of ground heat stored in individual systems, decreasing the spacing between ATES systems can also increase the total energy provided by those systems in a given area. In the Netherlands, it has been shown that controlled reduced thermal interference allows 30–40% more thermal energy to be provided in an area compared to a scenario where thermal interference is completely avoided (Sommer et al. 2015).

8.2 Geochemical Impacts

Potential geochemical impacts caused by shallow geothermal systems in the terrain can be primary or secondary. Primary impacts are those related to inappropriate design or construction of geothermal heat exchangers. Secondary impacts are those triggered by the generation of thermal plumes or hydraulic anomalies in the underground geological environment during the operation of shallow geothermal installations.

One of the major risks directly related to shallow geothermal systems is the poor insulation of boreholes drilled for the construction of geothermal heat exchangers (open and closed). Poor insulation is responsible for hydraulic interconnection between aquifers separated by aquitards or aquicludes (low permeability layers). Poorly sealed leaking wells are considered a major problem in groundwater protection strategies and have been linked to contamination problems in groundwater bodies exploited for drinking water supply (Koh et al. 2016; Lacombe et al. 1995). In the city of Munich, the supply of drinking water depends on groundwater reserves located in a semi-confined aquifer below a shallow aquifer partially isolated by an aquitard. The local authorities only allow geothermal exploitation of the shallow aquifer by means of open-loop geothermal heat exchangers in the shallow upper aquifer. Drilling into the aquitard is prohibited, in order to avoid possible leakage through poorly sealed BHE or geothermal wells (Böttcher et al. 2019). Another case of interest is the city of Ljubljana where, in almost all districts, the drilling of boreholes for exploiting shallow geothermal energy resources is prohibited, due to the use of its shallow urban aquifer to produce drinking water supply for the city (Janža 2017). Primary impacts are related to the establishment of a hydraulic connection between (1) groundwater bodies and the surface runoff or (2) different groundwater bodies. The generation of hydraulic connections creates a circulation pathway for pollutants to enter pristine groundwater bodies (EPA 1997).

In addition, the inappropriate construction of closed-loop geothermal heat exchangers is the main cause of heat carrier fluid losses, which are considered pollutants in some cases. The heat carrier fluids circulating in these exchangers consist of a mixture of water with an organic antifreeze, usually ethylene glycol, propylene glycol or betaine, which are considered potential groundwater pollutants (Klotzbücher et al. 2007). In open-loop exchangers, inappropriate design and/or construction of the operating wells or the pipe line of the secondary circuit (wells-plate exchanger) can lead to a lack of isolation of the subsurface conditions from the groundwater and expose it to atmospheric conditions. This situation causes different problems related to changes in the physicochemical properties of the groundwater and the rebalancing of CO_2 and O_2 partial pressures, changes that are responsible for secondary impacts on the groundwater environment when these waters are re-injected back into the aquifers from which they were pumped (Abesser 2010). Inappropriate design of geothermal wells can result in groundwater inundation phenomena (García-Gil et al. 2015c), ground subsidence (García-Gil et al. 2016a), overpressures in tunnels, etc. Geothermal wells even have the potential to interfere with groundwater users. Although these systems have a non-consumptive use, the extracted water is subsequently injected, and hydraulic levels can be locally modified.

Secondary impacts are related to the response of the groundwater environment to the increase (positive or negative) of groundwater temperatures, modification of physicochemical properties and variations of groundwater flow paths produced by shallow geothermal installations. Geochemical processes are clearly affected by temperature changes, not only causing a perturbation in the thermodynamic balance of homogeneous and heterogeneous reactions, but also affecting the kinetics of the reactions involved (Appelo and Postma 2005; Langmuir 1997).

The thermal impact of open systems on sorption processes of trace elements, including heavy metals, has been investigated by identifying limited impacts on endothermic and exothermic sorption reactions (García-Gil et al. 2016b). Other field studies have described minimal impacts of geothermal wells on major elements and isotopic composition (Park et al. 2015), trace metals (Saito et al. 2016) and emerging organic pollutants (García-Gil et al. 2018a). Column (Bonte et al. 2014) and *batch* (Griebler et al. 2016) experiments have been performed. ATES-induced thermal impacts have been related to modifications of the mobility of organic compounds in sediments (Brons et al. 1991).

8.3 Ecological Impacts

Groundwater ecosystems provide extensive and complex biotopes for a variety of indigenous microbial biocenoses including bacteria, fungi, viruses and protozoa (Griebler and Lueders 2009). In addition to these indigenous microbial communities, microbes adapted to the environment of the intestinal tract of homeothermic animals can survive in the subsoil. These allochthonous communities pose a public health risk, as several groups of enteric bacteria, viruses and protozoa that are included in these communities are commonly considered to be disease-causing pathogens in humans (Buffie and Pamer 2013; Snow 1855). In addition, non-enteric pathogens such as *Legionella pneumophila*, *Staphylococcus aureus* and *Pseudomonas aeruginosa*, as well as free-living amoebae, need to be controlled in aquifers used for human consumption (Anaissie et al. 2002; Cateau et al. 2014). In most urban aquifers there is clear evidence of vulnerability to microbial contamination (Powell et al. 2003). The US Environmental Protection Agency (EPA) was the first governmental organisation to adopt a drinking water regulation that provides standards for maximum allowable levels of various microorganisms in drinking water. The persistence of enteric pathogens in aquatic environments depends on temperature, pH, predation or competition with indigenous microorganisms and other pathogenic bacteria, dissolved oxygen, dissolved organic matter, particle attachment, salt content and other solutes (John and Rose 2005; Madigan and Martinko 2006). Temperature is a crucial physical variable controlling microbial activity. While indigenous microorganisms are adapted to different temperature ranges (psychrophilic, mesophilic and thermophilic microorganisms), enteric pathogens are highly adapted to warm temperatures of approximately 37 °C, which corresponds to the body temperature of *warm-blooded* animals (West et al. 1991). Existing and increasing anthropogenic pressure on urban groundwater resources in densely populated areas, where sewage pollution is ubiquitous, has allowed the transport and survival of sewage-derived microorganisms to groundwater (Rutsch et al. 2008) to expand. Relevant release of nutrients and pathogens from leaking urban sewage networks, together with an increase in groundwater temperature caused from an intensive exploitation of shallow geothermal resources (cooling), may lead to a significant increase in pathogen occurrence and counts, which could imply a public health problem. However, recent studies by García-Gil et al. (2018b) assessed the occurrence of pathogenic bacteria, including coliforms, faecal streptococci, *Escherichia coli, Clostridium perfringens, Salmonella spp. Staphylococcus aureus*, *Legionella pneumophila* and *Pseudomonas aeruginosa*, as well as free-living amoebae, in order to investigate the change of pathogen content in groundwater in relation to thermal plumes generated by shallow geothermal installations. The authors identified a significant and widespread decrease in pathogen content in groundwater affected by thermal plumes. In addition, a trend of recovery of pathogen contents was observed with increasing sampling distance from the nearest GWHP system. This decrease, together with the recovery with distance to the GWHP systems, allowed discussion of the possible existence of thermal shock on bacteria (pseudopasteurisation) inside plate heat exchangers, in the secondary circuit of GWHP systems.

Temperature changes also have an impact on the indigenous groundwater biocenosis. These indigenous biocenoses contribute to groundwater purification and filtration, thus becoming crucial components of subsurface ecosystems (Hahn 2006; Hancock et al. 2005; Hunt and Wilcox 2003). Thermal plumes from shallow geothermal systems have been experimentally demonstrated in the field to affect the composition of indigenous microbial communities in groundwater and their diversity in an oligotrophic aquifer (Brielmann et al. 2009; Lienen et al. 2017). However, although a modification in microbiological composition and diversity is identified in an aquifer heated by a thermal plume, the authors indicate that the integrity of the aquifer as an ecosystem is not modified, and the respective ecosystem functions remain unchanged. In their case study, the authors observed significant diversity in the aquifer studied prior to the operation of the geothermal installation, which increased with thermal impacts. They also point out that the lack of scientific studies investigating the thermal impact of shallow geothermal systems on groundwater ecosystems and makes it impossible to define threshold thermal impacts to ensure the integrity of groundwater ecosystem functions. However, they indicate that temperature changes between captation and injection geothermal wells of $\pm$ 6 ºC, as considered in some governmental good practice guidelines, are acceptable (VDI4640/4, 2004) and could be higher, although scientific studies are lacking. For planktonic micro-organisms in a pond ecosystem, it has been shown that the direct effects of temperature increase are much less important than the effects of nutrients, and that a combination of temperature and nutrient changes triggered very complex dynamics (Christoffersen et al. 2006). Therefore, the ecologically admissible thermal impact for a groundwater ecosystem cannot yet be established.

Other problems resulting from the injection of *hot* water into aquifers is the modification of bacterial growth rates and the clogging of injection boreholes due to the accumulation of gels of biological origin. Although the problem is not always clearly visible, Garnier (2012), through his experience in the French city of Lyon, or Burté (2018), from the study of numerous geothermal farms located in the alluvial areas of the Rhine, Saône or Rhône rivers, demonstrate the high degree of affection of bacterial growth induced by the implementation of open geothermal systems.

Uncontrolled growth of microorganisms can have negative consequences on the performance of geothermal systems. It is possible to observe biological precipitates, which considerably reduce the flow through well casing screens (Bezelgues-Courtade et al. 2013) and the permeability of the aquifer itself. In addition, bacteria are also responsible for many of the corrosion processes that frequently occur in heat exchanger circuits. Psychrophilic, mesophilic and thermophilic bacteria can be found in cooling circuits (Kim and Lee 2019), but two main groups of bacteria are mainly detected from the point of view of their interaction with the installation:

(a) Bacteria that form gelatinous capsules that trap very fine particles and end up causing clogging problems in filters. Among other genera we can find: *Flavobacterium, Proteus, Pseudomonas, Serratia or alkaligen.*
(b) Corrosion-causing bacteria. This group includes:

- Acid-producing bacteria of the genus *Thiobacillus*, which obtain energy by oxidising sulphur (S^{2-}) to elemental sulphur (S^0) and oxidising it to produce sulphuric acid.
- Iron bacteria (ferrobacteria). Bacteria which require the presence of oxygen to develop, they are able to obtain energy by oxidising ferrous iron (Fe^{2+}) into ferric iron (Fe^{3+}). Oxygen consumption leads to reducing conditions that favour the proliferation of sulphate-reducing bacteria.
- Sulphate-reducing bacteria. These are anaerobic organisms that use sulphate as an electron acceptor, producing sulphuric acid as metabolic waste. Eighteen bacterial genera with this capacity have been described (*Campylobacter, Desulfuromonas, Desulfovibrio, Desulfobacterium*, etc.). They have a very wide distribution. They find their optimum pH between 5.5 and 9.5 units. The optimum temperature range, very dependent on the specific species, is very wide, between 10 and 60 °C. They are also very adaptable to varying environmental conditions.

8.4 Geotechnical Impacts

The operating activity of shallow geothermal installations can also generate different geomechanical problems that are ultimately likely to cause structural damage to buildings and infrastructure. Most materials expand or contract under a change in temperature. When differential expansion/contraction behaviour of different materials occurs, temperature changes can be responsible for structural changes of the sediment and can even cause structural damage (Somerton 1992). Constrained bodies that are unable to expand experience internal stress as their temperature changes (Sect. 2.2.3). The biggest thermal expansion problems occur in BTES installations (Sect. 5.6) as they operate at higher temperatures of more than 60 °C (Sanner and Knoblich 1998). Shallow geothermal systems operating at high temperatures near the surface in the unsaturated zone, especially those with horizontal heat exchangers, are at risk of generating free convection air flows with vapour migration, causing progressive drying and even shrinkage of clay materials in the subsurface (Gabrielsson et al. 2000). Installations working at sub-zero temperatures can generate ice in the pores of rocks and sediments causing expansion and gelifraction, which can cause ground uplift with all the geotechnical implications for nearby infrastructure (Penner 1962). When temperature changes are of lesser magnitude, there is a periodic expansion and contraction of the subsoil and, when it comes to geostructures, there is a risk of structural damage (Dalla Santa et al. 2016; Svec and Palmer 1989).

The activity of pumping water and its re-injection in geothermal well installations poses different additional geotechnical risks. Groundwater pumping and injection produce changes in the hydrostatic pressure in the pores that can modify the stress tensor of the solid matrix, which is susceptible to rearrangement or collapse, producing settlements, even uplift, in the case of injection. Groundwater pumping can generate fine and sand entrainment. In addition to damaging the impellers of the submersible pumps, plate heat exchangers and valves of the secondary circuit pipe line, the fines entrainment can clog the grids of the geothermal collection and injection wells, and can even clog the aquifer itself in the vicinity of the wells. Fines entrainment can be significant, several kilograms per year, and even more in wells with poorly designed gravel packs (Fig. 6.2). When the process of fines entrainment is very advanced, significant subsidence phenomena can occur, leading to ground collapse and major structural damage (Shu and Ma 2015).

Where groundwater interacts with evaporite materials during the captation and injection of groundwater for thermal use by shallow geothermal systems, serious geotechnical problems may occur. Evaporite materials are composed of highly soluble minerals such as halite (NaCl), sylvinite (KCl), gypsum ($CaSO_4$-$2H_2O$) and anhydrite ($CaSO_4$). In general, these materials are not in direct contact with groundwater under natural regional flow regime conditions, as they are too soluble and would be dissolved if present. If, during the construction of a closed or open geothermal heat exchanger, these evaporite materials are perforated, the groundwater flow is susceptible to modification (García-Gil et al. 2016a, b) and will generate significant dissolution voids, with high risk of dissolution subsidence. If groundwater comes into contact with anhydrite, the mineral can hydrate to form gypsum, expanding and causing ground uplift, which can lead to structural damage (Goldscheider and Bechtel 2009).

References

Abesser C (2010) Open-loop ground source heat pumps and the groundwater systems: a literature review of current applications, regulations and problems. NER Council, British Geological Survey Nottingham, p 31

Alcaraz M, García-Gil A, Vázquez-Suñé E, Velasco V (2016) Advection and dispersion heat transport mechanisms in the quantification of shallow geothermal resources and associated environmental impacts. Sci Total Environ 543(Part A):536–546. https://doi.org/10.1016/j.scitotenv.2015.11.022

Anaissie EJ, Penzak SR, Dignani MC (2002) The hospital water supply as a source of nosocomial infections: a plea for action. Arch Intern Med 162(13):1483–1492. https://doi.org/10.1001/archinte.162.13.1483

Andrews CB (1978) The impact of the use of heat pumps on ground-water temperatures. Groundwater 16(6):437–443. https://doi.org/10.1111/j.1745-6584.1978.tb03259.x

Appelo CAJ, Postma D (2005) Geochemistry, groundwater and pollution, 2nd ed. CRC Press, Boca Raton

Arthur S, Streetly HR, Valley S, Streetly MJ, Herbert AW (2010) Modelling large ground source cooling systems in the Chalk aquifer of central London. Q J Eng Geol Hydrogeol 43(3):289–306. https://doi.org/10.1144/1470-9236/09-039

ASHRAE (2016) Ashrae handbook 2016: HVAC systems and equipment: SI edition. ASHRAE

Attard G, Rossier Y, Winiarski T, Eisenlohr L (2016) Deterministic modeling of the impact of underground structures on urban groundwater temperature. Sci Total Environ 572:986–994. https://doi.org/10.1016/j.scitotenv.2016.07.229

Bayer P, Attard G, Blum P, Menberg K (2019) The geothermal potential of cities. Renew Sustain Energy (in press)

Benz SA, Bayer P, Menberg K, Jung S, Blum P (2015) Spatial resolution of anthropogenic heat fluxes into urban aquifers. Sci Total Environ 524–525:427–439. https://doi.org/10.1016/j.scitotenv.2015.04.003

Bezelgues-Courtade S, Durst P, Garnier F, Ignatiadis I (2013) ImPAC Lyon: evaluer l'impact environnemental et thermique de l'exploitation des aquifers superficiels pour la climatisation. Aspectos Tecnológicos e Hidrogeológicos de la Geotermia. AIH-GE, Barcelona, pp 51–58

Bonte M, Stuyfzand PJ, van Breukelen BM (2014) Reactive transport modeling of thermal column experiments to investigate the impacts of aquifer thermal energy storage on groundwater quality. Environ Sci Technol 48(20):12099–12107. https://doi.org/10.1021/es502477m

Böttcher F, Casasso A, Götzl G, Zosseder K (2019) TAP—thermal aquifer potential: a quantitative method to assess the spatial potential for the thermal use of groundwater. Renew Energy 142:85–95. https://doi.org/10.1016/j.renene.2019.04.086

Brielmann H, Griebler C, Schmidt SI, Michel R, Lueders T (2009) Effects of thermal energy discharge on shallow groundwater ecosystems. FEMS Microbiol Ecol 68(3):273–286. https://doi.org/10.1111/j.1574-6941.2009.00674.x

Brons HJ, Griffioen J, Appelo CAJ, Zehnder AJB (1991) (Bio)geochemical reactions in aquifer material from a thermal energy storage site. Water Res 25(6):729–736. https://doi.org/10.1016/0043-1354(91)90048-U

Buffie C, Pamer E (2013) Microbiota-mediated colonization resistance against intestinal pathogens. Nat Rev Immunol 13:790–801. https://doi.org/10.1038/nri3535

Burté L (2018) Study of clogging phenomena and treatment optimisation of geothermal operations on shallow aquifers. Université Rennes 1

Cateau E, Delafont V, Hechard Y, Rodier MH (2014) Free-living amoebae: what part do they play in healthcare-associated infections? J Hosp Infect 87(3):131–140. https://doi.org/10.1016/j.jhin.2014.05.001

Christoffersen K, Andersen N, Søndergaard M, Liboriussen L, Jeppesen E (2006) Implications of climate-enforced temperature increases on freshwater pico- and nanoplankton populations studied in artificial ponds during 16 months. Hydrobiologia 560(1):259–266. https://doi.org/10.1007/s10750-005-1221-2

Dalla Santa G, Galgaro A, Tateo F, Cola S (2016) Modified compressibility of cohesive sediments induced by thermal anomalies due to a borehole heat exchanger. Eng Geol 202:143–152. https://doi.org/10.1016/j.enggeo.2016.01.011

EPA (1997) Manual on environmental issues related to geothermal heat pump systems. USEP Agency, National Service Center for Environmental Publications, p 98

Epting J et al (2017) The thermal impact of subsurface building structures on urban groundwater resources—a paradigmatic example. Sci Total Environ 596–597:87–96. https://doi.org/10.1016/j.scitotenv.2017.03.296

Fasci ML, Lazzarotto A, Acuna J, Claesson J (2018) Thermal influence of neighbouring GSHP installations: relevance of heat load temporal resolution. IGSHPA Research Track. International Ground Source Heat Pump Association, Stockholm. https://doi.org/10.22488/okstate.18.000019

Fascì ML, Lazzarotto A, Acuna J, Claesson J (2019) Analysis of the thermal interference between ground source heat pump systems in dense neighbourhoods. Sci Technol Built Environ 25:1–21. https://doi.org/10.1080/23744731.2019.1648130

Ferguson G (2006) Potential use of particle tracking in the analysis of low-temperature geothermal developments. Geothermics 35(1):44–58. https://doi.org/10.1016/j.geothermics.2005.11.001
Ferguson G (2009) Unfinished business in geothermal energy. Ground Water 47(2):167–167. https://doi.org/10.1111/j.1745-6584.2008.00528.x
Ferguson G, Woodbury AD (2006) Observed thermal pollution and post-development simulations of low-temperature geothermal systems in Winnipeg, Canada. Hydrogeol J 14(7):1206–1215. https://doi.org/10.1007/s10040-006-0047-y
Fujii H, Itoi R, Fujii J, Uchida Y (2005) Optimizing the design of large-scale ground-coupled heat pump systems using groundwater and heat transport modeling. Geothermics 34(3):347–364. https://doi.org/10.1016/j.geothermics.2005.04.001
Gabrielsson A, Bergdahl U, Moritz L (2000) Thermal energy storage in soils at temperatures reaching 90 °C. J Sol Energy Eng 122(1):3–8. https://doi.org/10.1115/1.556272
Galgaro A, Cultrera M (2013) Thermal short circuit on groundwater heat pump. Appl Therm Eng 57(1–2):107–115. https://doi.org/10.1016/j.applthermaleng.2013.03.011
García-Gil A, Vázquez-Suñe E, Schneider EG, Sánchez-Navarro JÁ, Mateo-Lázaro J (2014) The thermal consequences of river-level variations in an urban groundwater body highly affected by groundwater heat pumps. Sci Total Environ 485–486:575–587. https://doi.org/10.1016/j.scitotenv.2014.03.123
García-Gil A, Vázquez-Suñé E, Sánchez-Navarro JA, Lázaro J (2015a) Recovery of energetically overexploited urban aquifers using surface water. J Hydrol 1(1):111
García-Gil A, Vázquez-Suñé E, Sánchez-Navarro JÁ, Mateo Lázaro J, Alcaraz M (2015b) The propagation of complex flood-induced head wavefronts through a heterogeneous alluvial aquifer and its applicability in groundwater flood risk management. J Hydrol 527:402–419. https://doi.org/10.1016/j.jhydrol.2015.05.005
García-Gil A et al (2016a) A reactive transport model for the quantification of risks induced by groundwater heat pump systems in urban aquifers. J Hydrol 542:719–730. https://doi.org/10.1016/j.jhydrol.2016.09.042
García-Gil A et al (2016b) A city scale study on the effects of intensive groundwater heat pump systems on heavy metal contents in groundwater. Sci Total Environ 572:1047–1058. https://doi.org/10.1016/j.scitotenv.2016.08.010
García-Gil A et al (2018a) Occurrence of pharmaceuticals and personal care products in the urban aquifer of Zaragoza (Spain) and its relationship with intensive shallow geothermal energy exploitation. J Hydrol 566:629–642. https://doi.org/10.1016/j.jhydrol.2018.09.066
García-Gil A et al (2018b) Decreased waterborne pathogenic bacteria in an urban aquifer related to intense shallow geothermal exploitation. Sci Total Environ 633:765–775. https://doi.org/10.1016/j.scitotenv.2018.03.245
Garnier F (2012) Contribution to the biogeochemical evaluation of the impacts related to the geothermic exploitation of the near-surface aquifers: experiments and simulations on a pilot and real installation scale. Université d'Orléans
Goldscheider N, Bechtel TD (2009) Editors' message: the housing crisis from underground—damage to a historic town by geothermal drillings through anhydrite, Staufen, Germany. Hydrogeol J 17(3):491–493. https://doi.org/10.1007/s10040-009-0458-7
Griebler C, Lueders T (2009) Microbial biodiversity in groundwater ecosystems. Freshw Biol 54(4):649–677. https://doi.org/10.1111/j.1365-2427.2008.02013.x
Griebler C et al (2016) Potential impacts of geothermal energy use and storage of heat on groundwater quality, biodiversity, and ecosystem processes. Environ Earth Sci 75(20):1391. https://doi.org/10.1007/s12665-016-6207-z
Gringarten A, Sauty J (1975) A theoretical study of heat extraction from aquifers with uniform regional flow. J Geophys Res 80(35):4956–4962
Gultekin A, Aydin M, Sisman A (2016) Thermal performance analysis of multiple borehole heat exchangers. Energy Convers Manag 122:544–551. https://doi.org/10.1016/j.enconman.2016.05.086

Hahn HJ (2006) The GW-Fauna-Index: a first approach to a quantitative ecological assessment of groundwater habitats. Limnologica 36(2):119–137. https://doi.org/10.1016/j.limno.2006.02.001

Hancock PJ, Boulton AJ, Humphreys WF (2005) Aquifers and hyporheic zones: towards an ecological understanding of groundwater. Hydrogeol J 13(1):98–111. https://doi.org/10.1007/s10040-004-0421-6

Hecht-Mendez J, Molina-Giraldo N, Blum P, Bayer P (2010) Evaluating MT3DMS for heat transport simulation of closed geothermal systems. Ground Water 48(5):741–756. https://doi.org/10.1111/j.1745-6584.2010.00678.x

Herbert A, Arthur S, Chillingworth G (2013) Thermal modelling of large scale exploitation of ground source energy in urban aquifers as a resource management tool. Appl Energy 109:94–103. https://doi.org/10.1016/j.apenergy.2013.03.005

Hunt RJ, Wilcox DA (2003) Ecohydrology—why hydrologists should care. Groundwater 41(3):289–289. https://doi.org/10.1111/j.1745-6584.2003.tb02592.x

Janža M (2017) Management of the groundwater resource beneath the city of Ljubljana. Procedia Eng 209:100–103. https://doi.org/10.1016/j.proeng.2017.11.135

John DE, Rose JB (2005) Review of factors affecting microbial survival in groundwater. Environ Sci Technol 39(19):7345–7356. https://doi.org/10.1021/es047995w

Kim H, Lee J-Y (2019) Effects of a groundwater heat pump on thermophilic bacteria activity. Water 11(10)

Klotzbücher T, Kappler A, Straub KL, Haderlein SB (2007) Biodegradability and groundwater pollutant potential of organic anti-freeze liquids used in borehole heat exchangers. Geothermics 36(4):348–361. https://doi.org/10.1016/j.geothermics.2007.03.005

Koh E-H, Lee E, Lee K-K (2016) Impact of leaky wells on nitrate cross-contamination in a layered aquifer system: methodology for and demonstration of quantitative assessment and prediction. J Hydrol 541:1133–1144. https://doi.org/10.1016/j.jhydrol.2016.08.019

Kupfersberger H, Rock G, Draxler JC (2017) Inferring near surface soil temperature time series from different land uses to quantify the variation of heat fluxes into a shallow aquifer in Austria. J Hydrol 552:564–577. https://doi.org/10.1016/j.jhydrol.2017.07.030

Kurevija T, Vulin D, Krapec V (2012) Effect of borehole array geometry and thermal interferences on geothermal heat pump system. Energy Convers Manag 60:134–142. https://doi.org/10.1016/j.enconman.2012.02.012

Lacombe S, Sudicky EA, Frape SK, Unger AJA (1995) Influence of leaky boreholes on cross-formational groundwater flow and contaminant transport. Water Resour Res 31(8):1871–1882. https://doi.org/10.1029/95wr00661

Langmuir D (1997) Aqueous environmental geochemistry. Prentice Hall, Upper Saddle River, NJ

Law YLE, Dworkin SB (2016) Characterization of the effects of borehole configuration and interference with long term ground temperature modelling of ground source heat pumps. Appl Energy 179:1032–1047. https://doi.org/10.1016/j.apenergy.2016.07.048

Lienen T et al (2017) Effects of thermal energy storage on shallow aerobic aquifer systems: temporary increase in abundance and activity of sulfate-reducing and sulfur-oxidizing bacteria. Environ Earth Sci 76(6):261. https://doi.org/10.1007/s12665-017-6575-z

Lippmann MJ, Tsang CF (1980) Ground-water use for cooling: associated aquifer temperature changes. Groundwater 18(5):452–458. https://doi.org/10.1111/j.1745-6584.1980.tb03420.x

Luo J, Kitanidis PK (2004) Fluid residence times within a recirculation zone created by an extraction–injection well pair. J Hydrol 295(1):149–162. https://doi.org/10.1016/j.jhydrol.2004.03.006

Madigan MT, Martinko JM (2006) Brock biology of microorganisms. Pearson Prentice Hall, Upper Saddle River, NJ

Milnes E, Perrochet P (2013) Assessing the impact of thermal feedback and recycling in open-loop groundwater heat pump (GWHP) systems: a complementary design tool. Hydrogeol J 21(2):505–514. https://doi.org/10.1007/s10040-012-0902-y

Mueller MH, Huggenberger P, Epting J (2018) Combining monitoring and modelling tools as a basis for city-scale concepts for a sustainable thermal management of urban groundwater resources. Sci Total Environ 627:1121–1136. https://doi.org/10.1016/j.scitotenv.2018.01.250

Park Y, Kim N, Lee J-Y (2015) Geochemical properties of groundwater affected by open loop geothermal heat pump systems in Korea. Geosci J 1–12. https://doi.org/10.1007/s12303-014-0059-x

Penner E (1962) Ground freezing and frost heaving. Canadian Building Digest, No. CBD-26. https://doi.org/10.4224/40000788

Powell KL et al (2003) Microbial contamination of two urban sandstone aquifers in the UK. Water Res 37(2):339–352. https://doi.org/10.1016/S0043-1354(02)00280-4

Rivera JA, Blum P, Bayer P (2016) Influence of spatially variable ground heat flux on closed-loop geothermal systems: line source model with nonhomogeneous Cauchy-type top boundary conditions. Appl Energy 180:572–585. https://doi.org/10.1016/j.apenergy.2016.06.074

Rutsch M et al (2008) Towards a better understanding of sewer exfiltration. Water Res 42(10–11):2385–2394. https://doi.org/10.1016/j.watres.2008.01.019

Rybach L, Mongillo M (2006) Geothermal sustainability—a review with identified research needs. Geotherm Resour Council (GRC) Trans 30:8

Saito T et al (2016) Temperature change affected groundwater quality in a confined marine aquifer during long-term heating and cooling. Water Res 94:120–127. https://doi.org/10.1016/j.watres.2016.01.043

Sanner B, Knoblich K (1998) High temperature underground thermal energy storage. In: International geothermal conference, New Jersey, USA

Sciacovelli A, Guelpa E, Verda V (2014) Multi-scale modeling of the environmental impact and energy performance of open-loop groundwater heat pumps in urban areas. Appl Therm Eng 71(2):780–789. https://doi.org/10.1016/j.applthermaleng.2013.11.028

SGU (2016) Vägledning för att borra brunn. Sveriges Geologiska Undersökning, Geological Survey of Sweden, Uppsala

Shu B, Ma B (2015) Study of ground collapse induced by large diameter horizontal directional drilling in sand layer using numerical modeling. Can Geotech J 52:150226182447000. https://doi.org/10.1139/cgj-2014-0388

Snow J (1855) On the mode of communication of cholera. John Churchill, London

Somerton WH (1992) Chapter IV. Thermal expansion of rocks, developments in petroleum science. Elsevier, pp 29–38. https://doi.org/10.1016/S0376-7361(09)70024-X

Sommer W, Valstar J, Leusbrock I, Grotenhuis T, Rijnaarts H (2015) Optimization and spatial pattern of large-scale aquifer thermal energy storage. Appl Energy 137:322–337. https://doi.org/10.1016/j.apenergy.2014.10.019

Stauffer F, Bayer P, Blum P, Giraldo NM, Kinzelbach W (2013) Thermal use of shallow groundwater. Taylor & Francis, Boca Raton

Strack ODL (1989) Groundwater mechanics. Prentice Hall, Englewood Cliffs, NJ

Svec OJ, Palmer JHL (1989) Performance of a spiral ground heat exchanger for heat pump application. Int J Energy Res 13(5):503–510. https://doi.org/10.1002/er.4440130502

Urich C, Sitzenfrei R, Möderl M, Rauch W (2010) Einfluss der Siedlungsstruktur auf das thermische Nutzungspotential von oberflächennahen Aquiferen. Oesterr Wasser Abfallwirtsch 62(5–6):113–119. https://doi.org/10.1007/s00506-010-0188-z

VDI4640/4 (2004) Thermal use of the underground—direct uses. VDIRichtlinien 4640, Part 4 VDI/DIN Handbuck Energietechnik. Verlag, B., Berlin

Verda V, Guelpa E, Kona A, Lo Russo S (2012) Reduction of primary energy needs in urban areas trough optimal planning of district heating and heat pump installations. Energy 48(1):40–46. https://doi.org/10.1016/j.energy.2012.07.001

West JM, Grogan HA, McKinley IG (1991) The role of microbiology in the geological containment of radioactive wastes. Dev Geochem 6:205–215. https://doi.org/10.1016/B978-0-444-88900-3.50024-2

Zhu K, Blum P, Ferguson G, Balke K, Bayer P (2010b) The geothermal potential of urban heat islands. Environ Res Lett 5(4). https://doi.org/10.1088/1748-9326/5/4/044002

Chapter 9
Management and Governance of Shallow Geothermal Energy Resources

9.1 Management of Shallow Geothermal Energy Resources

The word *management* comes from the Latin *gestio*, a word composed of *gestus* from the verb *gerere* (to lead) and the suffix -tio (action and effect). The concept of management refers to the action of bringing or directing available resources towards a goal. Management of shallow geothermal energy resources can be understood as:

A complex set of measures aimed at (1) ***meeting the demands of socio-economic development*** *and (2) achieving* ***sustainable development*** *of the use of shallow geothermal energy resources in terms of their availability as renewable energy and sustainability, both technically in the operation of the facilities and environmentally with the associated ecosystems.*

The management of shallow geothermal energy resources consists of searching for the technical, economic and environmental balance between heat demand and the existing geothermal potential in an area. Management measures will primarily aim at modulating demand to find such a balance. While demand modulation is relatively easy to carry out by controlling access to the resource, resource modulation is generally not an option. If geothermal energy resource enhancement has been contemplated, it has only been considered in cases of overexploitation and as a remediation proposal from a theoretical point of view (Epting et al. 2017a). In general, the natural resource management strategy consists of quantifying the existing resources and then adjusting the sustainable demand (Lynch 2009).

In order to assess which demand is in balance with the available resources, it is absolutely necessary to quantify the shallow geothermal energy potential. On the basis of this knowledge, it is determined how much of this potential can be extracted in compliance with the criteria of technical, economic and environmental sustainability.

A. García Gil et al., *Shallow Geothermal Energy*, Springer Hydrogeology,
https://doi.org/10.1007/978-3-030-92258-0_9

9.1.1 Shallow Geothermal Energy Potential

The term *geothermal potential* is often used, but with different meanings. In most cases it refers to geothermal resource mapping (Kohl et al. 2003; Ondreka et al. 2007) and is quantified as thermal energy in absolute terms [J] or normalised by area [J m^{-2}], volume [J m^{-3}] or time [W]. In some cases, it is simply assessed qualitatively on a relative basis by zoning ranges of higher or lower geothermal potential (Muffler and Cataldi 1978; Noorollahi et al. 2008). Rybach (2015) introduced the definition of urban geothermal potential by proposing three hierarchical categories, used for other renewable energies (Angelis-Dimakis et al. 2011; Voivontas et al. 1998) applied to geothermal potential concept. The total thermal energy available from a physical point of view determines the *theoretical potential.* The fraction of the theoretical potential that can be exploited by the available technologies is called the *technical potential.*

9.1.1.1 Theoretical Potential

One way of calculating the theoretical geothermal potential is the so-called *heat in place* E_{TC} [J]. This approach consists of considering a reservoir of volume V [m^3] and calculating its internal energy U [J], and its total energy E_T [J]. Assuming that the considered body has no kinetic or potential energy, the heat in place can be calculated as (Muffler and Cataldi 1978):

$$E_{TC} = [\phi c_w(1 - \phi)c_m]V\Delta T \qquad (9.1)$$

where ϕ [–] is the porosity, c_w and c_m [J m^{-3} K^{-1}] are the volumetric heat capacity of the water and the solid, respectively, and $\Delta T = T_R - T_{min}$ [K] is the temperature change between the reservoir temperature T_R [K] and the minimum temperature at which the reservoir could be brought to during geothermal exploitation T_{min} [K]. Note that the internal energy is not an absolute value but a variation relative to the initial temperature of the system. In any case, this definition is a legacy of deep geothermal energy and is oriented to the absorption of heat from an exceptionally *hot* reservoir. In shallow geothermal energy, T_R is close to the annual mean atmospheric temperature (see Chap. 3) and geothermal heat exchangers are able to absorb heat but can also dissipate heat. Therefore, a distinction should be made between theoretical absorption and dissipation potential.

According to this definition, the geothermal reservoir is simplified to a closed adiabatic system without heat transfer to the surroundings. In reality this box behaviour does not occur and, although it is a simple method and can give a semi-quantitative idea of the potential, its use is questionable. A geothermal reservoir exploited by a geothermal installation on a permanent basis will experience at least a natural replenishment of thermal energy.

A somewhat more refined definition of the theoretical potential would include the thermal energy available from a geothermal reservoir over a period of time, including heat storage and recharge. The temperature change in the reservoir will tend to equilibrate back to its surroundings through a replenishment process, which will be limited by the heat transfer mechanisms existing in the subsurface, from the slow diffusion of heat by conduction in low permeability porous media, to the advection of heat caused by groundwater flow. Such heat transfer phenomena are transient and are not covered by the above definition of theoretical potential. The heterogeneity in the geology and the uncertainty in the description of its physical properties are crucial factors that make the calculation of the theoretical geothermal potential difficult and limit the estimates to an approximation. The thermal regime beneath cities (where the demand is located) can be very complex, complicating an accurate estimation of the potential even more. In the presence of uncertainty, most of the theoretical geothermal potential studies rely on simplifying assumptions to facilitate the mapping of thermal potential [J] or potential density [J m^{-2}] on a large scale.

In transmissive urban aquifers, where heat advection dominates, it was proposed that the theoretical geothermal potential is that supplied by groundwater (Epting et al. 2018). Knowing the Darcy velocity of groundwater, the heat advection per cubic metre of reservoir can be calculated as:

$$E_{adv} = Q_D c_w \Delta T t \tag{9.2}$$

where Q_D is the Darcy volumetric flow rate [m^3 s^{-1}], c_w [J m^{-3} K^{-1}] is the volumetric heat capacity of water, and $\Delta T = T_i - T_R$ [K] is the difference between the temperature of the water entering the considered reservoir volume T_R [K] and the temperature of the reservoir itself T_i [K].

9.1.1.2 Technical Potential

The technical potential refers to the fraction of the theoretical potential that can be utilised with a given technology. Most of the published works on shallow geothermal potential focus on the technical potential, for installations with open or closed-loop geothermal heat exchangers.

The technical potential is mainly calculated according to technological constraints such as the maximum drawdown caused by geothermal wells, the temperature increase caused in the ground or the drilling depth of boreholes (García-Gil et al. 2015a; Haehnlein et al. 2010). A distinction is made between the general technological potential, defined by the hypothetical application of a technology for full exploitation within predefined physical or regulatory limits, and the specific technological potential for an installation with a specific design.

Installations with open-loop geothermal heat exchangers

The most direct and simple estimation of calculating the technical potential in terms of energy E_{TEC} [J] is by defining the *recovery factor* δ [–] as the ratio between the extractable heat and the heat relative to the theoretical geothermal potential as heat in place [J] (Eq. 9.1). E_{TC} [J] (Eq. 9.1):

$$\delta = \frac{E_{TEC}}{E_{TC}} \Rightarrow E_{TEC} = \delta E_{TC} \tag{9.3}$$

The recovery factor δ is commonly considered for deep geothermal aquifers and reservoirs. It is a function of the geothermal exploitation technology considered, the operating temperature range, the petrophysical properties and the type of aquifer exploited. Experimental values in the range of 0.05–0.5 are often suggested (Bodvarsson 1974; Iglesias 2003; Muffler and Cataldi 1978). In a shallow geothermal installation, provided with a well doublet the extraction factor is given by Lavigne (1977):

$$\delta = \frac{1}{3}\frac{T_w - T_i}{T_{suf} - T_i} \tag{9.4}$$

where T_w [K] is the temperature of the groundwater at the top of the aquifer (at the water table in unconfined aquifers), T_{suf} [K] is the ground surface temperature, and T_i [K] is the temperature of the water in the geothermal injection well.

The technical potential in terms of energy (Eq. 9.3) does not give any idea of the availability of energy over time. To express the potential as power [W], and not as energy [J] in shallow geothermal installations with open-loop heat exchangers, it is possible to consider the flow rate [$m^3\ s^{-1}$] of extractable groundwater, Q [$m^3\ s^{-1}$] of extractable groundwater. Thus, the thermal power P_{TEC} [W] for the calculation of the technical potential in systems with open-loop geothermal heat exchangers would be:

$$P_{TEC} = \mathrm{Q}c_w \Delta T = \mathrm{Q}c_w (T_w - T_i) \tag{9.5}$$

Equation 9.5 is similar to Eq. 9.2, the difference being that Q, unlike Q_D, refers to the exploitation technology, in particular, the flow rate extractable by a geothermal groundwater well. The technical geothermal potential in this case will depend on the extractable flow rate Q extractable from an aquifer. It is therefore a hydrogeological factor that depends on the hydraulic properties of the ground. One way of estimating Q is the application of *Thiem's* analytical solution for steady-state wells, commonly used in pumping tests (García-Gil et al. 2015a):

$$\mathrm{Q} = s_{max}(r) 2\pi \mathrm{T}\left[\ln\left(R \cdot r^{-1}\right)\right]^{-1} \tag{9.6}$$

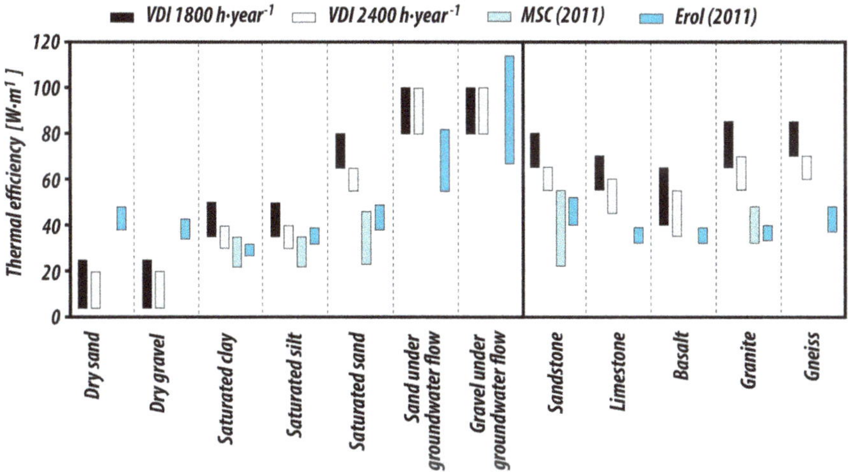

Fig. 9.1 Thermal efficiency rates η_I [W m^{-1}] for borehole heat exchangers in single U-shaped closed-loop with a length between 40 and 100 m, in heating mode and for an annual operating time depending on the lithology according to the VDI (1800 h year^{-1}, 2400 h year^{-1}) (VDI4640/1 2000), MCS (2400 h year^{-1}) (MCS 2011), Erol (2400 h year^{-1}) (Erol 2011) guidelines. Modified from Bayer et al. (2019)

where R [m] is the radius of influence, $s_{max}(r)$ [m] is the maximum allowable descent at a radial distance r [m] to the extraction well and T [m^2 s^{-1}] is the transmissivity.

Installations with borehole heat exchangers

The heat flow transferred per linear metre [W m^{-1}] of a BHE, referred to in this context as *thermal efficiency* η_I [W m^{-1}], is tabulated for different ground materials, both for rock and sediment, and different ranges of pore volume water content (Stauffer et al. 2013; VDI4640/2 2001). Tabulated values are provided by different shallow geothermal guides (Fig. 9.1). Using these tabulated values and a geological map, it is possible to map the technical potential P_{TEC} [W] for installations with closed-loop geothermal heat exchangers as follows:

$$P_{TEC} = \sum z_i \eta_{I_i} \tag{9.7}$$

where the technical potential P_{TEC} [W] of the BHE shall be the sum of the i-thermal potentials of each of the sections of geological material traversed with a thickness of z_i [m] and the associated thermal efficiency η_{I_i} [W m^{-1}]. Some applications of this technique can be found in the specialized literature (De Filippis et al. 2015; Gemelli et al. 2011; Ondreka et al. 2007; Schiel et al. 2016).

9.1.2 Existing Management Approaches

The granting of geothermal discharge rights for new shallow geothermal systems requires the assessment of the impacts on the groundwater environment. This assessment can only be possible through an advanced understanding of the thermal regime in the subsurface and especially in urban groundwater bodies. The most appropriate tools to adequately describe and capture the most relevant thermal boundary conditions include monitoring and numerical modelling. The recent availability of simple and reliable temperature measurement devices and the advancement of three-dimensional heat transport models facilitate the investigation of groundwater heat transport at different scales. Fujii et al. (2007) proposed to develop thematic site suitability mapping of GSHP systems based on local geological and hydrogeological information. The authors demonstrated the possibility of using field data and numerical models of groundwater flow and heat transport for such mapping. Zhu et al. (2010) and Zhu et al. (2011) evaluated the possibilities of utilising heat stored in urban aquifers by SUHI effect for indoor space heating of buildings. In Cologne (Germany) and Winnipeg (Canada), detailed groundwater temperature measurements revealed high groundwater temperatures in the city centres and indicated a warming trend of up to five degrees Celsius (Zhu et al. 2011). Elevated subsurface temperatures are also observed in other cities, such as Shanghai and Tokyo, where it was estimated that the associated shallow geothermal energy potential could supply heating demand even for decades. Allen et al. (2003) proposed that overheated urban aquifers in northern European countries could use the SUHI effect to provide low-enthalpy geothermal energy for indoor space heating in buildings.

Epting et al. (2013) describe the use of thermal groundwater for cooling purposes, the river-aquifer interaction affecting groundwater temperature patterns, and the main natural and anthropogenic thermal processes, including the effect of buildings with heated basements reaching the aquifer, with emphasis on the progressive increase in building density and its relation to the SUHI effect. Epting et al. (2013) defined the concept of a *potential natural state* in undisturbed conditions prior to exploitation. García-Gil et al. (2014) investigated the thermal consequences of river level variations in an urban groundwater body intensively affected by the intensive exploitation of shallow geothermal energy resources. Their results illustrate the thermal impact of surface water recharge from rivers during flood events on an urban alluvial aquifer. A spatial and temporal assessment of *cold* recharge events during river floods and the interaction with associated shallow geothermal installations demonstrated how such knowledge allows for improved thermal management of the urban aquifer. This work highlighted the importance of having an understanding of the varying influence of hydraulic and thermal boundary conditions due to specific geological and hydrogeological conditions in urban environments.

9.1.3 Management Concepts

In order to assess the anthropogenic changes that have already occurred on the thermal regime of an urban groundwater body, it is necessary to reproduce its *current thermal state* and how it has reached this point, according to the different existing hydraulic and thermal boundary conditions, and then compare it with a hypothetical *potential natural state*. The comparison gives a better understanding of (1) the role of river boundaries on the thermal states of cities, (2) the effect of high temperature injection introduced into urban aquifers, and (3) the important thermal effect of building structures on the subsurface, which is commonly underestimated due to lack of reliable temperature data and information.

Currently, many regional environmental and public water authorities are confronted with an increasing number of applications for groundwater use for indoor air-conditioning of buildings. It is common practice to authorise such activity, including thermal discharge, on the basis of a definition of a *potential natural state* of groundwater thermal resources. However, there is no universal standard procedure for estimation of the *potential natural state*. Consequently, there is no consolidated concept for the future management of urban groundwater resources for thermal use. In this sense, the *relaxation factor* concept (García-Gil et al. 2015c) can be applied as a science-based criterion for a more equitable and sustainable geothermal use of shallow energy resources. This concept consists of reserving part of the available shallow geothermal resources for future installations and preventing a monopolisation of the resource. It is also necessary to consider the thermo-hydraulic impacts on third installations, as well as in the definition of maximum and minimum injection temperature thresholds. These practices should be considered as possible steps towards a more sustainable thermal use of shallow urban aquifers.

For effective management of urban aquifers, the subdivision of the urban aquifer into groundwater bodies as management units for groundwater and thermal water resources can be useful. The delineation of water bodies is not only relevant for optimisation of water resources management but also allows definition of boundaries and deducing groundwater and heat flows, including possible pollutants in the groundwater across these boundaries. Therefore, these system inflows and outflows represent important boundaries that can explain the main water quality issues for groundwater resources management.

The *current thermal state* of a groundwater body is composed of the main natural (hydrological and hydrogeological) and anthropogenic boundary conditions, as well as the different users of shallow geothermal installations. It is on these contours where it is necessary to centre the monitoring activity, focused on reproducing the groundwater thermal regime in cities by means of numerical models. These strategies stand out for (1) comparing a current thermal state with a potential natural state of the investigated urban groundwater bodies, (2) calculating the relaxation factor, and (3) contemplating possible remediation measures if necessary.

9.1.3.1 Shallow Geothermal Monitoring Networks

Details of the various monitoring systems as well as the configuration and calibration of the groundwater flow and heat transport models can be found as a case study example in Chap. 11 for the city of Zaragoza and in other examples in the literature (Epting et al. 2013; Epting and Huggenberger 2013; García-Gil et al. 2014). Geothermal control networks do not cover the whole territory but focus on groundwater bodies where the modification of ground temperature can generate environmental and technical problems for other installations. In areas where there is no groundwater flow, there are no general monitoring networks on the scale of cities. In piezometers, continuous measurements of piezometric level and groundwater temperature are taken. In order to make more detailed interpretations of the vertical temperature distribution, it is possible to find piezometers with several multilevel sensors at different depths, with depth intervals from 0.5 to 1 m. Multilevel observation piezometers are designed to capture various processes such as (1) the effects of the interaction between groundwater and the subsurface thermal regime, (2) the influence of groundwater thermal use on the downward geothermal gradient, and (3) the thermal impacts of underground constructional structures reaching an aquifer and their influence on the groundwater thermal regime, including seasonal heating phases. In addition, piezometers can be located 20–100 m downstream of major open-loop geothermal installations or in urban areas not thermally affected by any geothermal installations to monitor the SUHI effect.

9.1.3.2 Thermal Groundwater Regimes

In order to assess the *current thermal state*, it is necessary to analyse how this state has been reached, according to the different hydraulic and thermal boundary conditions. This state can be obtained by studying the historical development of cities from historical maps and information on relevant events that have occurred in the groundwater bodies of interest. To understand the thermal regime of the aquifer, it is necessary to evaluate the Darcy velocities (Sect. 2.4.1) of groundwater flow, the quantitative information on the dominant heat transport mechanism (conduction vs. advection) provided by the dimensionless Peclet number, Pe [–] (Sect. 2.3.10) and the thermal lag coefficients (M_{th}). The thermal lag coefficients are obtained by Bear (1988):

$$M_{th} = \frac{1 + (1 - \phi)c_s\rho_s}{\phi c_w \rho_w} \tag{9.8}$$

where ϕ [–] is the porosity, c [J kg^{-1} K^{-1}] and ρ [kg m^{-3}] are the specific heat capacity and density of the solid phase (s) of the porous matrix and pore water (w), respectively. Finally, it may be appropriate to evaluate the *thermal memory* effect, which expresses the time that elapses from the occurrence of a change in the thermal regime until a new thermal equilibrium is reached in the groundwater body.

9.1.3.3 Relaxation Factor and Remediation Strategies

There is currently a widespread practice worldwide in the use of shallow geothermal resources that follows the simple rule of *first come, first served.* To move away from this unsustainable practice, it is necessary to develop technical solutions with a balanced use of resources, where heating and cooling demand are equal throughout the year. In practice, such a balance cannot be achieved in most urban areas. Concepts for temporary heat storage in deeper geological formations, or in the unsaturated zone, are ongoing research topics. One solution to this problem is the application of the *Indirect Relaxation Factor* (*IRF*) concept (Garcia-Gil et al. 2015c). The efficiency of the heat pump is highly dependent on the relative temperature between the geothermal reservoir and the domestic system (Sect. 4.3). This dependence allows relating the relative temperature difference to the shallow geothermal potential in the subsurface. In the case of a cooling demand scenario, the operating temperatures for the installations are restricted within the range between the aquifer background temperature (T_B), and the maximum temperature (T_{max}) at the reinjection site, which is defined by the competent authority. The temperature change (ΔT) induced by a shallow geothermal installation on the extraction well of another installation must be in the range of the operating temperatures. The IRF is then defined as the geothermal potential reserved for third party installations, taking into account the range of allowed operating temperatures and can be expressed as:

$$IRF = \frac{\Delta T - T_B}{T_{max} - T_B} \tag{9.9}$$

Since the ratio representing the IRF is normalised by the subsurface background temperature and the operational temperature ranges, the subsurface temperature deviation (ΔT) of the initial geothermal resource can be utilized as an indicator of the thermal state of the shallow geothermal reservoir. $0 < IRF < 1$ illustrates the degree of thermal influence that has already occurred. Values tending to zero correspond to an *uninfluenced state.* $IRF > 1$ corresponds to a highly impacted state, indicating inappropriate use of the resource requiring corrective actions.

In urban areas, subsurface thermal regimes generally cannot be characterised by a constant T_B. In addition, for urban groundwater bodies, two different regimes can be distinguished by defining two T_B; one T_B influenced by anthropogenic thermal changes that have already occurred (*current thermal state*) and a natural T_B which represents an anthropogenically uninfluenced potential. This distinction of T_B allows to promote different types of management strategies for urban shallow geothermal energy resources, with a choice between a more restrictive and conservative strategy or an exploitation-oriented development strategy.

Possible remediation measures for groundwater bodies declared at qualitative risk include using groundwater for heating purposes and re-injecting relatively *cold* water into the aquifer (Epting and Huggenberger 2013), and artificially recharging the aquifer with *cold* surface water (García-Gil et al. 2015b).

9.2 Governance Policies

In order to achieve a holistic governance system for shallow geothermal energy resources, it is necessary to define polices to guide the resource management process. In the following, the four key guiding principles on which the management structure of shallow geothermal energy resources could be based are exposed (García-Gil et al. 2020a).

First, the intensive and biased exploitation of shallow geothermal energy resources for heat production in domestic and industrial urban infrastructure may lead to a reduction of shallow geothermal resources. Heat production is understood as the transfer of thermal energy between different thermodynamic systems, including the shallow geothermal energy reservoir and the buildings and infrastructure involved in the exploitation of shallow geothermal energy. In this respect, the loss of geothermal resources associated with mismanagement could compromise the renewability of the thermal resource and thus restrict access to it for decades. To avoid this, the first policy, ***Sustainable development and exploitation of shallow geothermal energy resources***, envisages management focused on preventing the problems of (1) geothermal overexploitation and unsustainable development, (2) negative thermal interference between installations and (3) inefficient use of geothermal energy resources.

Second, uncontrolled exploitation of shallow geothermal energy resources can result in a reduction of environmental quality and, in the worst case, although unlikely, in a threat to human health. The management problems that arise around this issue include (1) hydrodynamic remobilisation of pre-existing contamination of aquifers due to groundwater pumping and injection (hydrogeological regime change), drilling and installation of wells. In areas where contaminated terrains exist, the existence of geothermal wells can trigger the movement of contaminant plumes due to induced groundwater flow. The exploitation of shallow geothermal energy resources may also trigger homogeneous and heterogeneous geochemical reactions that could eventually increase the groundwater content of pre-existing contaminants in the medium such as (2) inorganic trace metals, (3) organic and (4) microbiological contaminants. Furthermore, (5) the thermal discharge of thermal plumes coming from shallow geothermal systems into surface water bodies may induce harmful consequences for groundwater-dependent ecosystems. Finally, shallow geothermal energy exploitation could contribute (VI) to the SUHI effect. To avoid these management problems, management decisions should be guided by the second policy, ***Environmentally friendly use of shallow geothermal energy resources***.

The sustainable development of shallow geothermal energy must address the potential conflict between shallow geothermal energy exploitation and other uses of the subsurface, both pre-existing and future, especially in an urban context. Management issues arising from this potential conflict of subsurface uses include alteration of groundwater quality for human consumption, as well as other secondary uses such as irrigation, industrial, recreational, among others. In addition, the exploitation of geothermal energy resources can generate, in certain situations, geochemical impacts that can trigger terrain subsidence or generate different impacts on subsurface infrastructures. Potential conflicts with other urban subsoil uses must be coordinated and, therefore, the ***exploitation of shallow geothermal energy resources in coordination with other subsurface uses*** is considered to be the third policy.

Finally, the sustainable use of shallow geothermal resources depends on the successful implementation of planned management measures. Therefore, the fourth policy, ***Effective management of shallow geothermal energy resources***, refers to the effectiveness of the specific management approach adopted. The different management problems that compromise the successful implementation of management plans are: (1) management in a context of limited information, (2) conflicts of interest, (3) inefficient management of shallow geothermal energy resources, (4) management measures inappropriate to the specific conditions of the thermal reservoir, (5) disabling environment, (6) uncertainty, and (7) illegal activity and high implementation costs.

9.3 Overall Structure of the Management Framework

The overall structure of the shallow geothermal energy resource management framework proposed here is derived from the conceptual development of each of the *governance policies* described above using a hierarchical system (Fig. 9.2). As a second level of hierarchical ranking, potential *management problems* related to the respective defined governance policies are defined. Since the management problems that may occur are different for each locality, decision-makers are expected to establish their own governance policies once they have identified the existing management problems. Having defined the governance policy to be followed according to the existing problems, managers can propose *management objectives* following the adopted governance policies. Considering that one or more objectives can be

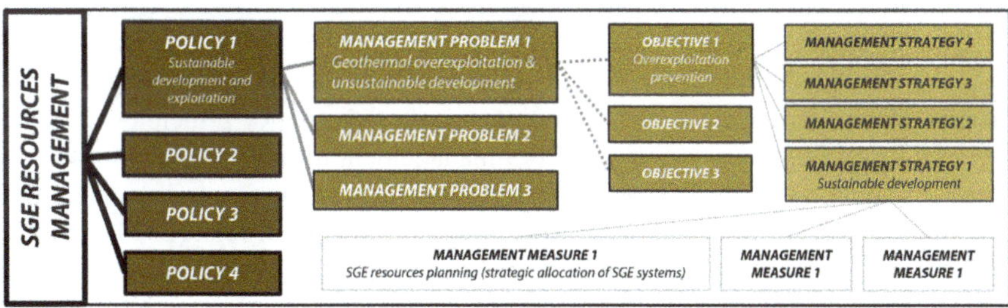

Fig. 9.2 Simplified structure of the proposed management framework for shallow geothermal energy (SGE) resources, showing five levels of governance. In this example, only one of the management measures belonging to one of the four management strategies is shown. This simplification applies to the rest of the hierarchical management levels. Only the extended governance levels are named in the diagram. An extended version of this structure shown is provided in tables for each governance policies set out in this chapter

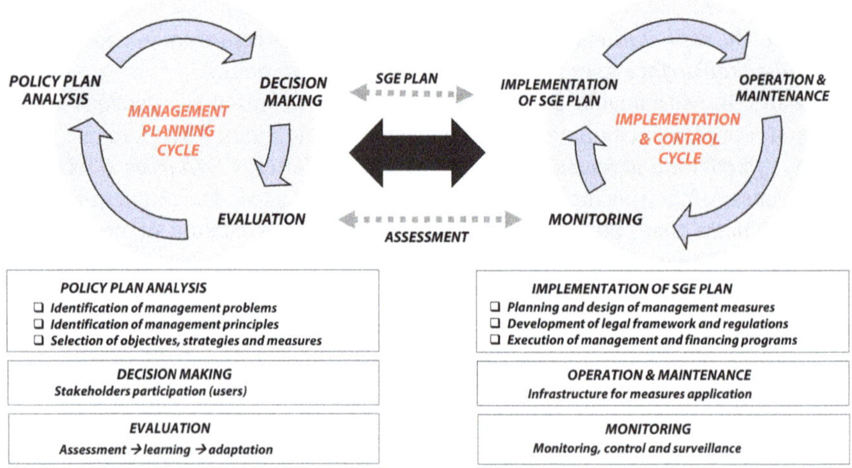

Fig. 9.3 Diagram of the dual-cycle adaptive management concept for shallow geothermal energy resource management from García-Gil et al. (2020a)

assigned to address a management problem, the overall structure of the management framework presented here considers management objectives as a third hierarchical level. In addition, to achieve each management objective, there are different *management strategies* (fourth level) for which specific *management measures* (fifth level) can be proposed. As an example, following the branch developed in Fig. 9.2, the strategic siting of shallow geothermal installations, the discharge authorisation procedure and the planning of district networks fed by shallow geothermal installations are three possible management measures to follow a sustainable development strategy. This strategy can be adopted to meet the objective of preventing overexploitation. Achieving this objective will contribute to improving the management problem of geothermal overexploitation and its unsustainable development.

A comprehensive review of all shallow geothermal energy resource management concepts has resulted in 289 management elements or concepts, organised into five hierarchical management levels: 4 guiding management policy principles; 21 problems; 27 objectives; 58 strategies and 179 management measures. The complete list of all management concepts and their hierarchy make up the overall management structure for shallow resources and can be found in García-Gil et al. (2020a). A thorough review of the measures proposed here would reveal that many are shared with several strategies of different objectives of the same management problem or even different problems. The more objectives the adoption of a measure helps to meet, the more important is that management measure, especially if the objectives are those considered as priorities by managers.

9.3.1 Sustainable Development and Exploitation of Shallow Geothermal Energy Resources

The emergence of new shallow geothermal installations in a limited urban environment can lead to a concentration of installations in certain areas. If, in these areas, the exploitation of geothermal resources far exceeds the natural renewability of the resource, the phenomenon of overexploitation occurs. In other words, an unsustainable development of shallow geothermal energy in a limited environment is likely to generate a situation of intensive exploitation of the resource that, over time, ends up being unsustainable, giving rise to a situation of geothermal overexploitation. Although, in an overexploitation scenario there is thermal interference between installations in a generalised way, the poor location of one installation next to another can also generate negative thermal interference between installations, ultimately leading to situations of overexploitation with low geothermal development in the area.

In addition, poor resource use and low efficiency of shallow geothermal installations can accelerate the process of overexploitation by increasing the exploitation of resources compared to an efficient system. All these problems are interrelated and affect the first governance policy exposed above. In the following, the management concepts necessary to achieve this guiding principle are developed.

9.3.1.1 Overexploitation and Unsustainable Development

Perhaps the most relevant problem endangering the sustainable development and exploitation of shallow geothermal energy resources is the overexploitation of these resources and the unsustainable development of shallow geothermal energy (Table 9.1). This problem can be solved by setting two management objectives: (1) the prevention of overexploitation and (2) the long-term sustainable use of shallow geothermal resources.

First, the objective of avoiding overexploitation of shallow geothermal resources must consider the most appropriate management strategies. A first strategy would be sustainable development. This may require management measures such as controlling the authorisation of thermal discharge (operating permits) of shallow geothermal installations according to an established strategic plan. Another measure would be to limit access to the resource, preferably during the processing of the operating permits. In this regard, access control should be considered, taking into account the size of the installation and the number of existing shallow geothermal installations, as well as the type of geothermal heat exchanger used (open or closed-loop). These measures suggest the use of an approach based on a system of exploitation rights to manage the limited resources in a particular city area or geological volume. Considering controlled access to the resource would limit the number of users with rights and responsibilities to exploit the resources and prevent overexploitation. In addition to these measures, the future of shallow geothermal seems to lie in the smart

Table 9.1 Management problems (PROB), management objectives (OBJ) and management strategies (STG) for the sustainable development and exploitation of shallow geothermal energy (SGE) resources

Management policy I	
Sustainable development and exploitation of shallow geothermal energy resources	
PROB	Geothermal overexploitation and unsustainable development
OBJ	Preventing overexploitation
STG	Sustainable development
STG	Identification of areas at risk of overexploitation
STG	Control of exploitation efforts
STG	Management of growing demand for SGE to pursue sustainability
OBJ	Long-term sustainable use of SGE resources
STG	Understanding of heat and hydraulic regimes in the subsurface
STG	Prioritization of SGE demands
STG	Sustainable development
STG	Enforcement/compliance for a rights-based system
STG	Control of exploitation efforts
STG	Long-term stability of production temperatures in SGE systems
STG	Promotion of a balanced use of the resources
STG	Stand-still principle: Maintenance of COP of SGE systems at its current level, at minimum
OBJ	Recovery of sustainability in areas under overexploitation
STG	Characterization of overexploited areas
STG	Increase of SGE supply in areas under overexploitation (remediation)
STG	Reduction of overexploitation (mitigation)
PROB	Negative thermal interference between/in installations
OBJ	Reduction of negative thermal interference
STG	Precautionary measures
STG	Limitation of the number of participants with rights and responsibilities
STG	Reduction of thermal interferences between/within exploitation
STG	Allocation of limited rights to net annual heat transfer into the aquifer
STG	Prevention of unbalanced heat transfer in peak demands
OBJ	Minimization of thermal shortcut (self-interference)
STG	Adequate SGE systems design
OBJ	Minimization of thermal interference between SGE systems
STG	Reduction of unbalanced energy transfer of neighboring installations
OBJ	Efficient use SGE resources
STG	Efficiency principle

(continued)

Table 9.1 (continued)

Management policy I	
Sustainable development and exploitation of shallow geothermal energy resources	
PROB	Inefficient use of geothermal resources
OBJ	Efficient use SGE resources
STG	Efficiency principle

interconnection of shallow geothermal installations through *District Heating and Cooling* (DHC) networks that allow transferring thermal energy between users before producing thermal discharge, thus reducing resource demand. 5G DHC networks fed by shallow geothermal installations (ultra-low temperature networks) are the ultimate expression of the concept of smart heat recovery at the district or neighbourhood scale of a city.

A second management strategy, which can also be considered relevant, is the identification of areas at risk of overexploitation. This can be carried out through measures such as the cartographic delimitation of areas with a high density of geothermal installations already under a overexploitation situation and areas at risk of becoming under overexploitation. In support of this measure, it would be necessary to establish a system for monitoring, surveillance and control of underground temperatures, regimes and operating performance coefficients of thermal installations with geothermal production systems.

The third strategy would be to control the exploitation efforts on the managed system. It would be advisable to adopt measures to limit access to the resource for new shallow geothermal installations in areas of conflict, to take into account areas of overexploitation during the granting of exploitation licences, and adopt measures for monitoring, surveillance and control of impacts on the subsurface and the exploitation regime of the installations. The fourth strategy would be to act on the growing demand for geothermal energy through measures to limit the installed capacity of geothermal heat pumps, to identify factors responsible for increasing demand, to identify current demand and trends, and to establish incentives in unexploited areas.

The second management objective is the long-term sustainable use of shallow geothermal energy resources, where resource renewability and exploitation are close to long-term equilibrium. To achieve this objective, the preferred strategy is characterisation of the existing thermal and hydrogeological regimes in the area, which requires R&D measures and the provision of advisory services to facility users and stakeholders. A second strategy would be to prioritise essential facilities (e.g., hospitals) over lower priority facilities to ensure the long-term availability of geothermal resources in the former. Therefore, priority use concept should be considered during the processing of operating licences.

It is clear that sustainability will only be achieved if a sustainable development strategy is followed, where the management measures considered include the assessment and quantification of the long-term sustainability of each of the installations during the processing of their operating authorisation, as well as planning the location of installations where the conditions of renewability of the resource are fulfilled. Wherever possible, sustainability is most likely to be achieved through the adoption of incentive measures for the use of shallow geothermal installations interconnected to DHC grids.

Another interesting strategy is to adopt a system of exploitation rights that takes into account the authorisation procedures for thermal discharge, thus granting exploitation rights to users for a defined period of time close to the average lifetime of a shallow geothermal installation (e.g., 25 years). In any case, it is recommended that the operating rights be reviewed every four years and that the competent authority be entrusted with the task of authorising its extension depending on whether the initially authorised operating conditions are met. In order to ensure sustainability, it will be necessary to impose a limit on the thermal energy transferred with the underground thermal reservoir in the operating licences. These conditions could be compensated to users by granting access rights to the resource and establishing protection perimeters to guarantee its sustainability.

Limiting the thermal energy transferred to the ground has sufficient relevance to be considered as a strategy, in general, for controlling the intensity of exploitation of shallow geothermal installations. Thus, it is advisable to take into account measures to control the access of new installations in conflict areas, to establish a monitoring, surveillance and control system of the exploitation regimes, coordinated with measures to impose mandatory exploitation limits, including flow rates, temperatures and thermal changes in the geothermal heat exchangers, both in open and closed-loop, during the authorisation procedure of the thermal discharge. In order to assess whether the operation is sustainable, it will be necessary to establish a monitoring, surveillance and control system of the temperature in the subsoil (mainly groundwater), of the production temperatures (input from the geothermal heat exchangers), of the aforementioned mandatory operating limits, as well as of the coefficients of performance of the installations.

A shallow geothermal installation will be sustainable over time if, after an exploitation cycle, the reservoir returns to the initial conditions. The temperature of the geothermal reservoir cannot be measured directly, only indirectly through the production temperatures, i.e., the temperatures of the heat carried fluid of the geothermal heat exchanger (open or closed) feeding the heat exchanger of the geothermal heat pump. Another strategy that can contribute to a sustainable use of resources consists of keeping the production temperatures of shallow geothermal installations stable over time. In a shallow geothermal environment where production temperatures over time tend to increase (or decrease) after three or four annual operating cycles, this will indicate a medium- to long-term problem of unsustainability. In this context it may be useful to establish a monitoring, surveillance and control system of production temperatures, operating regimes and coefficients of performance to identify undesirable trends as an indicator of unsustainability.

Promoting energy-balanced resource exploitation annually could be considered a smart strategy to reduce the thermal impacts (heat plumes) of shallow geothermal installations and achieve approximate initial conditions each year. Balanced heat transfer for heating and cooling has been identified as a good sustainability indicator for shallow geothermal installations (García-Gil et al. 2019). The installations with the largest biases in heat demand are those that produce thermal plumes that can be used by other installations, known as nested installations (García-Gil et al. 2020b), and their controlled promotion can be an interesting management measure in this strategy. Other measures could include the incentivisation of balanced annual net balances and taxation of geothermal installations with operating regimes heavily biased towards cooling or heating through increasing tax rates. Finally, following the principle of nondegradation as a strategy will allow for the least possible modification of the underground thermal reservoir without long-term degradation. Measures to avoid degradation include carrying out management actions to maintain or reduce coefficients of performance of geothermal heat pumps, developing a monitoring, surveillance and control system of heat pump coefficients of performance to identify peak values and trying to identify the problem. In cases where there is a real problem in this respect, a thermal impact assessment based on the coefficients of performance may be required.

The third major management objective for the problem of overexploitation and unsustainable development of shallow geothermal energy is the restoration of sustainability in overexploited areas. To achieve such recovery, the first strategy to be considered is the characterisation and delimitation of overexploited areas. This strategy involves the identification of overexploited areas and their mapping. This requires a monitoring, surveillance and control system to identify unacceptable values of production temperatures, exploitation regimes and yields. In addition, the identification of abandoned installations due to resource depletion would represent the worst possible situation and would be the main indicator of overexploited areas. Another recovery strategy for overexploited areas would be their remediation. As a first approximation, incentive measures should be adopted for installations with biased energy balances in order to promote recovery. Other active measures on the field have been proposed from a theoretical point of view (Epting et al. 2013; García-Gil et al. 2015b). Another similar strategy for the recovery of overexploited areas would be the mitigation of overexploitation by demand modulation towards less intensive exploitation. Overexploited areas can be recovered by considering the support of nested installations, considering the revocation or additional limitation of pre-existing licences (licensing procedure should consider this situation), as well as providing incentives at conflicting geothermal installations to reduce unbalanced exploitation.

9.3.1.2 Negative Thermal Interferences Between/at Shallow Geothermal Installations

The second most important problem for achieving sustainable development in the exploitation of shallow geothermal energy resources is thermal interference between geothermal installations. As demonstrated in Chaps. 2 and 4, the performance of a geothermal heat pump depends on the temperature of the thermal reservoir being exploited. If the reservoir is affected by the activity of another installation nearby, i.e., by a thermal plume from another installation, the coefficient of performance could be reduced, and the affected installation would end up making more aggressive use of the resource to meet its demand. This effect may lead to a non-renewable use of the resource, thus contributing to a potential overexploitation situation.

The management objective to be taken into account is the reduction of thermal interference through the strategy of following the precautionary principle, a widespread principle in environmental policies in the European Union (TFEU 2010). This strategy requires the adoption of important measures such as the definition of a minimum distance between geothermal heat exchangers of the different installations. Another relevant measure is the limitation of the maximum circulation temperature of the heat transfer fluid in geothermal heat exchangers. In addition, it can be useful to limit the temperature changes, both in the ground at a certain distance from the BHEs and between the water pumped and injected in the case of geothermal wells. Several precautionary measures covering this issue are already considered in international regulations (Haehnlein et al. 2010). If thermal interference occurs between two open-loop geothermal installations, a significant precautionary measure to be adopted is to monitor groundwater temperatures at an equidistant location between the impacted geothermal capture wells, and the injection wells responsible for the interference, by means of a piezometer control point. Limiting and controlling the flow rates and injection temperatures of the impacting geothermal wells is crucial.

Other strategies to be followed to reduce negative thermal interference include limiting the number of users with the right to operate, implementing measures to control access to the affected areas, and reducing interference between installations and self-interference. The latter strategy includes measures already mentioned, such as distance restrictions between shallow geothermal installations, restrictions on working temperatures in geothermal heat exchangers, implementation of a groundwater monitoring, surveillance and control system, restrictions on the depth of geothermal heat exchangers, restrictions on groundwater bodies that can be exploited by shallow geothermal energy, monitoring, surveillance and control of coefficients of performance of geothermal installations in order to detect interference, as well as delimitation of temporary and/or spatial areas to prevent the appearance of new exploitation systems in order to protect existing installations.

The allocation of exploitation rights in terms of annual net energy balance would be another strategy to follow. This would require the definition of an TUTE quota as the *Total Unbalanced Thermal Energy* transferred to the subsurface per year. A distinction can be made between the establishment of a soft TUTE, when it is indicative, and a hard TUTE, when it is mandatory for the user of a shallow geothermal installation. In order to guarantee this measure, it would be necessary to adopt a monitoring, surveillance and control system for the transfer of thermal energy to the ground of the shallow geothermal energy systems.

The last strategy considered here for reduction of negative thermal interference is the prevention of unbalanced heat transfer during peak demand periods. If thermal interference is identified, it is likely to be due to a bias in the heat demand of the installation responsible for the interference, especially produced by seasonal peaks in demand, such as during the summer period. Therefore, it is interesting to consider controlled episodes of thermal point discharge to an urban collection system (e.g., sewerage network).

A second management objective is the minimisation of thermal short-circuit (thermal self-interference and/or thermal recycling), by designing an appropriate configuration of geothermal heat exchangers. It requires the adoption of hydrogeological characterisation measures prior to the design of the installation (head contour maps, pumping tests, etc.) and assessment of the thermal self-interference process during the thermal discharge authorisation procedure.

Once the management objective related to coping with thermal self-interference has been addressed, it is necessary to consider the objective of thermal interference with other installations. In order to try to reduce this effect, it may be appropriate to set a strategy to reduce the energy imbalance of the installations with the terrain. Knowing the operating regime of the installation responsible for the thermal interference, it will be necessary to limit the flow rates and operating temperatures until the impact is assumable.

The last management objective against negative interference is the efficient use of available shallow geothermal energy resources. Following the principle of efficiency, negative thermal interference can be minimised. The operation will be more efficient if the possible effect of thermal self-interference is considered during the evaluation process. It is necessary to maximise the coefficient of performance of shallow geothermal installations, and a minimum coefficient of performance value for installations can required by authorities, as well as minimum utilisation of the permitted operating quota or a compulsory TRT for larger installations.

9.3.1.3 Inefficient Use of Geothermal Resources

Resource efficiency contributes to achieving many management objectives, but the underlying problem is of sufficient importance to be considered a management problem. A geothermal installation that maximises resource use will be closer to using the annually renewed resources, as well as being more economically beneficial to the user, and will promote the use of this technology to other users. Thus, the objective of efficient use of resources could be established and the principle of efficiency considered as a strategy to be followed. The measures to be adopted in this regard include evaluation of the thermal self-interference process during the thermal discharge authorisation procedure, maximising the coefficient of performance of shallow geothermal installations, the requirement of a minimum coefficient of performance, the minimum use of the authorised exploitation quota and the obligation to carry out TRTs in relevant installations.

Table 9.2 Management problems (PROB), management objectives (OBJ) and management strategies (STG) for the environmentally friendly use of shallow geothermal energy (SGE) resources

Management policy II Environmentally friendly use of shallow geothermal energy resources	
PROB	Hydrodynamic remobilisation of pre-existing contamination
OBJ	Shallow geothermal operations using groundwater of good quality
STG	Operate geothermal wells outside contaminated areas
PROB	Reduction of the overall environmental quality of the subsurface
OBJ	Reduction of environmental impacts
STG	Precautionary measures
STG	Understanding how SGE exploitation impact the ecosystem function
OBJ	Establishment of cause-effect relationships for environmental impacts
STG	Use of the best available science for decision-making
STG	Study of physical, biological and chemical processes triggered by SGE use
OBJ	Identification of potential subsurface quality deterioration
STG	Environmental monitoring, surveillance and control system
PROB	Contribution to the subsurface urban heat island (SUHI) effect
OBJ	Preventing the contribution to the SUHI effect in sensitive areas
STG	Control of shallow geothermal installations in areas sensitive to the SUHI effect
PROB	Enhancement of existing microbiological contamination
OBJ	Prevention of the potential enhancement of microbiological contamination
STG	Control of SGE activities in microbiologically-contaminated areas
PROB	Thermal groundwater discharge to surface water bodies
OBJ	Prevention of potentially negative environmental impacts on hyporheic zones
STG	Control of thermal groundwater discharge to surface water bodies
PROB	Enhancement of existent (emergent) organic contamination
OBJ	Prevention of the potential enhancement of emergent organic contamination
STG	Control of SGE activities in emergent organic contamination areas
PROB	Enhancement of existent inorganic trace metals contamination
OBJ	Prevention of possible enhancement of inorganic trace metals contamination
STG	Control of SGE activities within areas affected by inorganic trace metals contamination

9.3.2 *Environmentally Friendly Use of Shallow Geothermal Energy Resources*

With regard to the second guiding principle (Table 9.2), maintaining an environmentally friendly use of shallow geothermal energy resources requires addressing threats to human health or to the environment arising from their exploitation.

In general, it can be stated that the exploitation of shallow geothermal resources is an environmentally friendly activity and the impacts on the environment are minimal or negligible. In the vast majority of cases, it does not pose a risk to human health or the environment. However, there may be different management problems in specific situations that need to be considered by managers. It is also necessary to understand that the underground environmental conditions are not always the best. Shallow geothermal activity, especially from shallow geothermal installations with open-loop geothermal heat exchangers, can potentially enhance existing subsoil contamination. However, it is also true that temperature rise can also lead to thermal remediation of contamination (García-Gil et al. 2016b; Vidonish et al. 2016).

9.3.2.1 Hydrodynamic Remobilisation of Pre-existing Contamination in Aquifers

It is generally agreed that hydrodynamic remobilisation of pre-existing contamination in urban aquifers, due to the operation of geothermal wells, represents the most relevant management problem when water supply wells exist in the groundwater body managed. Geothermal installations with open-loop geothermal heat exchangers operating in contaminated areas may cause the mobilisation of pre-existing contamination to other locations by groundwater advection, thus contributing to a possible decrease in groundwater quality in critical pristine areas. Persistent pollutants such as heavy metals and polycyclic aromatic hydrocarbons (PAHs) are commonly found in urban groundwater environments as point source pollution (Bonneau et al. 2017; Schirmer et al. 2013), posing a real threat to urban water resources intended for human consumption. In addition, hydrodynamic mobilisation could also contribute to the release of pollutants from groundwater into hydraulically connected surface water bodies (Engelhardt et al. 2011). The management objective linked to this problem is to develop shallow geothermal exploitation in areas not affected by contamination. This objective is not entirely realistic in shallow aquifers in urban areas, given the high probability of contamination. Nevertheless, shallow urban aquifers are not often used for drinking water supply. It follows that, in aquifers not exploited for human consumption, this limitation of shallow geothermal exploitation could not be taken into account. Conversely, in groundwater bodies under cities exploited for drinking water supply, the use of shallow geothermal will be very limited (Koren and Janža 2019). The strategy to follow, only in urban shallow aquifers with drinking water developments, is to operate geothermal wells outside contaminated areas. Such contaminated areas should be mapped and consulted during the discharge permitting process.

9.3.2.2 Reduction of the Overall Environmental Quality of the Subsoil

The second most relevant issue facing the second management policy proposed is the shallow geothermal activity leading to a reduction in the overall environmental quality of the subsurface. Management objectives include reducing the environmental impacts of geothermal heat exchangers, establishing cause-effect relationships between thermal impacts and environmental quality deterioration.

Regarding the first objective of minimising the environmental impact of geothermal heat exchangers, it is considered a basic management strategy to follow the precautionary principle and establishing different measures. One of the most important is to carry out leakage (or tightness) tests on closed-loop geothermal heat exchangers. Even if the closed loop of an exchanger is completely insulated, there is a risk of friction and breakage of the exchanger piping. An additional important measure is to consider the specific regulation of the heat carrier fluids that will circulate inside it and thus ensure the least possible environmental impact in the event of breakage. In any case, an environmental impact assessment must be carried out during the processing of operating licences, always proportional to the size of the installation (depth and number of heat exchangers). Another useful measure in cases of multi-layer aquifers, where the aim is to protect a groundwater body at a certain depth, is to limit the depth of the geothermal heat exchangers, whether in an open or closed-loop. In order to prevent surface contamination from reaching groundwater bodies or circulating between multilayer aquifers, it is recommended that boreholes be sealed off after the cessation of exploitation of shallow geothermal resources. The foundation process of shallow geothermal heat exchangers must be specifically regulated, additionally guaranteeing the above measure and the leakage of heat transfer fluid that may occur in case of leakage of the closed geothermal heat exchanger piping. All these measures can be implemented during the processing of thermal discharge authorisations, their review and decommissioning. As mentioned in other strategies, it is also necessary here to carry out tightness tests for the pipe line in closed-loop geothermal heat exchangers. Another relevant measure is to precisely control the depth of the geothermal heat exchangers (open and closed). In exceptional cases it may be necessary to establish temporary and/or spatially protected zones.

A second strategy is to understand the impacts of shallow geothermal energy on the functioning of ecosystems. To this end, it would be advisable to coordinate R&D&I activities between institutions and to establish a consultancy service for users of installations, especially for installers of geothermal heat exchangers.

The second management objective for the environmental subsurface degradation problem is the identification of cause-effect relationships of the thermal impacts generated by shallow geothermal energy systems on the environment. This objective can consider the use of scientific knowledge in decision making and the study of the physical, biological and chemical processes related to shallow geothermal energy exploitation. In order to use the available scientific knowledge, it is necessary to establish measures for monitoring and evaluation of the environmental status during the exploitation of shallow geothermal resources, as well as to coordinate R&D&I activity with other research institutions and to establish advisory services. This last measure is also necessary for the successful study of processes responsible for environmental degradation.

The third and final objective for this management problem is the identification of the deterioration in the environmental quality of the subsurface. The basic strategy is the implementation of a system of environmental monitoring, surveillance and control, especially adapted to groundwater quality.

9.3.2.3 Contribution to the Subsurface Urban Heat Island (SUHI) Effect

The thermal impacts caused by shallow geothermal energy exploitation on the subsoil are added to those caused by other human infrastructures in the subsoil. Intensive geothermal exploitation could amplify the SUHI effect and, if this occurs in sensitive areas, it can lead to economic and welfare damage. For example, a tunnel through which road traffic circulates needs to be at a certain temperature; if the ground heats up, the design conditions of the tunnel's air-conditioning system are altered and can lead to serious problems. If geothermal exploitation is to be carried out, it may be necessary to establish a management objective related to the prevention and mitigation of the contribution of shallow geothermal energy to the SUHI effect in sensitive areas. In this case, the strategy to be considered is the control of shallow geothermal installations in areas sensitive to the SUHI effect. For this purpose, it would be necessary to consider management measures to map the areas affected by SUHI. In these areas, the potential impacts of geothermal energy on sensitive areas will have to be taken into account during the authorisation procedure for thermal discharge. If sensitive areas are related to human health/environmental comfort, the existing risk should be assessed. In exceptional cases, measures may be taken to establish protected areas (temporary and/or spatial) as well as restrictions on the depth of exploitation of shallow geothermal resources.

9.3.2.4 Enhancement of Existing Microbiological Contamination

Thermal use of groundwater in urban aquifers can induce groundwater heating. If microbiological contamination is present, prior to or during the thermal use of groundwater, this type of contamination is believed to increase and lead to further problems. However, sometimes the opposite is true, and thermal use can pseudopasteurise the groundwater and reduce existing microbiological loads. If there are indications of previous problems of this concern in the managed water body, it may be advisable to set a management objective related to the enhancement of microbiological contamination, especially if there are wells for drinking water supply or wells for irrigation of parks and gardens. The latter may be connected to sprinkler irrigation systems and are capable of misting groundwater and holding microbiological contaminants, including pathogenic organisms, in colloidal suspension. The most advisable strategy to follow would be the control of shallow geothermal installations in microbiological contaminated areas and environmental control.

If this strategy is considered, the thermal discharge authorisation procedure should consider whether there is microbiological deterioration of the soil in the area of operation. If the answer is affirmative, the operation may not be authorised or measures for monitoring, surveillance and control of the microbiological content of the groundwater may be required. Such a measure requires the establishment of protected areas (temporary and/or spatial) through the mapping of microbiological contamination zones. This will allow the resource management authority to assess risks to humans and the environment. The resource managers may also adopt measures restricting the authorised exploitation in terms of depth of exploitation of shallow geothermal energy resources.

9.3.2.5 Thermal Groundwater Discharge into Surface Water Bodies

Thermal plumes generated by shallow geothermal installations, and the possible enhancement of existing pollution from them, can end up being discharged into surface water bodies. In the case of urban areas, usually in rivers where groundwater is discharged to rivers through the hyporheic zone. If thermal and/or enhanced pollution plumes affect this hyporheic zone, its ecosystems, which are of vital importance for the environmental quality of rivers (Hancock et al. 2005), could deteriorate. At present, there is no scientific evidence of the magnitude of impacts on the ecology of the hyporheic zone from the above described thermal discharge. Even if degradation of the hyporheic zone was produced, the thermal discharge zones would be limited to the river section passing through a metropolitan area. The length of the affected river section in an urban area would be infinitesimal compared to the length of the river. In any case, managers can apply the precautionary principle and set an objective of preventing potentially negative environmental impacts in hyporheic areas, given that there is no scientific evidence to either effect. In any case, the recommended strategy is to control thermal discharge of groundwater into surface water bodies. The measures to be adopted would consist of assessing the risks associated with thermal discharge into surface water bodies. The first step would be the mapping of groundwater thermal discharge zones in urban areas. The second step would be to implement a monitoring, surveillance and control system of groundwater discharge temperature and ecological quality. In the event of any type of environmental risk, measures can be adopted to restrict installations during the thermal discharge authorisation procedure and even the establishment of protected areas (temporary and/or spatial) or restrictions on the depth of exploitation, if deemed necessary.

9.3.2.6 Enhancement of Existing Contamination by Regulated and Emerging Organic Compounds

In addition to regulated organic pollutants (hydrocarbons, herbicides, etc.), there is organic trace element pollution of an anthropogenic nature, consisting of new compounds used in industry, food, pharmacology, agriculture, etc., capable of altering the physiology of target receptors with possible harmful consequences, i.e., emerging organic contaminants. These contaminants enter the environment with a high resistance to degradation, resulting in the detection of emerging organic contaminants that are increasingly common in the environment and especially in urban aquifers (Jurado et al. 2012), where shallow geothermal energy is exploited.

In the context of the management of shallow geothermal resources, there may be situations where organic contaminants exist, thus posing a health hazard from drinking water supply wells, irrigation wells, etc. In situations where the competent authority considers that a risk exists, it can adopt the management objective of preventing the potential enhancement of trace organic contamination. The strategy to be followed would be the control of shallow geothermal installations in areas presenting organic contaminants. Problems of trace organic pollution enhancement can also be avoided during the thermal discharge authorisation procedure. In that case, a monitoring, surveillance and control system of organic pollutants in groundwater can be established. Once contaminated areas have been identified, protected areas (temporary and/or spatial) can be established. The mapping of organic contaminant zones can be a useful tool. In addition, as in the previous cases, restricting the depth of exploitation of shallow geothermal energy resources can avoid the presence of organic pollutants.

9.3.2.7 Enhancement of Existing Inorganic Trace Element Contamination

Temperature changes in the soil can lead to sorption reactions of inorganic trace elements, compounds potentially toxic for humans and the environment. These sorption reactions can be endothermic or exothermic, where a temperature change in the subsurface environment will tend to fix the pollutants in the solid phase or its release from the solid phase to groundwater. In situations where the soil is contaminated by inorganic trace elements, the most appropriate approach may be to set a target to prevent the potential enhancement of trace element contamination. The strategy to follow, as in the previously discussed cases of contamination enhancement, would be to control shallow geothermal installations in areas of inorganic contamination. The most important measure to be considered would be the mapping of areas affected by inorganic trace element contamination, where identified and defined areas would be considered during the thermal discharge authorisation procedure. Depending on these areas, protected areas (temporary and/or spatial) would be established. In addition, for safety reasons, a monitoring, surveillance and control system of groundwater quality should be implemented. In the case of multi-layered aquifers, depth restrictions could be imposed on the exploitation of shallow geothermal energy resources. Finally, a risk assessment for humans and the environment may be necessary if the presence of these contaminants in the managed area is unknown.

9.3.3 *Exploitation of Shallow Geothermal Resources in Coordination with Other Subsoil Uses*

The third guiding principle (Table 9.3) refers to the use of shallow geothermal energy resources in coordination with other subsurface uses. Where the subsurface has different uses, a conflict of interest may arise between them.

9.3.3.1 Modification of the Chemical Quality of Groundwater Intended for Human Consumption

There is a general consensus among 13 European geological surveys (García-Gil et al. 2020a) that the biggest conflict of interest that may arise is between shallow

Table 9.3 Management problems (PROB), management objectives (OBJ) and management strategies (STG) for the exploitation of shallow geothermal energy (SGE) resources in coordination with other subsurface uses

Management policy III	
Shallow geothermal energy coordination with other urban subsurface uses	
PROB	Groundwater quality as water supply
OBJ	Maintenance of groundwater quality standards
STG	Precautionary principle
PROB	Groundwater use conflicts (irrigation, industrial, recreational, etc.)
PROB	Urban subsurface use conflicts (general approach)
OBJ	Prevention/control of crosscutting conflicts
STG	Prevention and mitigation of crosscutting issues
PROB	Geotechnical impacts (subsidence)
OBJ	Prevention of fines migration into groundwater heat pumps systems
STG	Ensurance of laminar flow in pumping/injection wells
OBJ	Prevention of dissolution subsidence
STG	Groundwater isolation from atmospheric conditions
PROB	Impacts on underground structures
OBJ	Reduction of thermal impacts in tunnels (ventilation design)
STG	Consideration of temperature-sensible subsurface infrastructures in thermal impact assessment during the licensing process

geothermal energy use and the use of groundwater for human consumption. The main problem of this crosscutting issue is modification of the chemical quality of groundwater extracted for human consumption. The management objective, in coordination with the strategies of the third management policy proposed, would be maintenance of the minimum chemical quality required for human consumption. Given the difficulty of controlling the geochemical impacts triggered by thermal plumes, it could be necessary to adopt the precautionary principle as a key strategy to be followed. The most useful measures in this regard would be to identify or define groundwater protection areas for water supply, so that shallow geothermal energy activity does not take place in such sensitive areas of general interest. It is recommended to consider the elaboration and use of groundwater management maps defining the areas of priority use for shallow geothermal energy.

9.3.3.2 Conflict with Other Groundwater Uses (Irrigation, Livestock, Industrial, Recreational, Etc.)

A second management problem is the conflict of shallow geothermal energy use with other groundwater uses (irrigation, livestock, industrial, recreational, etc.). Although this second problem is separated from the first by its importance, the objective, strategy and management measures are analogous to the first management problem. Other issues to consider are cases of specific conflicts of interest with the geotechnical use of subsoil and underground infrastructure.

9.3.3.3 Conflict with Urban Subsurface Uses (General Case)

The third problem is the general approach dealing with conflicts related to other urban subsurface uses. It is necessary to understand subsurface uses in urban areas, identify potential conflicts and adopt a management objective based on the avoidance of conflict of interest situations. One possible strategy to achieve this objective is to prevent and mitigate conflict of interest situations in urban subsurface use. This would require inventories and mapping of all subsurface uses and, in cases of conflict, consideration of this in the authorisation procedure for the thermal discharge of the geothermal installations. In very specific cases, where there is a risk, exceptional monitoring, surveillance and control measures can be adopted in conflict areas, and restrictions placed on the depth of exploitation of shallow geothermal resources and/or the establishment of protected areas.

9.3.3.4 Geotechnical Impacts (Subsidence)

In the case of geotechnical impacts, the exploitation of shallow geothermal energy resources can induce a subsidence problem when specific geological conditions exist. Two main management objectives can be established. The first is the prevention and control of fines washing related to geothermal wells. This objective intends to avoid the subsidence phenomenon due to subtraction of unconsolidated material by creep during the operation of geothermal wells. To achieve this objective, it would be necessary to ensure laminar flow of groundwater around geothermal wells, so that the turbulent flow responsible for fines entrainment does not occur. This strategy can be implemented by establishing regulations to guarantee quality in the design, construction and maintenance of geothermal wells. There is also the problem of subsidence by dissolution (García-Gil et al. 2016a; Garrido et al. 2016b). If evaporite materials (gypsum, halite, anhydrite, etc.) are involved in the shallow geothermal energy exploitation process, it would be appropriate to establish a management objective to prevent subsidence by dissolution. The main strategy to be followed is isolation of the atmospheric conditions of the circulating groundwater by open-loop geothermal heat exchangers. In this way, it would be possible to regulate the mandatory pressurisation of the secondary circuit through which groundwater circulates. The main reason for this measure would be to prevent the loss of carbon dioxide from the groundwater to the atmosphere and to avoid calcite scaling that would force the flow of water injection towards a possible soluble evaporite basement.

9.3.3.5 Impacts on Underground Structures

Regarding the problem of subsurface use conflict with underground infrastructures, the objective of reducing thermal impacts on tunnels and their ventilation design conditions can be considered. It would be necessary to assess whether there are thermal impacts on sensitive underground structures, and it would require the adoption of measures for monitoring, surveillance and control of sensitive underground structures and an inventory and mapping of these structures. In cases of conflict, there would be the possibility of establishing protected areas that would have to be considered during the thermal discharge authorisation procedure. If the structure is located at a certain depth, restrictions on the depth of exploitation of shallow geothermal energy resources could be adopted.

9.3.4 Effective Management of Shallow Geothermal Resources

The effective management of shallow geothermal energy resources is the fourth management policy proposed (Table 9.4). The management of renewable resources requires taking into account a multitude of complex and dynamic issues that must be guided by the principle of efficiency.

Table 9.4 Management problems (PROB), management objectives (OBJ) and management strategies (STG) effective management of shallow geothermal energy (SGE) resources

Management policy IV	
Effective management of shallow geothermal energy resources	
PROB	Managing in the context of data-poor urban subsurface body
OBJ	Providing an efficient management of SGE while improving a data-poor context
STG	Improvement of the overall SGE data system
STG	Simple management approaches (low information gathering)
STG	Use of simple statistics to manage the SGE resources
STG	Relying on users' knowledge of their own SGE systems
PROB	Conflicts of interest
PROB	Inefficient management of the SGE resources
OBJ	Diminishing of enforcement problems and compliance
STG	Providing legal certainty (economic stability)
STG	Co-management approach
STG	Maximization of economic profits for SGE users
STG	Establishment of a SGE market
STG	Adoption of a rights-based system
STG	Increase of investment security
OBJ	Flexible iterative management approach
STG	Adaptive management
PROB	Management measures dependence to site-specific conditions
OBJ	Adaptation of management measures to local boundary conditions
STG	Decentralization of SGE resources management
PROB	Disabling environment
OBJ	SGE capacity development (building)
STG	Development of appropriate policy and legal frameworks
STG	Institutional training courses
STG	Establishment of institutions responsible for the governance of shallow geothermal resources
PROB	Uncertainty
OBJ	Coping with uncertainty
STG	Adaptive approach (adjustments and improvements mid-stream)
STG	Management measures applicable to a wide range of scenarios
PROB	Illegal activity and heavy enforcement costs
OBJ	Implementation of an integrative and inclusive approach
STG	All the parties involved need a voice in the decisionmaking
STG	Ensuring an inclusive and participatory approach
STG	Co-management approach

9.3.4.1 Management in a Data-Poor Environment

Given that subsurface ground data are generally very limited and economically expensive to obtain, and that shallow resource energy management is an emerging branch, it is necessary to provide an efficient management approach while efforts are made to improve the information gaps. To achieve this goal, shallow geothermal energy resource management is still a pioneering activity in many countries in Europe and worldwide. Arguably, many of the competent authorities seeking to manage shallow geothermal energy resources will face this challenge in the context of limited information on the resource and on existing demand. This problem may be the most common and the one that concerns managers the most. To address it, it may be useful to consider providing efficient management during data acquisition as an objective. The strategy to follow is the improvement of the overall shallow geothermal data system, through data acquisition requirements, data analysis and the construction of geo-referenced databases. In addition, a simplistic management approach can be acquired with low initial data collection, making use of simple statistics and temporarily taking over existing knowledge acquired by users of geothermal installations.

9.3.4.2 Conflict of Interest Between the Stakeholders Involved in the Management Process

A second important management problem would be the conflict of interest between stakeholders, i.e., all parties involved in the management process. To ensure the objective of reducing the number of conflict of interest cases, it is proposed to consider co-management of shallow geothermal energy resources as a possible solution. This would make the resources partly self-regulating, thus decreasing the imposition of possible management measures taken, and thus increasing the chances of compliance. In addition, co-management can be implemented by including affected parties in decision-making throughout the planning process. If all parties are involved in decisionmaking and understand the management issues, it will be easier for the interests of the parties to be respected in the event of a conflict.

9.3.4.3 Inefficient Management of Shallow Geothermal Resources

The third problem for effective management of shallow geothermal resources is the inefficiency of the management measures adopted. These measures may be inefficient either because they are not accepted or not complied with by the parties that should adopt them. If measures are complied with, they may not achieve the desired effect, causing the established objectives to be missed.

The first objective is to reduce the problems of legal compliance and acceptance. To this end, six management strategies can be proposed. The first would consist of providing legal security to users, so that they acquire economic stability as insurance against any eventuality. If the user complies with the regulations, the manager will ensure that the rest of the users comply with them, avoiding damage to their interests. If another user were to cause damage with economic consequences to his operation, the regulations would imply financial compensation from the offender. In other words, legal protection of the rights and benefits of the user of a shallow geothermal installation is necessary during the resource management process. The second strategy would be to adopt a joint co-management approach; it would entail a sharing of authority and responsibility between stakeholders, mainly between the users of shallow geothermal installations and the competent authority. It is also advisable to involve all stakeholders in all management decisions taken. This will ensure that the parties accept the measures as their own.

A third strategy is to consider management measures that tend to maximise the economic profits of users with shallow geothermal installations. Regulated access through a discharge permit procedure tends to protect shallow geothermal energy resources and guarantee stable background ground temperatures of the subsurface through shallow geothermal operations. Another strategy to reduce conflicts of interest is to establish a market for shallow geothermal energy resource exploitation rights, where measures for temporary or permanent transfer of resource exploitation rights, and the inclusion of individual exploitation quotas in the thermal discharge permit can be considered. Another possibility is to establish a strategy based on a controlled allocation system of exploitation rights through a thermal discharge permit procedure. These thermal discharge permits will consider 20 to 30-year duration, in order to guarantee long-term rights with the possibility of renewal/revise for shorter periods (e.g., 4 years) depending on the fulfilment of controlled operation requirements. The final strategy is to increase investment security. This can be obtained through the thermal discharge authorisation procedure and by guaranteeing installations that the background temperature will be reasonably stable and close to the background temperature of the subsurface under a natural/pristine thermal regime.

The second management objective for the efficient management of shallow geothermal energy resources would be to adopt a flexible approach through the measures to be taken in an iterative management approach. This means establishing an adaptive management strategy according to the effectiveness of the measures, modulating the intensity of the measures until the expected results are achieved. This requires the development of standardised indicators to assess the effectiveness of management measures adopted in the management planning of shallow geothermal energy resources.

9.3.4.4 Inappropriate Management Measures to Specific Geothermal Reservoir Conditions

An ineffective management problem may result from the adoption of management measures that are inappropriate for the specific characteristics of the shallow geothermal reservoir managed. Adaptation of management measures to local boundary conditions is necessary. This entails decentralisation of the management of shallow geothermal resources by transferring part of the competencies from the central government to local entities at the urban scale and a procedure of thermal discharge authorisation by the local competent authority.

9.3.4.5 Disabling Environment

Lack of knowledge of shallow geothermal energy technology and the ground thermal regime by stakeholders, including managers and users, creates a disabling environment that leads to ineffective management of shallow geothermal resources. There is a need to set an objective for capacity building (training) in shallow geothermal energy problems. This can be done through strategies of developing management policies and a clear and scientifically motivated regulatory framework, conducting training courses at institutional level and creating specific administrative bodies responsible for the governance of shallow geothermal energy resources within the regional or local administration.

9.3.4.6 Uncertainty

In some cases, uncertainty associated with managing shallow geothermal resources can be a major problem. For example, there may be situations where the heat demand of the installations is not well known, or how the subsurface will respond to long-term temperature changes in terms of chemical quality of the subsurface, among many other issues. Uncertainty is inevitable and will always exist. The solution is to cope with uncertainty in the best possible way. To this end, useful strategies to be considered include the adaptive management approach, which includes adjusting and improving during the management process, establishing a cyclical process and measures applicable to a wide range of possible scenarios. In the latter strategy, scenario assessment is recommended.

9.4 Governance Model

The adaptive management approach (Holling 1978; Walters 2001) is the most widely accepted and applicable method for the governance of renewable natural resources in highly dynamic and complex environments. Adaptive management provides a framework for dealing with complex non-linear relationships between planning, implementation and control. It is based on cyclical and iterative activity during the management process. Given the high complexity and uncertainty associated with management of shallow geothermal energy resources, it is necessary to follow an adaptive management model (Fig. 9.3), as is done in the management of other natural renewable resources, e.g., water resources (Savenije and Van der Zaag 2008).

In the case of shallow geothermal energy resource management, two fundamental management processes are considered: (1) the planning process and the (2) implementation and control process. In order for the planning process to acquire the property of adaptability, it can be conceived as a dual-cycle adaptive management process (Fig. 9.3).

The first activity to be carried out in the planning cycle is identification and assessment of the existing management problems in the shallow geothermal reservoir to be managed, and the definition of possible management objectives, strategies and measures that best fit with the management policies established.

The second activity proposed here consists of decision-making to prepare and adopt a strategic action plan, considering specific and detailed management measures. The management plan constitutes the basic reference for actual implementation and, thus, is a first link between the planning cycle and the implementation and control process.

The third and last activity before restarting the planning cycle is review of the effectiveness of the adopted management plan. This involves comparing the results obtained from the implementation of specific management measures with the expected objectives. The review phase is the cornerstone of adaptive management, whereby decision makers learn about the possible shortcomings of some adopted management measures. The review process also includes adaptation of inefficient measures and their justified inclusion in the update of the management plan.

The management cycles are proposed to be separated to facilitate the whole management process. Once the first planning cycle is completed, each cycle can evolve independently as information continues to flow between cycles. This system allows for greater flexibility in management. For example, two implementation and control cycles can be carried out with the same management plan defined in a first planning cycle, or two planning cycles can be carried out without modifying the initial implementation and control cycle.

References

Allen A, Milenic D, Sikora P (2003) Shallow gravel aquifers and the urban 'heat island' effect: a source of low enthalpy geothermal energy. Geothermics 32(4–6):569–578. https://doi.org/10.1016/S0375-6505(03)00063-4

Angelis-Dimakis A et al (2011) Methods and tools to evaluate the availability of renewable energy sources. Renew Sustain Energy Rev 15(2):1182–1200. https://doi.org/10.1016/j.rser.2010.09.049

Bayer P, Attard G, Blum P, Menberg K (2019) The geothermal potential of cities. Renew Sustain Energy (in press)

Bear J (1988) Dynamics of fluids in porous media. Dover, USA

Bodvarsson G (1974) Geothermal resource energetics. Geothermics 3(3):83–92. https://doi.org/10.1016/0375-6505(74)90001-7

Bonneau J, Fletcher TD, Costelloe JF, Burns MJ (2017) Stormwater infiltration and the 'urban karst'—a review. J Hydrol 552:141–150. https://doi.org/10.1016/j.jhydrol.2017.06.043

De Filippis G, Margiotta S, Negri S, Giudici M (2015) The geothermal potential of the underground of the Salento Peninsula (Southern Italy). Environ Earth Sci 73(11):6733–6746. https://doi.org/10.1007/s12665-014-4011-1

Engelhardt I et al (2011) Comparison of tracer methods to quantify hydrodynamic exchange within the hyporheic zone. J Hydrol 400(1–2):255–266. https://doi.org/10.1016/j.jhydrol.2011.01.033

Epting J, Huggenberger P (2013) Unraveling the heat island effect observed in urban groundwater bodies—definition of a potential natural state. J Hydrol 501:193–204. https://doi.org/10.1016/j.jhydrol.2013.08.002

Epting J, Händel F, Huggenberger P (2013) Thermal management of an unconsolidated shallow urban groundwater body. Hydrol Earth Syst Sci 17(5):1851–1869. https://doi.org/10.5194/hess-17-1851-2013

Epting J, García-Gil A, Huggenberger P, Vázquez-Suñe E, Mueller MH (2017) Development of concepts for the management of thermal resources in urban areas—assessment of transferability from the Basel (Switzerland) and Zaragoza (Spain) case studies. J Hydrol 548:697–715. https://doi.org/10.1016/j.jhydrol.2017.03.057

Epting J, Müller MH, Genske D, Huggenberger P (2018) Relating groundwater heat-potential to city-scale heat-demand: a theoretical consideration for urban groundwater resource management. Appl Energy 228:1499–1505. https://doi.org/10.1016/j.apenergy.2018.06.154

Erol S (2011) Estimation of heat extraction rates of GSHP systems under different hydrogeological conditions. University of Tubingen

Fujii H, Inatomi T, Itoi R, Uchida Y (2007) Development of suitability maps for ground-coupled heat pump systems using groundwater and heat transport models. Geothermics 36(5):459–472. https://doi.org/10.1016/j.geothermics.2007.06.002

García-Gil A, Vázquez-Suñe E, Garrido E, Sánchez-Navarro JA, Mateo-Lázaro J (2014) The thermal consequences of river-level variations in an urban groundwater body highly affected by groundwater heat pumps. Sci Total Environ. https://doi.org/10.1016/j.scitotenv.2014.03.123

García-Gil A et al (2015a) GIS-supported mapping of low-temperature geothermal potential taking groundwater flow into account. Renew Energy 77:268–278. https://doi.org/10.1016/j.renene.2014.11.096

García-Gil A, Vázquez-Suñé E, Sánchez-Navarro JA, Lázaro J (2015b) Recovery of energetically overexploited urban aquifers using surface water. J Hydrol 1(1):111

García-Gil A, Vázquez-Suñe E, Schneider EG, Sánchez-Navarro JÁ, Mateo-Lázaro J (2015c) Relaxation factor for geothermal use development—criteria for a more fair and sustainable geothermal use of shallow energy resources. Geothermics 56:128–137. https://doi.org/10.1016/j.geothermics.2015.04.003

García-Gil A et al (2016a) A reactive transport model for the quantification of risks induced by groundwater heat pump systems in urban aquifers. J Hydrol 542:719–730. https://doi.org/10.1016/j.jhydrol.2016.09.042

García-Gil A et al (2016b) A city scale study on the effects of intensive groundwater heat pump systems on heavy metal contents in groundwater. Sci Total Environ 572:1047–1058. https://doi.org/10.1016/j.scitotenv.2016.08.010

García-Gil A et al (2019) Sustainability indicator for the prevention of potential thermal interferences between groundwater heat pump systems in urban aquifers. Renew Energy 134:14–24. https://doi.org/10.1016/j.renene.2018.11.002

García-Gil A et al (2020a) Governance of shallow geothermal energy resources. Energy Policy 138:111283. http://doi.org/10.1016/j.enpol.2020.111283

García-Gil A et al (2020b) Nested shallow geothermal systems. Sustainability 12(12):5152

Garrido EA, García-Gil A, Vázquez-Suñè E, Sánchez-Navarro JÁ (2016) Geochemical impacts of groundwater heat pump systems in an urban alluvial aquifer with evaporitic bedrock. Sci Total Environ 544:354–368. https://doi.org/10.1016/j.scitotenv.2015.11.096

Gemelli A, Mancini A, Longhi S (2011) GIS-based energy-economic model of low temperature geothermal resources: a case study in the Italian Marche region. Renew Energy 36(9):2474–2483. https://doi.org/10.1016/j.renene.2011.02.014

Haehnlein S, Bayer P, Blum P (2010) International legal status of the use of shallow geothermal energy. Renew Sustain Energy Rev 14(9):2611–2625. https://doi.org/10.1016/j.rser.2010.07.069

Hancock PJ, Boulton AJ, Humphreys WF (2005) Aquifers and hyporheic zones: towards an ecological understanding of groundwater. Hydrogeol J 13(1):98–111. https://doi.org/10.1007/s10040-004-0421-6

Holling CS (1978) Adaptive environmental assessment and management. International Institute for Applied Systems Analysis

Iglesias E (2003) First assessment of Mexican low-to medium-temperature geothermal reserves. Energy Sources 25(2):161–173. https://doi.org/10.1080/00908310390142226

Jurado A et al (2012) Emerging organic contaminants in groundwater in Spain: a review of sources, recent occurrence and fate in a European context. Sci Total Environ 440:82–94. https://doi.org/10.1016/j.scitotenv.2012.08.029

Kohl T, Andenmatten N, Rybach L (2003) Geothermal resource mapping—example from northern Switzerland. Geothermics 32(4):721–732. https://doi.org/10.1016/S0375-6505(03)00066-X

Koren K, Janža M (2019) Risk assessment for open loop geothermal systems, in relation to groundwater chemical composition (Ljubljana pilot area, Slovenia). Geologija 62:237–249. https://doi.org/10.5474/geologija.2019.011

Lavigne J (1977) Les resources géothermaiques françaises. Possibilités de mise en valeur. Ann Des

Lynch DR (2009) Sustainable natural resource management: for scientists and engineers. Cambridge University Press, Cambridge

MCS (2011) Microgeneration installation standard: MIS 3005. Department of Energy and Climate Change, Microgeneration Certification Scheme (MCS), London

Muffler P, Cataldi R (1978) Methods for regional assessment of geothermal resources. Geothermics 7(2):53–89. https://doi.org/10.1016/0375-6505(78)90002-0

Noorollahi Y, Itoi R, Fujii H, Tanaka T (2008) GIS integration model for geothermal exploration and well siting. Geothermics 37(2):107–131. https://doi.org/10.1016/j.geothermics.2007.12.001

Ondreka J, Rüsgen MI, Stober I, Czurda K (2007) GIS-supported mapping of shallow geothermal potential of representative areas in south-western Germany—possibilities and limitations. Renew Energy 32(13):2186–2200. https://doi.org/10.1016/j.renene.2006.11.009

Rybach L (2015) Classification of geothermal resources by potential. Geoth Energy Sci 3(1):13–17. https://doi.org/10.5194/gtes-3-13-2015

Savenije HHG, Van der Zaag P (2008) Integrated water resources management: concepts and issues. Phys Chem Earth Parts A/B/C 33(5):290–297. https://doi.org/10.1016/j.pce.2008.02.003

Schiel K, Baume O, Caruso G, Leopold U (2016) GIS-based modelling of shallow geothermal energy potential for CO_2 emission mitigation in urban areas. Renew Energy 86:1023–1036. https://doi.org/10.1016/j.renene.2015.09.017

Schirmer M, Leschik S, Musolff A (2013) Current research in urban hydrogeology—a review. Adv Water Resour 51:280–291. https://doi.org/10.1016/j.advwatres.2012.06.015

Stauffer F, Bayer P, Blum P, Giraldo NM, Kinzelbach W (2013) Thermal use of shallow groundwater. Taylor & Francis, London

TFEU (2010) Consolidated version of the treaty on the functioning of the European Union. European Parliament

VDI4640/1 (2000) Thermal use of the underground fundamentals, approvals, environmental aspects. Verein Deutscher Ingenieure

VDI4640/2 (2001) In: Ingenieure VD (ed) Thermal use of the underground: ground source heat pump systems, Berlin, p 41

Vidonish JE, Zygourakis K, Masiello CA, Sabadell G, Alvarez PJJ (2016) Thermal treatment of hydrocarbon-impacted soils: a review of technology innovation for sustainable remediation. Engineering 2(4):426–437. https://doi.org/10.1016/J.ENG.2016.04.005

Voivontas D, Assimacopoulos D, Mourelatos A, Corominas J (1998) Evaluation of renewable energy potential using a GIS decision support system. Renew Energy 13(3):333–344. https://doi.org/10.1016/S0960-1481(98)00006-8

Walters CJ (2001) Adaptive management of renewable resources. Blackburn Press, USA

Zhu K, Blum P, Ferguson G, Balke K-D, Bayer P (2010) The geothermal potential of urban heat islands. Environ Res Lett 5(4):044002. http://doi.org/10.1088/1748-9326/5/4/044002
Zhu K, Blum P, Ferguson G, Balke KD, Bayer P (2011) The geothermal potential of urban heat islands. Environ Res Lett 6(1). http://doi.org/10.1088/1748-9326/6/1/019501

Chapter 10
Legal Framework for Regulation

10.1 Policies, Strategies and Regulatory Standards in the European Union for the Promotion of Shallow Geothermal Energy

10.1.1 Policies and Strategies for the Promotion of Renewable Energies

The first EU strategy that comprehensively addressed energy and climate objectives was published in 2010 as *Energy 2020—A strategy for competitive, sustainable and secure energy* (COM/2010/0639). It aimed to reduce GHG emissions by at least 20%, increase the share of renewable energy by at least 20% of consumption, and achieve energy savings of 20% or more. By achieving these targets, the EU would help combat climate change and air pollution, decrease its dependence on fossil fuels from third countries and keep energy affordable for consumers and businesses. To meet these objectives, the strategy set out five priorities, including the *Strategic Energy Technology Plan* (*SET Plan*), which aims to accelerate the development and deployment of low-carbon technologies. It also sought to improve new technologies and reduce costs by coordinating national research efforts and helping to fund projects. The *SET Plan* promotes R&D&I efforts across Europe by supporting the most efficient technologies in the European Union's transformation to a low-carbon energy system. In order to support the implementation of the *SET Plan*, European Technology & Innovation Platforms (ETIPs) were created for joint action by EU countries, industry and researchers in key areas. In relation to geothermal energy, these comprise the *European Geothermal Energy Council* (EGEC) and the *Renewable Heating & Cooling (RH&C)* platforms.

The European Union has also set itself the long-term goal of reducing GHG emissions by 80–95% compared to 1990 levels by 2050. The *Energy Roadmap 2050*, also called the *2050 Energy Strategy* (COM/2011/885), explores the transition of the energy system in a way that is compatible with the GHG reduction target, while

A. García Gil et al., *Shallow Geothermal Energy*, Springer Hydrogeology,
https://doi.org/10.1007/978-3-030-92258-0_10

increasing competitiveness and security of supply. Through the *2050 Energy Strategy*, the European Commission set out four pathways for a more sustainable, competitive and secure energy system in 2050: energy efficiency, renewable energy (including shallow geothermal), nuclear energy, as well as carbon capture and storage.

In 2014, the European Council agreed on a new policy: *2030 Framework for Climate and Energy*, also called *2030 Energy Strategy*, which includes European targets and policies for the period between 2020 and 2030. The aim of this strategy was to help the European Union achieve a more competitive, secure and sustainable energy system, in order to meet its long-term GHG reduction target for 2050. The renewable energy and energy efficiency figures have subsequently been increased in the context of the *Clean Energy for all Europeans package*. The *2030 Energy Strategy* aims to help the European Union address the following issues: (1) implementation of the next step towards the target of reducing GHG emissions by 80–95% below the 1990 level by 2050, (2) dealing with high energy prices and the vulnerability of the EU economy to future price increases, especially for oil and gas, (3) increasing the EU's dependence on energy imports, often from politically unstable areas, (4) upgrading and replacing energy infrastructure and obtaining a stable regulatory framework for potential investors, and (5) agreeing on a GHG reduction target for 2030.

In addition, the *2030 Energy Strategy* proposed new targets and measures to make the EU economy and energy system more competitive, secure and sustainable. It included targets for reducing GHG emissions and increasing the use of renewable energy, and it proposed a new governance system and performance indicators. In particular, it proposed the following actions: (1) establishing a commitment to further reduce GHG emissions by setting a 40% reduction target by 2030 relative to 1990 levels, (2) setting a target for renewable energy use of at least 27% of energy consumption, with flexibility for member states to set national targets, (3) improving energy efficiency through possible amendments to the energy efficiency directive, (4) reforming the EU emissions trading scheme to include a market stability reserve, (5) establishing key indicators on energy prices, supply diversification, interconnections between member states and technological development in order to measure progress towards a more competitive, secure and sustainable energy system, and, finally, (6) designing and implementing a new governance framework for member state reporting, based on national plans coordinated and assessed at EU level.

The *Energy Union* strategy helped to provide secure, affordable and clean energy for EU citizens and businesses. It built further on the *2030 Energy Strategy* and the *European Energy Security Strategy*. It consisted of five closely related and mutually reinforcing policy areas: innovation and competitiveness, energy efficiency and decarbonisation of the economy. The European Commission's *Clean Energy for all Europeans package*, published in 2016, included a proposal for regulation on energy governance for the European Union as a whole.

Apart from energy policies, the European Union also sets air quality law and regulatory standards. Directive 2001/81/EC on national emission thresholds for certain atmospheric pollutants aims to limit emissions of ozone precursors, acidifying substances and fine particulate pollutants in order to improve the protection of the environment and human health. Directive 2008/50/EC of 21 May 2008 on ambient air quality and cleaner air for Europe defines the boundaries of the new air quality objectives for PM2.5, including exposure and limit values.

Mitigation of global warming is the main objective of the Paris Agreement (UNFCCC 2015), signed by 174 countries and the European Union in 2015. The agreement establishes a global framework to address climate change by limiting global warming to below two degrees Celsius and aiming to bring it down to 1.5 °C. To achieve this goal of the Paris agreement, countries must set climate targets every five years and increase the ambition of the targets over time. The agreement also aims to strengthen countries' capacity to address the impacts of climate change and support them in their efforts. Studies by the International Renewable Energy Agency (IRENA 2019) highlight cost-effective and available options for countries to meet climate commitments and contain rising global temperatures.

The EU's latest climate and energy policy is the *European Green Deal* (EC 2019), which assumes that the EU will gradually become climate and carbon neutral by 2050. To achieve that goal, a roadmap was proposed covering all sectors of the economy, including: (1) investing in environmentally friendly technologies, (2) supporting industry for innovation, (3) developing cleaner, cheaper and healthier private and public transport, (4) decarbonising the energy sector, (5) ensuring greater energy efficiency of buildings and (6) working with international partners to improve global environmental standards.

The European Union plans to provide financial support and technical assistance to help the people, businesses and regions most affected by the move towards a green economy. This is called the *Just Transition Mechanism*, which will help mobilise more than 100 billion euros over the period 2021–2027 in the most affected regions. Importantly, the mechanism will support the transition of coal mining areas in the EU towards the future decarbonised economy and society. Financial sources will be allocated according to specific criteria, and priority will be given to regions with a large number of employees in the coal, peat mining or oil and natural gas industry. The *European Green Deal* focuses on the need to decarbonise EU societies.

10.1.2 Regulatory Standards for the Increase of Renewable Energies

In support of *Energy 2020—A strategy for competitive, sustainable and secure energy*, the following EU directives entered into force: Directive 2009/28/EC on the promotion of the use of energy from renewable sources (RES Directive) and Directive 2012/27/EU on energy efficiency. The RES Directive specifies national renewable energy targets for each EU member state, taking into account its starting point and the overall potential for renewable energy. EU member states set out how they plan to meet these targets and the general course of their renewable energy policy in National Renewable Energy Action Plans (NREAPs), including the use of shallow geothermal energy for heating and cooling. Progress towards national targets is measured every two years, when EU countries publish national renewable energy progress reports.

Directive 2009/28/EC defines geothermal energy as *energy stored in the form of heat beneath the surface of solid earth.*

Directive 2012/27/EU establishes a set of binding measures to help the European Union reach its 20% energy efficiency target by 2020. According to the directive, all EU countries must use energy more efficiently at all stages of the energy consumption cycle, from production to final consumption. In 2016, the European Council proposed an update of the Energy Efficiency Directive 2012/27/EU which included a new energy efficiency target of 30% by 2030, and measures to update the directive to ensure that the new target is met. To help EU member states implement the Energy Efficiency Directive, the European Commission publishes guidance notes. In December 2018, the Renewable Energy Directive (RED II) 2018/2001/EU on the promotion of the use of energy from renewable sources entered into force, raising the overall EU target for the consumption of renewable energy sources by 2030–32%.

10.2 European Regulatory Legal Framework for the Use of Shallow Geothermal Energy

The first significant fact to be highlighted in relation to the regulation of shallow geothermal energy at the European level is that there is no specific directive with the aim of making sustainable use of shallow geothermal energy. The current state of affairs shows a situation where there are different directives dedicated to protecting the environment that can indirectly affect the thermal use of the subsurface.

It should be noted that shallow geothermal energy, except in very specific cases and as a derivative of other environmental problems, is essentially harmless to the environment. However, the lack of scientific knowledge of the effects of geothermal energy use on underground ecosystems and subsurface geochemistry means that existing regulations follow the precautionary principle and restrict the use of shallow geothermal energy, even in the absence of scientific evidence. In this respect, the legislative principles relevant to shallow geothermal thermal installation projects, as far as Europe is concerned, can vary considerably between countries. The legal framework is essentially determined by water legislation as well as natural resource regulations. Other legal areas that might be related in specific cases include nature and landscape conservation, energy, construction, soil conservation and protection against GHG emissions, among others.

10.2.1 Legal Framework at the National Level of Member States

The international legal framework for the regulation of shallow geothermal installations is still under development. The available literature shows the various regulatory approaches available for the use of shallow geothermal energy worldwide (Haehnlein et al. 2010; Jaudin 2013; Somogyi et al. 2017), highlighting the fact that this type of energy is more heavily regulated in European countries, due to the fact that its use is more common there (Francesco et al. 2016). In other countries, either deep geothermal energy is more commonly used or the standards regulating the use of shallow geothermal energy are not very robust and partially ambiguous. Some of the most common approaches consider that a minimum distance of 2.5 m (Austria) or 10 m (Finland and Sweden) should be set between the BHEs of geothermal installations and the boundaries of other users' plots (Haehnlein et al. 2010). For installations with open-loop geothermal heat exchangers, it has been proposed to set threshold temperature discharge values between 15 and 25 °C during operation in cooling mode, and between 2 and 5 °C for heating mode (Hähnlein et al. 2013; Tsagarakis et al. 2018). There is a varied approach across countries, with very different threshold values. A wide range of values has been defined empirically without validated scientific criteria but, at the same time, adapted to the locality-specific thermal regime on the subsurface (Haehnlein et al. 2010; Tsagarakis et al. 2018).

10.2.2 European Regulatory Legal Framework for the Protection of the Groundwater Public Domain

The European Union legislation that has the most significant influence on the use of shallow geothermal resources, especially in geothermal installations with open-loop geothermal heat exchangers (GWHPs), is the Water Framework Directive (WFD) 2000/60/EC (EU-WFD 2000). This directive committed the member states of the European Union to achieve good qualitative and quantitative status of all water bodies, including groundwater bodies, by 2015. The *framework* concept refers to the fact that it guides the steps to achieve the common objective without imposing a more traditional limit. The WFD objective was not achieved in 2015, as 47% of the EU water bodies affected by the regulation did not achieve the proposed objective. An important aspect of the WFD is the introduction of river basin districts and, based on these, the water management system. Directive 2006/118/EC (GWD) on the protection of groundwater against pollution and deterioration (EU-GWD 2006) refers to the protection of groundwater against pollution and deterioration and, as such, concerns shallow geothermal installations with open and closed-loop geothermal heat exchangers.

A groundwater abstraction for use in a geothermal installation is considered to extract an acceptable quantity when the quantity of groundwater abstracted does not exceed the recharge rate of the abstraction itself. The extraction of groundwater must be carried out without causing damage to the ecosystems directly dependent on it. The relevant laws of the EU member states have already transposed the provisions contained in the WFD and GWD Directives. Earlier regulations in many EU countries have been incorporated into the WFD. Key elements of the GWD include the description, assessment, classification and monitoring of the qualitative status of groundwater, as well as the determination and reversal of any significant increase in pollutant concentrations in groundwater bodies. The directive also specifies the measures that should be promoted in order to prevent or limit infiltration of pollutants into groundwater, as well as to prevent deterioration of groundwater conditions.

10.3 Legal Framework for Regulation in Spain

The legal framework for regulation of the exploitation of shallow geothermal energy resources in Spain is shown in this chapter as an example of the difficulties in many countries to find a clear and simple legal framework that facilitates the development of this renewable technology. In Spain, there is no clear and adequate legal framework regulating shallow geothermal installations, as there are no regulations for this purpose (Haehnlein et al. 2010; Somogyi et al. 2017). Previous studies suggest that general regulations (which are not specific to shallow geothermal systems) are limited to describing specifications in codes and technical regulations to indicate that such systems could be considered as potential solutions for reducing energy demand in buildings, leading to energy savings (Jaudin 2013). Recently, a review of the legal framework for shallow geothermal energy in selected European countries concluded that, in general, there are no specific procedures or regulations for geothermal systems in Spain (Tsagarakis et al. 2018).

10.3.1 Legal Definition of Shallow Geothermal Energy in Spain

According to the Spanish Geothermal Technology Platform (GEOPLAT 2010) shallow geothermal energy is defined as: *Energy stored in the ground or in groundwater at temperatures below 30 °C*. The associated geothermal resources depend on the temperature of the solid phase of the geological medium and the groundwater contained therein, which has temperatures below 30 °C, usually around the annual average surface temperature for the specific location under consideration. This definition is provided by the Spanish Geothermal Technology Platform (GEOPLAT 2010), a grouping of national stakeholders involved in this sector, including industry, scientific research and competent authorities. In the case of Spain, there is no legally binding definition of shallow geothermal energy.

10.3.2 Technical Guidelines for the Implementation of Good Practices

The first set of guidelines created, entitled *Design, implementation and monitoring of a shallow geothermal installation. Part 1: Vertical closed-loop systems.* (UNE 2014) describes the procedures necessary to design, plan and install closed-loop geothermal heat exchangers. The purpose of this standard is to define and promote the proper installation of closed-loop geothermal heat exchangers and their interconnections for heating, cooling or DHW production. In this way, the energy efficiency of the installations as a whole is guaranteed.

A second guide called *Technical guide for the design of closed-loop geothermal exchange systems* (ATECYR 2012) aims to establish the minimum technical conditions required for closed-loop geothermal systems used for space heating/cooling and DHW production by specifying design, installation and maintenance requirements.

10.3.3 Regulations for the Use of Shallow Geothermal Installations

Neither the installation nor the operation of systems using shallow geothermal energy are specifically regulated in Spain. The exploitation of shallow geothermal energy resources is within a complex legal framework (Fig. 10.1), which has two dimensions of complexity: (1) the multidisciplinary nature of the activities involved (natural resources, thermal installations and environmental aspects), which are affected by the four acts prior to the development of shallow geothermal energy systems, which were not specifically elaborated to consider this type of technology, and (2) in Spain, the legal competences of mining, environmental impact assessment and water have been transferred from the national to the regional level.

At the national level (Table 10.1), the Mining Act (L22/1973) and associated regulations (RD2857/1978, RD863/1985) define the legal concept of geothermal resources and thermal waters. Shallow and deep geothermal resources are not differentiated, but a future definition of these resources is taken into account, which requires the creation of a proposal by the competent ministries for energy and mining, with the approval of a prior report prepared by the Spanish Geological Survey (IGME).

In general, the production of electricity and the direct use of geothermal resources are subject to obtaining exploitation and research permits, as well as the approval of a licence, which is granted for a duration of 30 years. However, the Mining Act does not apply to geothermal installations where the extracted resources are used for the licence holder's own use and where the mining methods used do not have an exploitative purpose. If the thermal waters (groundwater with a temperature of four degrees Celsius above the annual average atmospheric ambient temperature of the specific location) are used for heating/cooling and the exploitation exceeds 500 th·hour^{-1} (581 kWh), the installation will then require permits and licences as

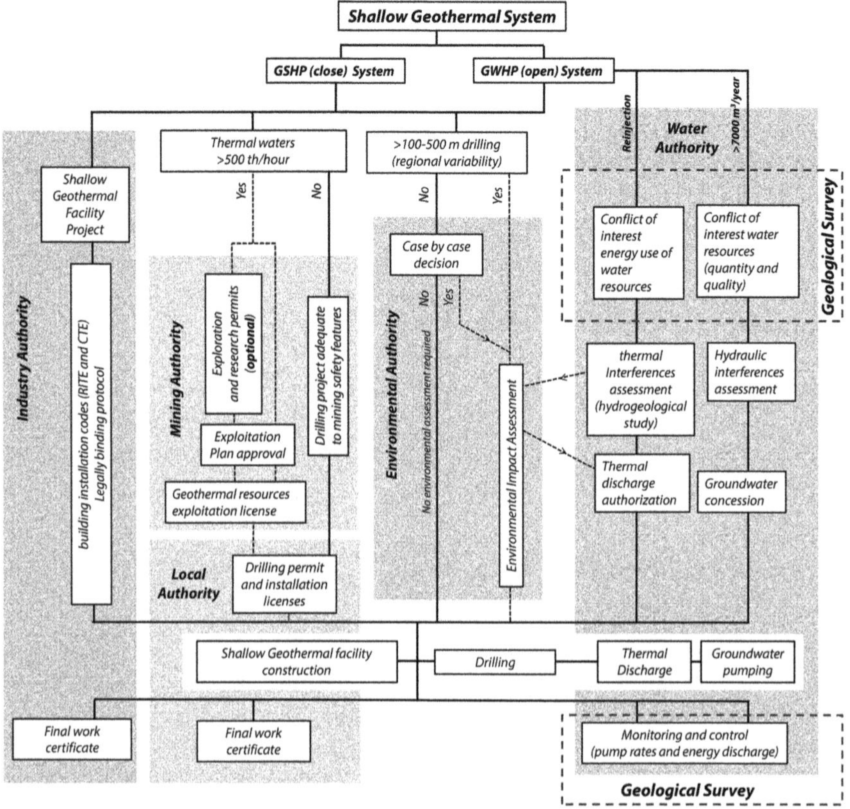

Fig. 10.1 Flow chart showing the procedures and regulations for the assessment of applications and granting of licences for the use of shallow geothermal energy

mentioned above. In conclusion, ordinary shallow geothermal installations are not affected by this law, as it only applies to cases where thermal waters are used, which is usually an exception. At the regional level (Table 10.1), local authorities are allowed to complement the Mining Act. However, only the region of Galicia has its own mining act, and it does not distinguish between the types of geothermal resources.

At the national level, the Act on Environmental Impact Assessment of projects (RDL1/2008 2008) and associated Regulations (L6/2010 2010) consider heat as a pollutant and suggest that assessment and monitoring of environmental impacts should potentially be required. However, there are no unified criteria to determine which geothermal systems would be affected by this regulation. At the regional level, only the regions of Andalusia (Fig. 10.2), Aragon (Fig. 10.2), Cantabria (Fig. 10.2) and the Balearic Islands (Fig. 10.2) have delimited depths where environmental impact assessment is mandatory (500, 500, 100 and 400 m, respectively). This fact indicates that environmental impact assessment is generally not mandatory for shallow geothermal systems.

Table 10.1 Documents that make up the legal framework for the use of shallow geothermal energy in Spain

Level	Legal documents	Type
National	Mining Act (L22/1973 1973; L54/1980 1980)	Mandatory
National	Regulations associated with the Mining Act (RD2857/1978 1978)	Mandatory
National	Design, implementation and monitoring of a shallow geothermal installation. Part 1: Vertical closed-loop systems. (UNE 2014)	Recommendation
National	Technical guide for the design of geothermal heat pump systems (ATECYR-IDAE) (ATECYR 2012)	Recommendation
National	Technical Building Code (CTE) and Regulation for Thermal Installations in Buildings (RITE) (RD238/2013 2013)	Mandatory
National	Act on Environmental Impact Assessment of projects (L6/2010 2010; L21/2013 2013; RDL1/2008 2008)	Mandatory
National	Regulations implementing the Aaw on Environmental Impact Assessment of projects (RD1131/1988, 1988)	Mandatory
National	Water Act (RDL1/2001 2001)	Mandatory
National	Regulations implementing the Water Act (RD638/2016 2016; RD849/1986 1986)	Mandatory
Regional	Regional environmental impact assessment acts of Andalusia, Aragon, Cantabria and the Balearic Islands (D16/2015 2015; L7/2007 2007; L11/2014 2014; ; L12/2016 2016; L17/2006 2006)	Mandatory
Regional	Mining Act for the Region of Galicia (L3/2008 2008)	Mandatory
Regional	Spanish Hydrological Plans (RD1/2016 2016)	Mandatory
Local	Municipal Ordinance on Means of Intervention in Town Planning Activity (ODZ99 2011)	Mandatory

The Water Act (RDL1/2001, 2001) regulates the use of groundwater and the discharge of thermal effluents in general, thus forcing users of shallow geothermal installations with open-loop geothermal heat exchangers to obtain a groundwater concession and authorisation for thermal discharge. The authorisation for thermal discharge requires a hydrogeological study and an environmental impact assessment, including the impacts of the discharge on the environment and on other shallow geothermal installations.

Closed-loop geothermal heat exchangers are generally excluded from these assessments. At the regional level, the River Basin Confederations, as competent authorities in the groundwater public domain, of the Duero (Fig. 10.2), Ebro (Fig. 10.2), Miño-Sil (Fig. 10.2), Tajo (Fig. 10.2) and the Cantabrian Domain (Fig. 10.2) basins have introduced different considerations in their respective water management plans (Garrido et al. 2016; RD1/2016 2016). This is in order to maintain good thermal quality of their groundwater bodies, including maximum acceptable values of relative temperature change between geothermal groundwater abstraction and re-injection wells (six to eight degrees Celsius) and maximum temperature thresholds (30 °C).

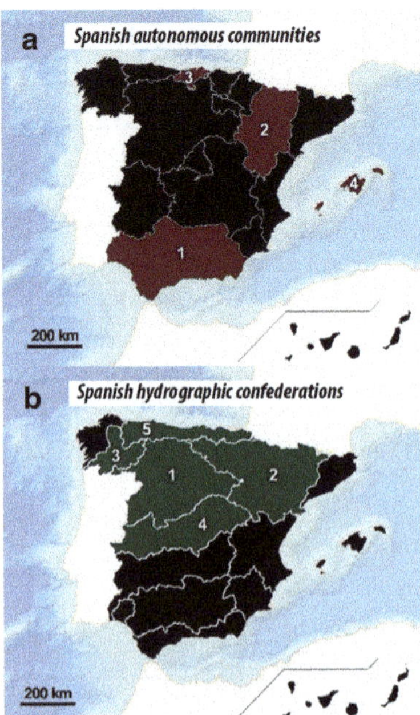

Fig. 10.2 **a** Autonomous Communities of Spain: Andalusia (1), Aragon (2), Cantabria (3) and Balearic Islands (4), where environmental impact assessment is mandatory for certain depths. **b** Spanish River Basin Confederations: Duero (1), Ebro (2), Miño-Sil (3), Tajo (4) and Cantabrian Domain (5), where there are considerations to maintain a good thermal quality of groundwater bodies. Base maps obtained from the Ministry for the Ecological Transition and the Demographic challenge

Finally, the efficiency, safety, control and responsibilities of thermal installations with geothermal heat pumps are regulated by the Technical Building Code (CTE) and the Regulation of Thermal Installations in Buildings (RITE, RD238/2013), created by the Spanish Ministry of Industry. These documents cover the requirements for design of the installation project, as well as execution of the works and the legalisation procedure.

10.4 Special Requirements for the Installation and Operation of Shallow Geothermal Installations

In Spain, an environmental impact assessment is only requested by the competent authority if the drilling work exceeds 500 m depth.

The national legal framework does not set a depth threshold value, so this responsibility has been transferred to the regional authorities. In pioneering areas, such as the Spanish city of Zaragoza, GSHP systems are allowed to inject groundwater at temperatures up to 30 °C, calculating the equivalent daily value of the temperature and integrating by the total energy dissipated during the daily activity. A relative change in temperature of eight degrees Celsius is allowed between extraction and injection geothermal wells. In addition, installations above 200 kW are continuously monitored and reported to the water authority by means of fortnightly operating data sets, including abstraction/injection flow rates and abstraction/injection temperatures at all geothermal wells of the installations.

The decommissioning of operational wells is only regulated by the Water Act when they affect the public hydraulic domain (RD849/1986), but there is no specific protocol (Guardiola-Albert et al. 2018). The Mining Law also encourages the re-establishment of natural conditions once exploitation has ended. There are no specific regulations by law for shallow geothermal installations.

It is also generally not considered necessary to report the operating regimes of shallow geothermal installations to the authorities. Depending on specific cases, if deemed relevant, the authorities may require reports in order to prevent overexploitation or environmental damage.

For further development of legislation in Spain, a framework is needed at the national level, in order to establish the needs for defining specific minimum requirements for the different local and regional authorities. First, it would be desirable to create a framework structure for the management of shallow geothermal energy resources, as well as a resource governance model. Among other aspects, it would be necessary to request the establishment of an inventory of shallow geothermal installations in order to define thermal affection perimeters around geothermal installations and implement them in software platforms based on geographic information systems (Alcaraz et al. 2016; Attard et al. 2020). This would allow defining a threshold value of installed capacity (e.g., 50 kW) as a criterion for carrying out the simplified or ordinary licensing process, which would assess whether a full environmental impact analysis is needed. The threshold value of 50 kW is based on the typical thermal load value for residential/domestic systems and marks the boundary with larger shallow geothermal installations for commercial, industrial and public sector use (Feuvre and Cox 2009).

A special feature of Spain that sets it apart from other northern European countries is its climatology. According to the State Meteorological Agency (AEMET), the average annual temperature in Spain is 15.1 °C (1981–2010) and rising (AEMET 2018), which is significantly higher than in the rest of the northern European countries. The implication is basically that the cooling demand is higher than the heating demand, so it is necessary to impose a relatively high limit for the injection temperature (30 °C or more), considering the typical equivalent daily injection temperature in shallow geothermal installations with open-loop geothermal heat exchangers in Spain (Muela Maya et al. 2018). The operating temperature limits and allowable temperature changes in urban areas, where shallow geothermal systems are concentrated should be different from non-urban areas, where there is no concentration of installations.

10.5 Future Need for Adaptation of the Spanish Regulatory Framework

The Spanish legal framework does not directly differentiate shallow geothermal resources within general geothermal resources, except for the delimitation of the maximum drilling depth (100–500 m) at the regional level, where regulations, depending on the region where the installation is located, will require an environmental impact assessment study. Therefore, these regulations would indirectly and partially define shallow geothermal resources that are not regulated in a general and direct way. In the case of shallow geothermal installations with open-loop geothermal heat exchangers (geothermal wells), a groundwater concession and authorisation for thermal discharge into the public water domain are required. Shallow geothermal installations with closed-loop geothermal heat exchangers are therefore unregulated in most regions of Spain. In practice, this legal regime is not entirely inadequate, as it gives freedom to develop shallow geothermal energy in areas where there are no defined groundwater bodies. However, the lack of a specific regulation explicitly stating this reality may generate uncertainty among potential investors in future shallow geothermal installations. The inclusion of specific aspects of shallow geothermal energy in river basin management plans is a major step forward (Garrido et al. 2016). Future updates of these hydrological management plans should include guarantees for the tightness of closed-loop geothermal heat exchangers operating in groundwater bodies.

There is robust evidence of the need for the elaboration of a specific legal framework for shallow geothermal energy utilization. Shallow geothermal resources should be defined and differentiated from deep geothermal resources in the Mining Act, including the former at the maximum threshold of 400 m depth for heat generation (heat understood as thermal energy transfer). Conventional drilling technology used up to 400 m does not pose a risk to the environment if a standardised drilling protocol with basic safeguards is followed. Furthermore, geothermal installations with an installed capacity of less than 50 kW should only require a simplified dossier procedure including essential information. For systems between 50 and 200 kW, an extended project report is recommended, including a hydrogeological report assessing possible thermal interference and possible interaction with contaminated groundwater. Systems exceeding 200 kW, in urban areas with defined groundwater bodies, should be forced to carry out an environmental impact study and be monitored, both their operating regime (García-Gil et al. 2020) and in piezometric networks of geothermal control in urban areas (Garrido et al. 2017). Another important point would be the fact that coordinated efforts between administrations are needed to unify the administrative procedure, so that interested parties can carry out the licence request in a single-step application. Any effort to promote this renewable, economically efficient and environmentally friendly technology will be fruitless if potential users are not offered a simple, fast and specifically developed legal procedure.

References

AEMET (2018) Resumen anual climatologico 2018. In: (AEMET) SMA (ed) Ministry for the Ecological Transition, Madrid

Alcaraz M, García-Gil A, Vázquez-Suñé E, Velasco V (2016) Use rights markets for shallow geothermal energy management. Appl Energy 172:34–46. https://doi.org/10.1016/j.apenergy.2016.03.071

ATECYR (2012) Guía técnica Diseño de sistemas de intercambio geotérmico de circuito cerrado. Instituto para la Diversificación y Ahorro de la Energía (IDAE), Madrid, p 45. 978-84-96680-60-9

Attard G, Bayer P, Rossier Y, Blum P, Eisenlohr L (2020) A novel concept for managing thermal interference between geothermal systems in cities. Renew Energy 145:914–924. https://doi.org/10.1016/j.renene.2019.06.095

D16/2015 (2015) Decreto 16/2015, de 10 de abril, de ordenación sanitaria territorial de la comunidad autónoma de las Illes Balears. In: Gobierno Cd (ed)

L54/1980 (1980) Ley 54/1980, de 5 de noviembre, de modificación de la Ley de Minas, con especial atención a los recursos minerales energéticos. In: Spain So (ed) Official state bulletin

RD2857/1978 (1978) Real Decreto 2857/1978, de 25 de agosto, por el que se aprueba el Reglamento General para el Régimen de la Minería. In: Energía MdIy (ed) Official state bulletin, p 10

EC (2019) Communication from the Commission to the European Parliament, the European Council, the Council, the European Economic and Social Committee and the Committee of the Regions the European Green Deal. In: Commission E (ed) Brussels

EU-GWD (2006) La Directiva 2006/118/CE relativa a la protección de las aguas subterráneas contra la contaminación y el deterioro (GWD) In: Europeo C (ed) Brussels

EU-WFD (2000) European water framework directive 2000/60/EC. European Parliament

Feuvre PL, Cox CSJ (2009) Ground source heating and cooling pumps in England and Wales: State of Play and Future Trends (Environment Agency Report). In: Directorate E (ed) Environment Agency United Kingdom, p 81

Francesco T et al (2016) How to boost shallow geothermal energy exploitation in the adriatic area: the LEGEND project experience. Energy Policy 92:190–204. https://doi.org/10.1016/j.enpol.2016.01.041

García-Gil A et al (2020) Defining the exploitation patterns of groundwater heat pump systems. Sci Total Environ 710:136425. https://doi.org/10.1016/j.scitotenv.2019.136425

Garrido E, García-Gil A, Arrazola Martínez C, Escayola Calvo O, Sánchez Navarro JA (2017) Usefulness of a groundwater temperature baseline monitoring network for the identification of thermal interferences between shallow geothermal exploitation systems in urban environments. In: Calvache ML, Duque C, Pulido-Velazquez D (eds) Congress on Groundwater and Global Change in the Western Mediterranean International Association of Hydrogeologists, Granada

Garrido E et al (2016a) Aspectos normativos de los aprovechamientos geotérmicos someros. La experiencia del acuífero urbano de Zaragoza, Las aguas subterráneas y la planificación hidrológica. Congreso hispano-luso. AIH-GE, Madrid. ISBN: 978-84-938046-5-7

Guardiola-Albert C et al (2018) Los pozos abandonados como fuente puntual de contaminación de los acuíferos. Ideas para

Haehnlein S, Bayer P, Blum P (2010) International legal status of the use of shallow geothermal energy. Renew Sustain Energy Rev 14(9):2611–2625. https://doi.org/10.1016/j.rser.2010.07.069

Hähnlein S, Bayer P, Ferguson G, Blum P (2013) Sustainability and policy for the thermal use of shallow geothermal energy. Energy Policy 59:914–925. https://doi.org/10.1016/j.enpol.2013.04.040

IRENA (2019) Global energy transformation: a roadmap to 2050 (2019 edn), International Renewable Energy Agency, Abu Dhabi

Jaudin F (2013) D2.2: General Report of the current situation of the regulative framework for the SGE systems. REGEOCITIES, p 50

L11/2014 (2014) Ley 11/2014, de 4 de diciembre, de Prevención y Protección Ambiental de Aragón. In: Aragón CAd (ed)

L12/2016 (2016) Ley 12/2016, de 17 de agosto, de evaluación ambiental de las Illes Balears. In: Balears CAdlI (ed). Official state bulletin

L17/2006 (2006) Ley 17/2006, de 11 de diciembre, de control ambiental integrado. Official state bulletin

L21/2013 (2013) Ley 21/2013, de 9 de diciembre, de evaluación ambiental. In: Estado Jd (ed) Official state bulletin

L22/1973 (1973) Ley 22/1973, de 21 de julio, de Minas. In: State S (ed) Official state bulletin

L3/2008 (2008) Ley 3/2008, de 23 de mayo, de ordenación de la minería de Galicia. In: Galicia PdlJd (ed)

L6/2010 (2010) Ley 6/2010, de 11 de junio, reguladora de la participación de las entidades locales en los tributos de la Comunidad Autónoma de Andalucía. In: Andalucía CAd (ed), BOE-A-2010-11492. Official state bulletin

L7/2007 (2007) Ley 7/2007, de 9 de julio, de Gestión Integrada de la Calidad Ambiental. Official Bulletin of the Andalusian Autonomous Government

Muela Maya S et al (2018) An upscaling procedure for the optimal implementation of open-loop geothermal energy systems into hydrogeological models. J Hydrol 563:155–166. https://doi.org/10.1016/j.jhydrol.2018.05.057

ODZ99 (2011) Ordenanza Municipal de Medios de Intervención en la Actividad Urbanística. In: Council ZC (ed), nº 99 03 mayo 2000. BOPZ, Zaragoza

RD1/2016 (2016) Real Decreto 1/2016, de 8 de enero, por el que se aprueba la revisión de los Planes Hidrológicos de las demarcaciones hidrográficas del Cantábrico Occidental, Guadalquivir, Ceuta, Melilla, Segura y Júcar, y de la parte española de las demarcaciones hidrográficas del Cantábrico Oriental, Miño-Sil, Duero, Tajo, Guadiana y Ebro. In: Ministerio de Agricultura AyMA (ed) Official state bulletin

RD1131/1988 (1988) Real Decreto 1131/1988, de 30 de septiembre, por el que se aprueba el Reglamento para la ejecución del Real Decreto Legislativo 1302/1986, de 28 de junio, de Evaluación de Impacto Ambiental. In: Urbanismo MdOPy (ed)

RD238/2013 (2013) Real Decreto 238/2013, de 5 de abril, por el que se modifican determinados artículos e instrucciones técnicas del Reglamento de Instalaciones Térmicas en los Edificios, aprobado por Real Decreto 1027/2007, de 20 de julio. In: Presidencia Mdl (ed) Official state bulletin

RD638/2016 (2016) Real Decreto 638/2016, de 9 de diciembre, por el que se modifica el Reglamento del Dominio Público Hidráulico aprobado por el Real Decreto 849/1986, de 11 de abril, el Reglamento de Planificación Hidrológica, aprobado por el Real Decreto 907/2007, de 6 de julio, y otros reglamentos en materia de gestión de riesgos de inundación, caudales ecológicos, reservas hidrológicas y vertidos de aguas residuales. In: Ministerio de Agricultura y Pesca AyMA (ed) Official state bulletin

RD849/1986 (1986) Real Decreto 849/1986, de 11 de abril, por el que se aprueba el Reglamento del Dominio Público Hidráulico, que desarrolla los títulos preliminar I, IV, V, VI y VII de la Ley 29/1985, de 2 de agosto, de Aguas. In: Urbanismo MdOPy (ed) BOE-A-1986-10638. Official state bulletin

RDL1/2001 (2001) Real Decreto Legislativo 1/2001, de 20 de julio, por el que se aprueba el texto refundido de la Ley de Aguas. In: Ambiente MdM (ed) Official state bulletin

RDL1/2008 (2008) Real decreto legislativo 1/2008, de 11 de enero, por el que se aprueba el texto refundido de la Ley de Evaluación de Impacto Ambiental de proyectos. In: Ambiente MdM (ed) BOE-A-2008-1405

Somogyi V, Sebestyén V, Nagy G (2017) Scientific achievements and regulation of shallow geothermal systems in six European countries—a review. Renew Sustain Energy Rev 68(Part 2):934–952. https://doi.org/10.1016/j.rser.2016.02.014

Tsagarakis KP et al (2018) A review of the legal framework in shallow geothermal energy in selected European countries: need for guidelines. Renew Energy. https://doi.org/10.1016/j.renene.2018.10.007

UNE (2014) UNE100715-1:2014 Guide for the design, implementation and monitoring of a geothermal system. Part 1: Vertical closed circuit systems. In: Normalización AEd (ed)
UNFCCC (2015) Adoption of the Paris Agreement. Proposal by the President (Draft decision). In: Change FCoC (ed) United Nations Office, Geneva (Switzerland), p 32

Chapter 11
Example of Application (I): The Management of Shallow Geothermal Energy Resources in the City of Zaragoza

The city of Zaragoza was chosen as a relevant case study of the application of a shallow geothermal energy resource management model for four reasons: (1) the city shows an important development of shallow geothermal energy systems, (2) Zaragoza shows a typology of geothermal installations focused on the use of open-loop geothermal heat exchangers, (3) the magnitude of the installed heating/cooling power in this type of installations, and (4) the high number of them in a reduced area of the city centre (Garrido and Sánchez Navarro 2009). The studies and research carried out by the Spanish Geological Survey (IGME) and the water authority, the Ebro Hydrographic Confederation (CHE), have provided detailed knowledge of shallow geothermal energy resources and a set of new criteria and decision-making tools to assist in the management of these resources. This chapter describes the different aspects of these resources and how they are exploited and managed, as well as the control mechanisms adopted by the competent authority to minimise the thermal impacts associated with the operation of shallow geothermal installations with open-loop geothermal heat exchangers.

11.1 The Early Exploitation of Shallow Geothermal Energy Resources

The exploitation of shallow geothermal energy resources in the city of Zaragoza arises in a particular way. The city has continental Mediterranean climate, which means that the demand for thermal installations is mainly focused on satisfying cooling demand, mainly in summer. The open-loop geothermal heat exchanger is the technology preferred in almost all thermal installations with geothermal heat pumps. This is not the case in most European cities, where both open and closed-loop geothermal heat exchangers are used, although the latter predominate. The use of closed-loop geothermal heat exchangers is usually associated with a simpler, more economical installation procedure that can be adapted to almost any terrain, making

A. García Gil et al., *Shallow Geothermal Energy*, Springer Hydrogeology,
https://doi.org/10.1007/978-3-030-92258-0_11

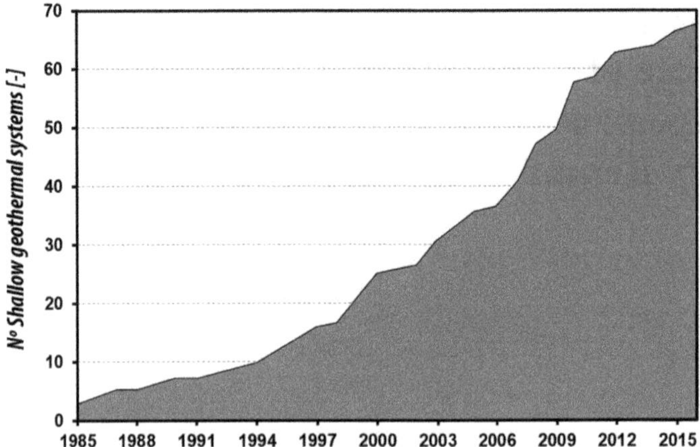

Fig. 11.1 Annual distribution of new shallow geothermal installations with open-loop geothermal heat exchangers in the city of Zaragoza

them highly suitable for small buildings and single-family houses, which are well established and widespread in other parts of Europe. In the case of Zaragoza, a city with a high building density and abundant groundwater resources that are easily accessible and available, it offers a perfect setting for installations with open-loop geothermal heat exchangers, i.e., geothermal wells.

The first known shallow geothermal installations in Zaragoza were built in the 1980s (Fig. 11.1). These early installations generally presented simple designs with geothermal wells (*standing column well*) such as the one shown in Fig. 11.2, which are implemented to meet small energy demands by capturing very shallow groundwater.

Subsequently, two episodes were responsible for the increase of shallow geothermal installations in the city: the first was at the end of the 1990s, coinciding with the Kyoto agreement in 1997; the second was after an outbreak of *Legionella pneumophila* in 2004 in cooling towers (aerothermal). The mortality associated with the latter event prompted city managers of large facilities to look for safe HVAC alternatives, finding an optimal solution to the health issue in water-water geothermal heat pumps. It is since these episodes that this type of thermal installations has proliferated, with a constant increase in the number of new installations every year.

The extension of Zaragoza's urban alluvial aquifer, the significant groundwater resources available and the advanced degree of technological development in heat pumps make it possible to design high-capacity thermal installations, which are used in emblematic buildings and infrastructures, in both public and private ownership. As a result, the number of shallow geothermal installations operating in the city centre is now close to 70, making Zaragoza a pioneer in Spain in the use of shallow geothermal energy resources and unique in Europe in utilizing large installations using open-loop geothermal heat exchangers.

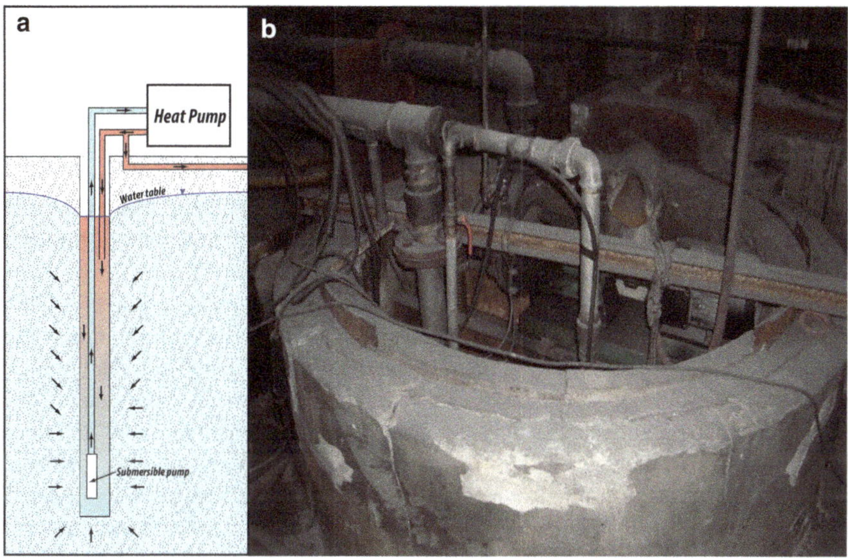

Fig. 11.2 **a** Diagram of the operation of a standing column well, modified from Banks (2012). **b** Example of a real standing column well located in the historic quarter of Zaragoza

Despite the fact that the oldest geothermal installations in the city may have been in operation for more than forty years, it has nevertheless been necessary to wait for the results of different research and studies carried out over the last two decades (García-Gil et al. 2015d; IGME-CHE 2014, 2019; Moreno et al. 2008) to obtain a detailed understanding of the level of development of shallow geothermal energy in Zaragoza. The results of these studies have made it possible to confirm the dimensions reached in the exploitation of shallow geothermal energy resources in urban areas, their relevance in the current context of the use of renewable energy sources in Spain and the degree of development of open-loop shallow geothermal heat exchanger technology at a European and international level.

In the last decade, the annual number of new requests for the commissioning of shallow geothermal installations processed by the water management body (CHE) has remained constant, either for the concession to exploit groundwater for the air-conditioning of buildings or for authorisation of the thermal discharge that this same process generates. The groundwater exploitation requested for air-conditioning with geothermal heat pumps corresponds to thermal installations in buildings that cover a wide range of activities and services, including the following: small neighbourhood communities, offices, banks, leisure premises, shopping centres, hotels, hospitals, industrial buildings and public buildings such as the University of Zaragoza, Zaragoza City Council and the Government of Aragon (regional government).

The level of development of shallow geothermal energy currently achieved would not have been possible without the firm commitment of both private and public initiatives. The initial implementation of this technology was in a framework lacking in specific management and regulation. However, with the current efforts of competent administration and with the understanding of the thermal and hydrogeological regimes of the urban aquifer, it has been possible to establish the bases for secure operation in the majority of installations. In short, all the actions that have occurred make the city of Zaragoza an urban centre with a high level of exploitation of shallow geothermal energy resources, as a pioneering spot in Spain in the massive implementation of large geothermal installations with open-loop geothermal heat exchangers and an example in adaptive management of shallow geothermal energy resources (Garrido and Sánchez Navarro 2009).

Knowing the dimensions and effects associated with exploitation of shallow geothermal energy resources on groundwater bodies is necessary information, which must come before the adoption of any management policy aimed at reducing undesired impacts and effects. This implies a significant human and economic effort to improve our knowledge of the physical groundwater environment in all its dimensions and then to investigate the mechanisms that govern groundwater dynamics and heat transport in this environment. This is an effective way to be in a position to predict, with some success, the future behaviour of the aquifer in the event of a change in the exploitation regime. Prior to the studies carried out within the IGME-CHE collaboration framework (IGME-CHE 2014, 2019), there was no history of specific work or research in other cities that addressed problems similar to those that could be found in the Zaragoza alluvial urban aquifer. Consequently, both organisations began to collaborate in order to initiate studies aimed at advancing knowledge of the exploitation model of the urban aquifer and the impacts caused. Some of the most relevant actions taken and the results obtained are described below.

The general objectives of the studies carried out in the city of Zaragoza included: (1) improvement of the knowledge of the groundwater body and the urban alluvial aquifer of Zaragoza, and (2) development of tools and criteria for the orderly and sustainable exploitation of groundwater and shallow geothermal resources, which can be useful to the competent authority.

The specific objectives were the following:

I. Improving knowledge of the hydrogeological functioning of the alluvial aquifer of the Ebro River in the urban sector of the city of Zaragoza.
II. Updating the inventory of geothermal installations in the city of Zaragoza and description of the technical characteristics of interest, in relation to groundwater and the urban aquifer.
III. Establishing an urban aquifer monitoring and observation network for thermal and hydraulic characterisation of the urban aquifer as a shallow geothermal energy reservoir.
IV. Identification and characterisation of thermal, hydrochemical and biological impacts on the environment, as well as the study of possible thermal–hydraulic interference between shallow geothermal installations.
V. Study of the hydraulic and thermal regime of the Zaragoza urban aquifer through the development of a mathematical model of groundwater flow and heat transport at city scale, capable of: (1) reproducing the functioning of the

aquifer, of (2) simulating the affection of thermal discharges, and (3) serving as a support tool for the management of groundwater resources and authorisation of thermal discharges (authorisation of exploitation of shallow geothermal energy resources).

VI. Establishing additional and complementary management criteria and indicators to the numerical simulation to facilitate decision-making by the competent authority.

The following sections of this chapter describe the most relevant results achieved for improvement of the existing knowledge on the intense shallow geothermal energy exploitation of the Zaragoza urban aquifer and its consequences.

11.2 Geological and Hydrogeological Framework

Zaragoza is located in the centre of a large sedimentary basin of Tertiary age. In its central sector, over a Palaeozoic basement and an incomplete Mesozoic cover, conglomerates, shales, sandstones and marls have accumulated, coming from the dismantling of the surrounding mountain ranges, as well as carbonate precipitates and evaporitic salts. During the Oligocene-Miocene, the basin shows endorheic facies and accumulates important thicknesses of sediments. In the Zaragoza vertical geological profile, gypsum and salts alternate with other fine-grained detrital materials, totalling nearly a thousand metres.

At the end of the Tertiary period, during the opening of the basin to the Mediterranean Sea, the materials accumulated during the endorheic phase began to be evacuated through a drainage network, with the Ebro River as the main collector. With the formation of a fluvial network, the deposit of clays, silts, sands and gravels also began, arranged in different levels of terraces. The different levels of terraces are generally found in contact with each other, facilitating the hydraulic connection between them and with the river Ebro, which is a determining factor for this group to be defined as the alluvial aquifer of the Ebro river. Their development is uneven, occupying large areas and gaining width on the right bank of the river Ebro, while on the left bank they are affected by a process of denudation due to migration of the riverbed towards the north, which constricts and reduces them considerably as they collide with a high front of impermeable tertiary materials. A similar process occurs in the two tributaries of the river Ebro, whose confluence to the Ebro river is in the east sector of the city: the river Gállego on the left bank, which has a highly developed system of stepped terraces and, to a lesser extent, the river Huerva, a tributary on the right bank.

Practically the entire city is overlaid on a vast system of terraces developed in the confluence area of these three rivers (Fig. 11.3). Due to hydrogeological and water management criteria, the hydrographic confederation (CHE) includes the Ebro terraces in the *groundwater body ES091MSBT058 Ebro Alluvial: Zaragoza* and the Gállego terraces in the *groundwater body ES091MSBT057 Gállego Alluvial.*

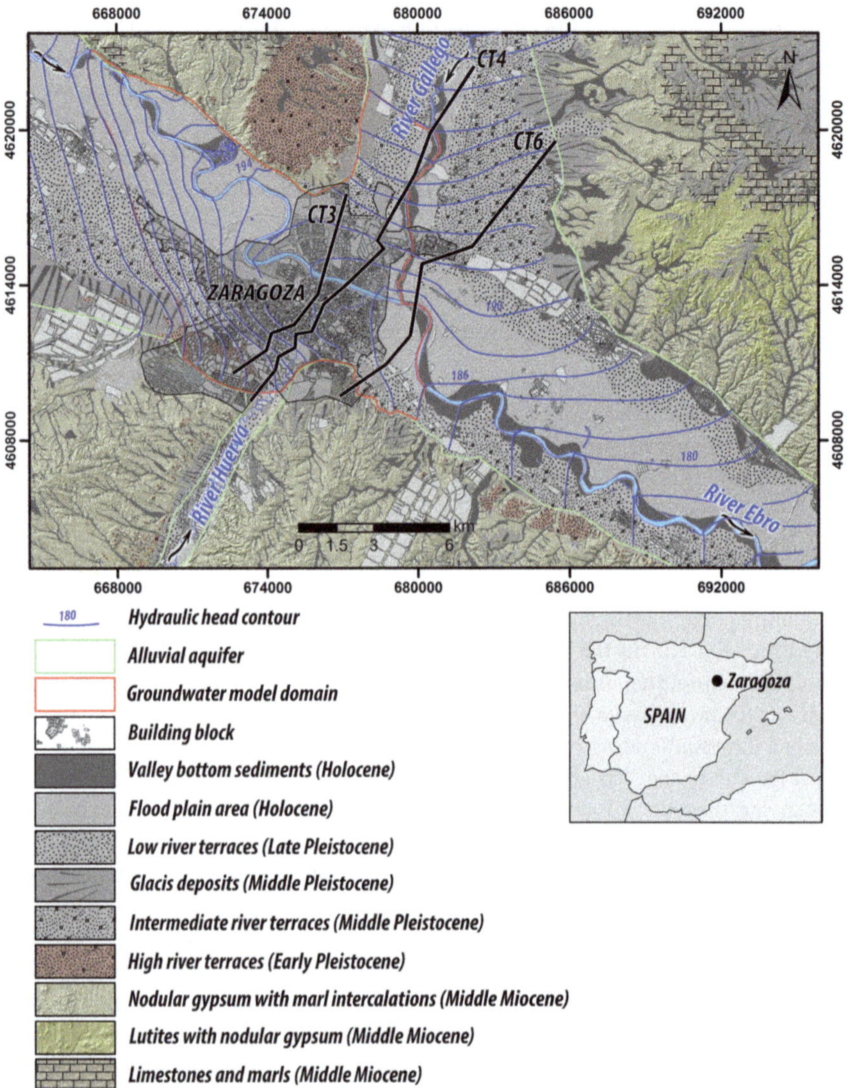

Fig. 11.3 Geological map and head contour map of the alluvial aquifer of Zaragoza. *Source* IGME (2017)

However, due to the particular and specific groundwater exploitation of these water bodies within the city, this particular sector is henceforth referred to as the *Zaragoza Urban Aquifer* (Garrido et al. 2006, 2010; Moreno et al. 2008).

Some authors have distinguished up to 12 terrace levels (Benito et al. 1998, 2010), but the most important levels, characterised by their extension and by maintaining the hydraulic connection, are the four most recent terraces T1 to T4, which rise between

3 and 73 m above the Ebro riverbed, and the current floodplain T0. They present fairly common granulometric and lithological features, with upper stretches where sands, shales and silts are abundant while, in the lower stretches, gravels are in the majority. Sometimes the clasts are cemented by carbonate, taking on the appearance of a caliche crust in the older and higher terraces. The sand content varies between 15 and 25% and is often arranged in lenses, in which this granulometric fraction is dominant.

Other Quaternary formations of relative interest are the deposits of sub-angular gravels, silts and sands with glacial morphology that link the alluvial materials and the peripheral Tertiary escarpments. Their thickness is variable and they are highly developed on the right bank of the Ebro, where they can be hydraulically connected with the higher terrace levels. Finally, dejection cones and flat-bottomed valleys or *vales* can be found frequently in the aquifer lateral boundaries but whose interest from a hydrogeological point of view is very limited and local.

Tertiary materials form the alluvial aquifer basement due to their very low permeability. There is, however, a close relationship between these two formations of very different permeability. The solubility of the salts is the main reason for the development in the tertiary gypsum and salts formations of a network of conduits and cavities linked to a poorly known karst aquifer. The underground flow is also the origin of other phenomena of collapse and subsidence by dissolution, which affects all the materials of the aquifer and which would explain the irregular morphology of the alluvial aquifer basement and the variations in aquifer thickness, but also the appearance of sinkholes on the surface, structures which are very common in the middle terraces around the city (Benito 1987; Soriano et al. 1994).

Current hydrogeological knowledge of the urban aquifer comes mainly from lithological, hydraulic tests and hydrochemical analyses extracted from more than 500 water wells, boreholes and reconnaissance piezometers, which commonly have depths of between 20 and 60 m. Recent works (Garrido et al. 2006; IGME-CHE 2014, 2019; Moreno et al. 2008) improve the description and characterisation of the aquifer, which is summarised below.

The urban aquifer of Zaragoza is a detritic alluvial aquifer, with a very high permeability and unconfined conditions, although very occasionally it may show semi-confinement behaviour where it accumulates significant thicknesses of silts and clays. The aquifer is crossed by three rivers; the Ebro, Gállego and Huerva. According to the gauging yearbooks of CEDEX (CEDEX 2019), the following historical average flow rates, respectively, of 230.23 $m^3\ s^{-1}$, 12.07 $m^3\ s^{-1}$ and 3.04 $m^3\ s^{-1}$, were measured at gauging stations located in the city.

It is difficult to identify the terrace boundaries from lithological borehole logs, and a single thickness is assumed for the Quaternary deposits as a whole, with highly variable thickness ranging between 5 and 60 m, being smaller on the left bank of the Ebro but reaching up to 80 m in depth near the river Gállego (Fig. 11.4).

The piezometric level is generally found between 7 and 34 m below the terrain surface, at elevations between 188 and 200 m m.a.s.l. Figure 11.3 summarises the geology of the Zaragoza area; it also shows the piezometric surface above the urban sector (IGME-CHE 2019). The piezometric level in the urban aquifer is conditioned

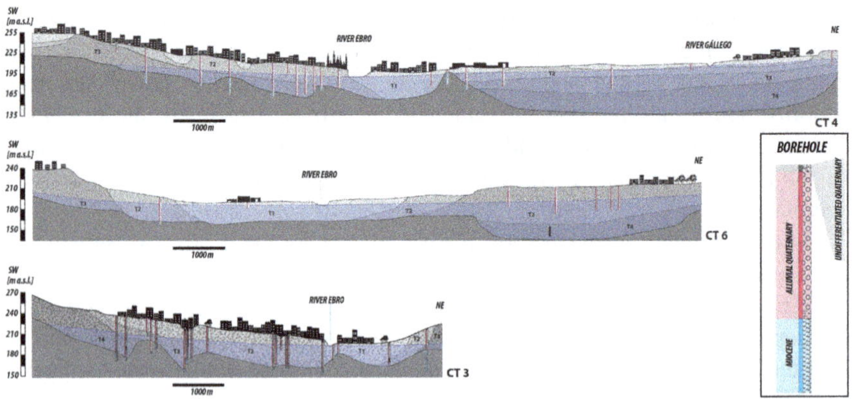

Fig. 11.4 Hydrogeological cross-sections of the Zaragoza urban alluvial aquifer. *Source* IGME-CHE (2019). Location of profiles in Fig. 11.3

by the hydraulic behaviour of the Ebro alluvial as a whole, by that of the Gállego alluvial in its lower part and confluence with the Ebro. The evolution of the level shows a significant seasonal oscillation related to the surface recharge of the water supplied for irrigation, with a maximum in summer and minimum levels in winter. However, the variation of the piezometric level also depends on the water level of the Ebro and Gállego rivers, the ordinary and extraordinary flood events, as well as the proximity or distance of the observation point to the respective riverbeds, which ultimately determines the drainage level of the underground flow (Fig. 11.5).

An exhaustive analysis of the groundwater head contour maps of the city reveals its complexity, with large changes in the hydraulic gradients in areas of lower permeability corresponding to high terraces, and modification of flow lines due to pumping and injections from the different geothermal wells in operation. The hydrogeological flow regime differs from one bank of the Ebro to the other. The left bank is the area where groundwater flows from the two alluvial deposits converge, with a clear influence of the flow from the river Gállego, much more noticeable further north and east of the city of Zaragoza. In the lower Gállego area, as can be deduced from interpretation of the piezometric data, the river shows an influent condition, i.e., it recharges the aquifer, and the groundwater flow adopts a component that runs approximately from north to south. In contrast, the river Ebro is mainly effluent and drains the aquifer. On its right bank, it drains the high terraces presenting SW to NE direction, then the groundwater flow paths change to a W-E direction and then, parallel to the river, the flow runs eastwards along the lower terraces of the city. The arrangement of the meanders of the river Ebro as it passes through the city of Zaragoza explains why the Ebro, along an urban section of some 9.5 km in length, is capable of draining some 30 hm^3 per year, of which 3 hm^3 $year^{-1}$ are underground contributions from the Ebro alluvial and 27 hm^3 $year^{-1}$ are provided by the Gállego alluvial aquifer, which means a linear flow rate of 0.09 L s^{-1} km^{-1} in the latter case (Garrido et al. 2006).

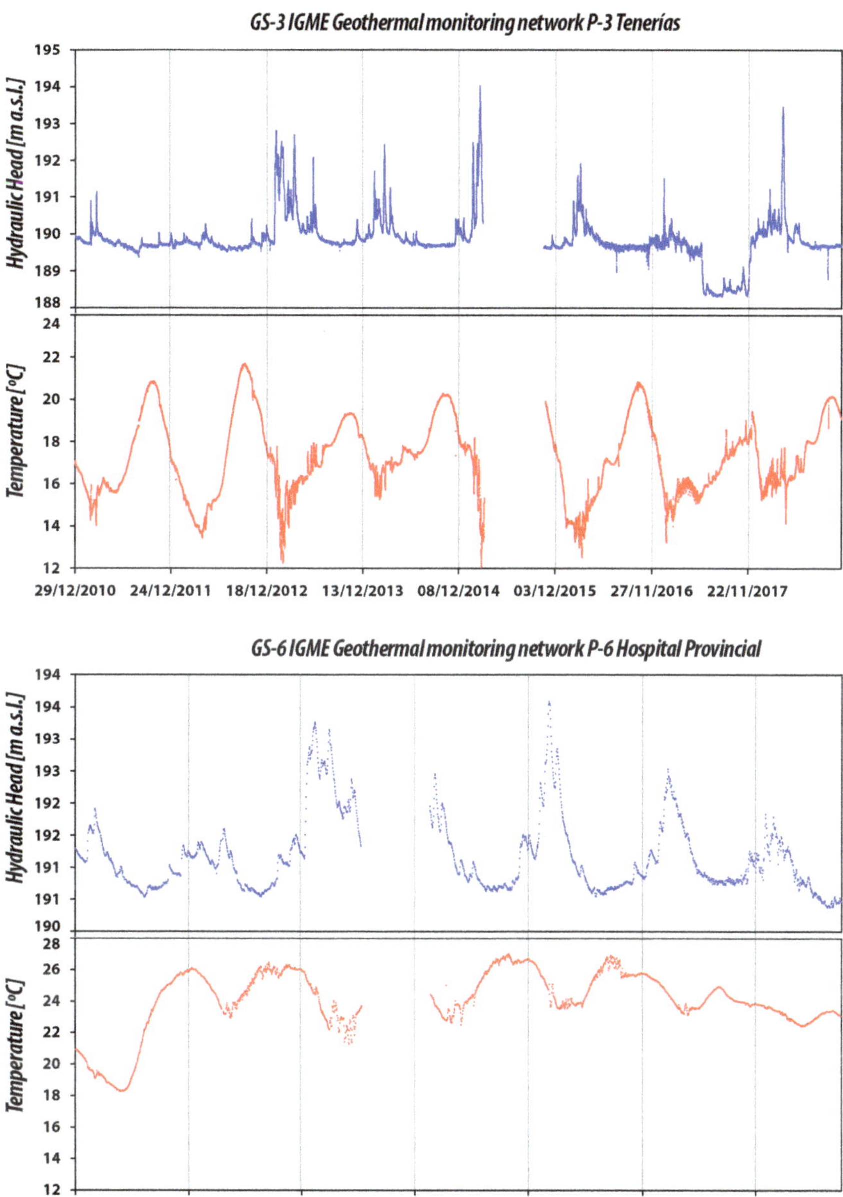

Fig. 11.5 Evolution of the piezometric level in 2 piezometers of the IGME geothermal control network, piezometers GS-3 Tenerías and GS-6 Hospital Provincial

The chemical composition of the groundwater has been exhaustively investigated on the basis of the research conducted since 2000 (Garrido et al. 2003; IGME-CHE 2014, 2019; IGME-DGA 2005). The composition differs considerably from one bank of the Ebro to the other as a result of the different origins of the waters that enter the aquifer, either from the Gállego or the Ebro basin, but also because of the tertiary formations that border the aquifer and through which the surface water circulates before infiltrating, dissolving large quantities of salts that end up being incorporated into the groundwater flow showing an increasing salinity.

Thus, the average electrical conductivity of groundwater on the right bank is around 1478 μS cm^{-1} at 20 °C, compared to an average value of 2656 μS cm^{-1} on the left bank, notably more saline. Both scenarios share generally mixed hydrochemical facies, sulphate-chloride and bicarbonate/calcium-sodium and magnesium, with contents between 30 and 50% for sulphates, 20–40% for chlorides, 40–50% for calcium ion and 30–50% in sodium.

The hydrogeological parameters obtained experimentally allow us to appreciate a very high transmissivity, with mean values around 2520 m^2 day$^{-1.}$ Transmissivity shows a heterogeneous spatial distribution in the urban aquifer within the range of 283 to 4121 m^2 day$^{-1.}$ Hydraulic conductivity is similarly high, with a mean value of 218 m day^{-1}, generally with values between 20 and 348 m day$^{-1.}$

About 300 boreholes are counted as active in the consolidated urban space of the city, and almost two thirds of them are directly related to geothermal installations. The last two decades have shown that there has been a remarkable transformation of historical groundwater use in Zaragoza (Garrido et al. 2012b). The opportunity offered by the aquifer to satisfy the high flows demanded by the new geothermal installations has led to a significant increase in demand for energy purposes, to the detriment of other uses traditionally attributed in the city, such as industrial, in a broad sense or, to a lesser extent, irrigation. However, of the significant volume of groundwater demanded by the city, which amounts to around 25 hm^3 per year for all uses, approximately 67% (16.6 hm^3) is for energy purposes, for HVAC of buildings and industrial *cooling* generation by heat pumps, the rest being divided between garden irrigation, recreational and other industrial uses.

It should be noted, however, that the net consumption of groundwater is comparatively very low, amounting to just over 8 hm^3 per year, i.e., one third of urban demand. Water abstraction for HVAC consumes barely one cubic hectometre per year, compared to the 3.3 hm^3 consumed annually for irrigation of parks and gardens. Lower consumption uses such as recreational and industrial uses reach 2.4 and 1.9 hm^3 year^{-1} respectively, representing 27.6% and 22.2% of total urban consumption, respectively.

11.3 Characterisation of Geothermal Exploitation

Until the end of the 1990s, in the city of Zaragoza, the greatest demand for groundwater in the city was destined to favour the supply for industrial uses in industrial sectors located in peri-urban areas and small facilities inside the metropolitan area. Growth experienced in recent decades, relocation of industry to the outskirts and urban redevelopment of the city centre have been key elements in the transformation of groundwater demand. Garden irrigation and HVAC of buildings by shallow geothermal systems are now the two uses with the highest demand in the city. The following distribution of urban groundwater use was assessed in 2010 (Garrido et al. 2012a), most probably still valid today: more than 68% of the extractions from the aquifer (16.4 hm^3 year^{-1}) are attributed to air geothermal wells, 14% (3.34 hm^3 year^{-1}) corresponding with pumping for irrigation, another 10% (2.4 hm^3 year^{-1}) is accounted for the sum of different recreational uses, while the industrial uses of the city as a whole was barely 2 hm^3 year^{-1}, i.e., 8% of the total. These groundwater extraction quantities contrast with the net consumption attributed to the different uses (Fig. 11.6). Open-loop geothermal systems account for only 11% of urban groundwater consumption (barely 1 hm^3 year^{-1}); thermal discharge water is commonly injected into the aquifer of origin and, in some cases, discharge into the sewerage network.

Recent actualization of the existing GWHP systems inventory bring the number of active installations in the city to 61. Adding other installations already decommissioned, the total number would exceed 70 known installations. Of the active installations, at least 11 do not return extracted groundwater to the aquifer but discharge it into the municipal sewerage system.

The GWHP systems investigated sowed a cooling-oriented design; however, at least 40% of the installations have a reversible heat pump, which gives them the additional capacity of working under a heating mode. Only one system is exclusively adapted for heating, and that is a thermal installation serving a sports complex with swimming pools. Heat pumps work in a regime that depends on the HVAC needs of each building, mainly marked by the activity that takes place inside the building and the availability of other *cooling-heating* generation systems powered by conventional energy sources (García-Gil et al. 2020a). It is common to find a biased cooling demand (and, consequently, the discharge of *hot* water into the aquifer) and concentrated in the summer season, with an average daily operating frequency of 6–8 h, except on public holidays when the facilities are closed. In certain installations, this schedule is extended and geothermal heat pumps operate throughout the whole year, providing

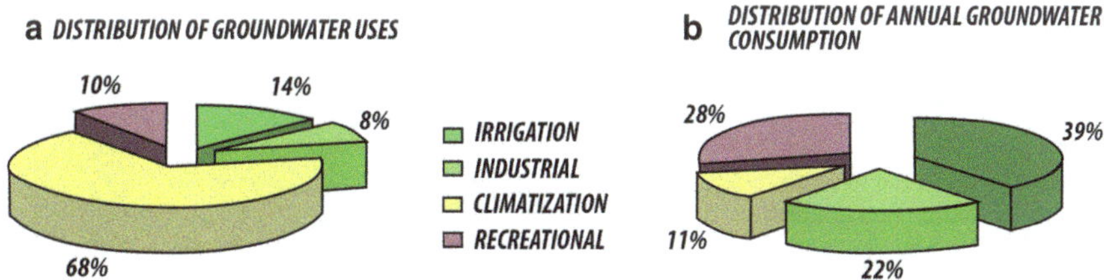

Fig. 11.6 Percentage distribution of groundwater exploitation and consumption in the city of Zaragoza according to its uses

continuous *cooling*, such as in theatres, TV studios, machine or computer rooms, and industrial *cooling* among others.

The heating demand is spread over the period from October to April, during which time thermal discharges can slightly lower the aquifer temperature. Even in winter, some geothermal systems may need to meet some of the cooling demand of buildings. Thus, heat pumps can simultaneously generate *heat* and *cold*, i.e., thermal recycling.

The total number of geothermal wells (open-loop geothermal heat exchangers) for thermal installations equipped with geothermal heat pumps is 202. Their location, nature and current status can be seen in Fig. 11.7. Almost 120 are groundwater captation wells, between 5 and 65 m deep, with an average depth of 33 m. The deepest geothermal wells are located at the highest coordinates of the city, on the T4 and T5 alluvial terraces. Each GWHP system presents one or two captation wells and may have up to four wells which operate alternately or reserving one in case maintenance work has to be carried out or if one of them is forced to stop. The number of injection wells is 82, and they present an average depth of 33 m. It is common for the GWHP systems to have a single injection well for returning the water. Nevertheless, GWHP systems may present three or four wells to distribute the discharge flow.

In addition to the number of active wells or installations in the city, the magnitude of the geothermal resources mobilised has been calculated from the power of all the heat pumps installed in the city of Zaragoza. Based on 2019 datasets covering the exploitation regimes of 75% of the inventoried installations, it is estimated that

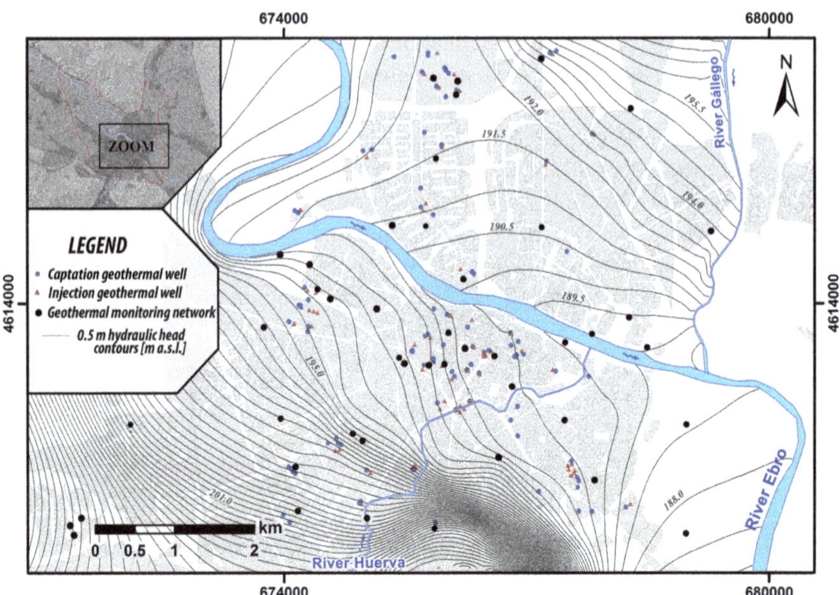

Fig. 11.7 Location of open-loop geothermal heat exchangers of shallow geothermal installations in the city of Zaragoza

the total installed capacity would be approximately 75–80 MW for cooling and 40–45 MW for heating, of which approximately two thirds would correspond to the power used in each installation. The most frequent nominal power of heat pumps is between 150 and 500 kW, but there are heat pumps of up to 1400 kW. Taking into account that geothermal systems usually have several heat pumps working simultaneously (Fig. 11.8), it can be deduced that the average value of the installed power is approximately 1700 kW, with extreme cases in which the installed power reaches high figures, such as the 8.5 MW in this case corresponding to a hospital (*Hospital Clínico Universitario Lozano Blesa*).

The groundwater demand of each system depends to a large extent on the energy needs of each building for air conditioning, as well as on the temperature of the water in the captation wells. Commonly, this demand is between 4 and 20 L s^{-1} of equivalent average flow rate. In some of the largest GWHP systems it can be well over 60 L s^{-1} which, in terms of annual volume, is over one cubic hectometre.

In a natural thermal regime, groundwater in Zaragoza has a reference temperature of 17 °C, but it is not uncommon to measure slightly higher values in different locations in the city, up to 2–3 °C more, which could be related to greater or lesser development of urbanisation in the surrounding area, i.e., the SUHI effect. Any deviation from these values should be understood in terms of thermal contamination of the aquifer, probably originating from thermal discharge from other nearby geothermal developments or self-interference. In both cases, the arrival of the thermal front can,

Fig. 11.8 Geothermal heat pumps in the old seminary building, Zaragoza City Hall

in extreme situations, compromise the thermal use of groundwater resources, the performance of the heat pumps and, ultimately, the technical sustainability of the geothermal systems.

In order to understand the temperature changes between captation and injection wells in the city of Zaragoza, during a first inspection in 2012 it was found that the geothermal systems applied average temperature jumps of 8.1 K which, in extreme situations, exceeded 25 K. This, in fact, meant finding discharges with up to 43.3 °C. Subsequently, taking as a reference a dozen GWHP systems monitored, the average temperature change observed was 4.1 K, reaching maximums of 20.3 K, which implies having discharges with temperatures of up to 36.2 °C.

In order to obtain more precise information, it is necessary to regulate and standardise the collection of exploitation data for subsequent study. To this end, the CHE, as the regional water authority, requests collaboration of the users of the GWHP systems so they can equip their installations with measuring and control equipment. The response has been positive and, thanks to this, around 45 facilities have installed flowmeters and temperature sensors which will contribute to the improvement of the systems management. The devices log 15-minutal measurements, 24 h a day and datasets are periodically sent to the CHE for verification.

The volume of information generated in the period from 2015 to 2017 was 1.56 million records. Therefore, applied upscaling methods provide daily equivalent data series containing essential information on the operating systems. In the case of flowrates, 15-minutal datasets measured in the city of Zaragoza are treated to provide daily or monthly abstraction volumes. When it comes to captation/injection temperatures, the 15-minutal datasets usually include very disparate extreme values treated as outliers and related mostly to room temperatures when the heat pump is not in operation. A method of integrating the daily transferred energy to calculate the daily equivalent discharge temperature from 15-minutal datasets is used (Muela Maya et al. 2018). The *equivalent temperature* (T_e) is defined as the integrated temperature value for a given time period (in this case daily), deduced from the energy transferred to the aquifer with a 15-minutal cadence:

$$T_e = \frac{\sum_0^{1\,dia} P_{15}}{Qc_w\rho_w} + \overline{T}_{ca} \tag{11.1}$$

being: $P_{15} = Q\ C_w\rho_w\Delta T$ [J s^{-1}] the thermal power transferred between each 15-minutal reading; Q [m^3 s^{-1}] the flow rate of water pumped between two 15-min readings; c_w [J kg^{-1} K^{-1}] the specific heat capacity of water; ρ_w [kg m^{-3}] the density of water;$\Delta T = (T_2 - T_1)$ [K] is the temperature change between two 15-min readings, where T_2 is the 15-min discharge temperature, and T_1 the 15-minutal captation; and $\overline{T}_{ca}$ [K] the daily arithmetic mean captation temperature.

As an example, Fig. 11.9 shows a time series of 15-minutal discharge temperature data for a hotel facility, representing the actual operating regime. The period of record is 273 days, in continuous operation, with an average flow rate of 661 m^3 day^{-1}. It can be seen that the temperature distribution varies according to the climate control

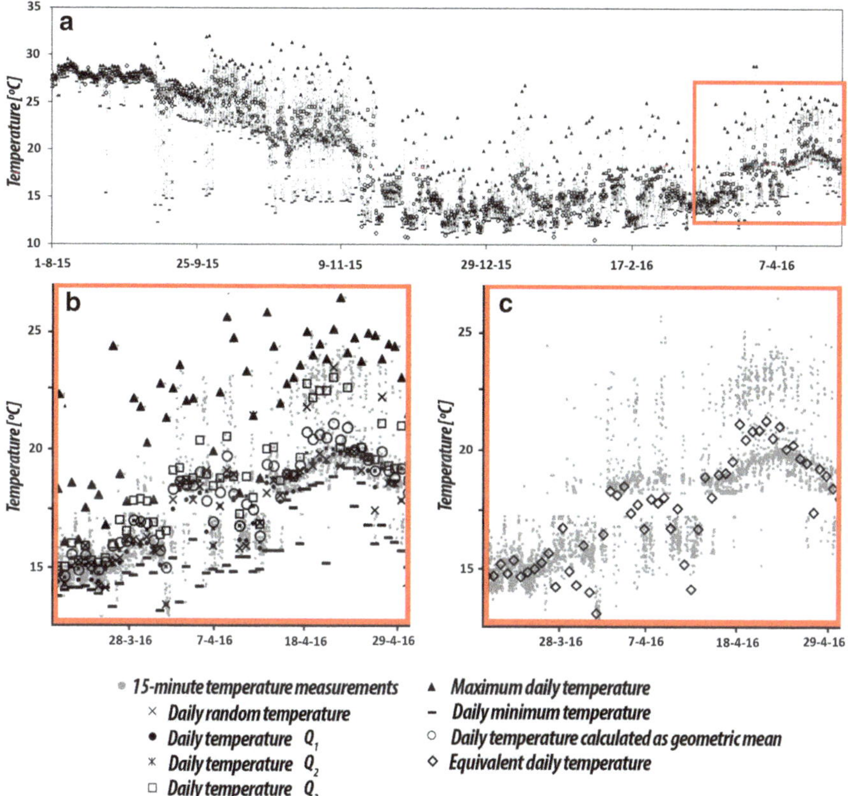

Fig. 11.9 **a** Graphical representation of the time series of the 15-minutal discharge temperature of the shallow geothermal system of a hotel installation in Zaragoza. **b** Detail of the distribution of the parameters calculated by daily cadence upscaling: interquartiles Q1, Q2 and Q3; maximum and minimum temperature; geometric mean of the temperature; and a daily random data. **c** Detail with calculation of the daily power equivalent temperature T_e

required at each moment, with the exception of small intervals of time without records in which the discharge temperature is adopted to be the same as the bottom temperature of the aquifer not thermally affected (17 °C).

The temperature series calculated by upscaling (geometric mean, interquartiles Q1, Q2, Q3; maximum, minimum and a random value of the series), have a daily cadence. It can be seen that the distribution of these daily values is close to the real temperature data, so that the maximum and minimum values are at the top and bottom of the graph, respectively. These indicators limit the range where the geometric mean and quartiles are shown.

The calculated daily time functions efficiently reproduce the operating regimes of these systems, with an error that is more than acceptable within a framework of sustainable resource management. This data treatment also makes it possible to (1) characterise the operating and shutdown periods of the installations, (2) verify that the flowrates and volumes actually used are in line with those of the administrative concession granted by the water authority, (3) confirm the maximum discharge temperatures, and (4) determine the temperature changes during systems operations.

11.4 The Zaragoza Geothermal Monitoring Network

An absolutely essential activity for the management of shallow geothermal energy resources consists of having direct access points to the subsurface geothermal reservoir, in this case to groundwater of the alluvial aquifer of Zaragoza. The access is obtained by means of observation and monitoring networks composed by different control points strategically distributed throughout the shallow geothermal reservoir. The first objective of this network is to obtain distributed values of the piezometric level and groundwater temperature, both in areas affected by thermal plumes and in areas not affected. A second objective is to have infrastructure for direct access to groundwater in order to collect groundwater samples and check the chemical composition and qualitative state of groundwater.

The location of the monitoring points of the shallow geothermal monitoring network of the city of Zaragoza is shown in Fig. 11.7. The location of the monitoring points is based on the existing infrastructure of wells and boreholes in the city, and the areas of particular interest near shallow geothermal installations. The network consists of 45 points; however, the number and position of the specific points that make up the network at any given time have been modified and adapted from the beginning to the situations and working conditions at each stage. The control points consist of publicly and privately owned boreholes and wells, with the majority belonging to the IGME, the CHE, the City Council or the University of Zaragoza. The IGME contributes 17 specifically constructed piezometers with 616 m of drilling in total. IGME piezometers provide ease of access and flexible constructive design according to the requirements needed. All the newly constructed piezometers fully penetrate into the aquifer, with an 80 mm diameter casing screened along the entire length of the water column inside. This allows the introduction of a submersible pump inside the network, thus facilitating control operations, sampling and enhancing the representativeness of the data obtained for all the piezometers.

The network has been equipped with autonomous devices that measure and store data on temperature, groundwater level and, in specific cases, electrical conductivity. A total of 37 datalogger-type devices are in operation and periodically require maintenance operations and retrieving of the stored data. Subsequently, collected data require filtering and processing to compensate the water pressure values with the atmospheric pressure values. The latter are measured and recorded in parallel with specific barometric devices.

The monitoring network has made it possible to observe oscillatory variation of groundwater temperature in the city with different amplitudes; from 0.3 °C in unaffected areas to almost 12 °C in areas affected by thermal plumes generated from shallow geothermal systems. This range of variation is still much lower than the thermal amplitude of more than 23 °C observed in the water of the Ebro river (Table 11.1). The river Ebro water temperature also shows a seasonal oscillation, as shown in Fig. 11.10. Piezometers next to the river tend to present a phase shift with respect to the river water temperatures, between 1 and 5 months depending on the distance of the piezometer to the river bed (García-Gil et al. 2014a).

Collected data from the geothermal monitoring network show that in the city of Zaragoza the aquifer temperature values range between 13.0 and 29.0 ºC, depending on the location of the monitoring point and proximity to the heat plumes generated by GWHP systems. Outside the influence of the geothermal installations, groundwater temperatures not affected by anthropogenic activities are found to be close to 17 ± 0.2 °C.

The shallow geothermal monitoring network also plays an important role in the execution of various studies for hydrochemical characterisation and quality control of groundwater in the urban aquifer. More than 400 groundwater samples have been

Table. 11.1 Maximum and minimum temperatures observed in the IGME piezometers of the shallow geothermal monitoring network (period 2010–2018) and of the Ebro river

Piezometer	Minimum T (°C)	Maximum T (°C)	Thermal amplitude (°C)	Seasonal minimum	Seasonal maximum
GS-1 Actur	16.09	22.93	6.84	December	May
GS-2 Tenerías	12.04	21.68	9.64	September	April
GS-3 Geology faculty	23.19	26.10	2.91	April	October
GS-4 José Sinués	18.95	26.95	8.00	January	August
GS-5 Albareda	16.22	29.00	12.78	February	September
GS-6 H. Provincial	18.32	27.10	8.78	May	October
GS-7 Aragonia	12.99	23.49	10.50	December	April
GS-8 Seminario	16.19	19.49	3.30	August	February
GS-9 Clavé	17.08	17.67	0.59	September	February
GS-10 Schindler	19.79	23.85	4.06	August	February
GS-11 Helios	17.73	19.74	2.01	July	January
GS-12 Boggiero	16.56	17.36	0.80	August	February
GS-13 Pº de La Mina	19.85	20.31	0.46	July	November
GS-14 Rosalía Castro	18.82	20.27	1.45	August	January
GS-15 Antonio Saura	17.60	17.75	0.15	October	April
GS-16 Bosco	26.16	28.20	2.04	July	November
River Ebro	3.60	27.40	23.80	July	January

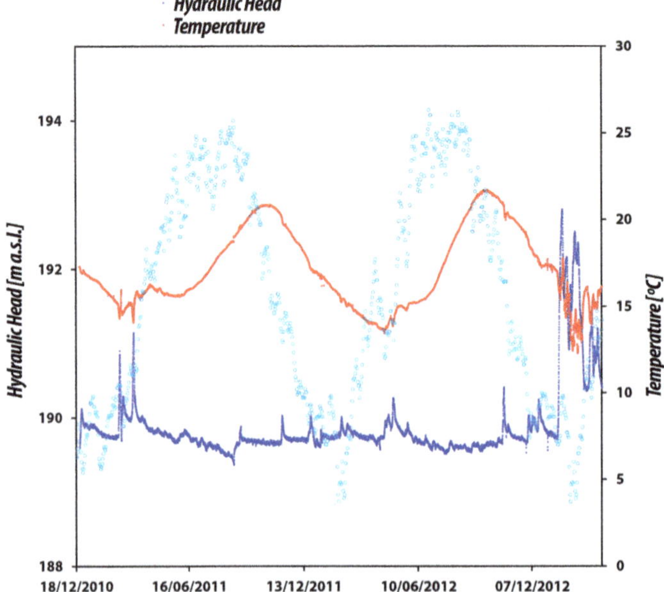

Fig. 11.10 Groundwater temperature (red) and piezometric level (dark blue) in the GS-3 Tenerías piezometer compared with the evolution of the river Ebro water temperature (light blue). River Ebro water temperature obtained from gauging station 909 in Zaragoza

taken so far, distributed over 12 field campaigns between 2011 and 2012, plus a last campaign in October-November 2016 that incorporates monitoring in the latest piezometers built by IGME.

Monitoring points are also a means of obtaining other geological or hydrogeological information useful in the process of modelling groundwater flow and heat transport or in improving knowledge about the operation of shallow geothermal installations. Thus, data can be obtained on hydraulic parameters (hydraulic conductivity, transmissivity, hydraulic gradient), thermal parameters (groundwater and unsaturated zone temperature, thermal conductivity), or the response of the aquifer to the injection of thermal discharges can be monitored. Significant examples of how the groundwater background temperature is affected and what the thermal response is as the exploitation regime or discharge conditions vary can be described from the time evolution of temperature records obtained from piezometers. This is the case of piezometers GS-5 (Fig. 11.11) and GS-9 (Fig. 11.12), located respectively next to the shallow geothermal installations of the *Hospital Provincial* and *Aragonia Shopping Centre.*

In the first case, Garrido et al. (2017) analyse the interference of the thermal discharge of a shallow geothermal installation A1 on the catchment wells C1 and C2 of the shallow geothermal installation A2 (Provincial Hospital), located 160 m downstream of the former installation in the same flow line. From the background temperature of the aquifer obtained in the piezometer PZ-9 and the thermal control

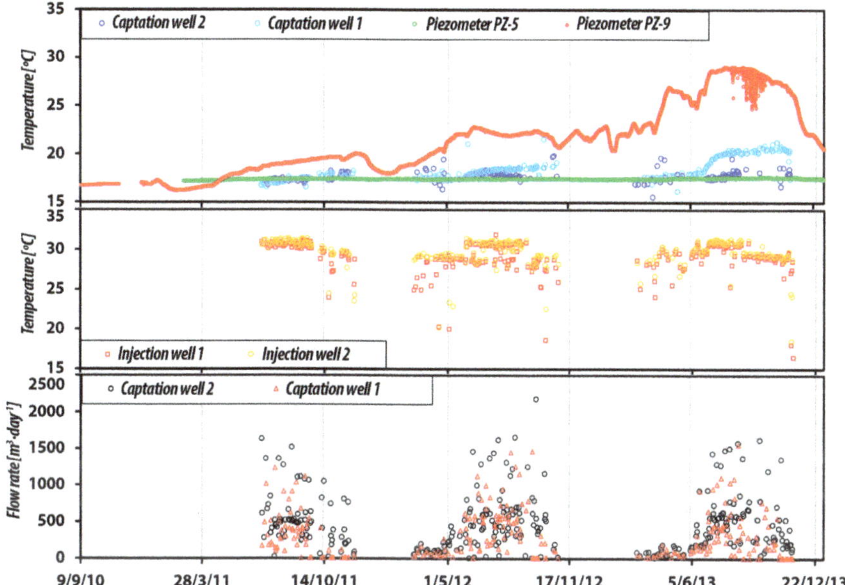

Fig. 11.11 Upper: evolution of the temperatures of captation wells C-1 and C-2 of shallow geothermal installation A2 compared to those of piezometers GS-5 (PZ-5) and GS-9 (PZ-9). Middle: evolution of the discharge temperatures of the shallow geothermal installation A2. Bottom: evolution of the operating flow rates in boreholes C-1 and C-2 of the shallow geothermal installation A2

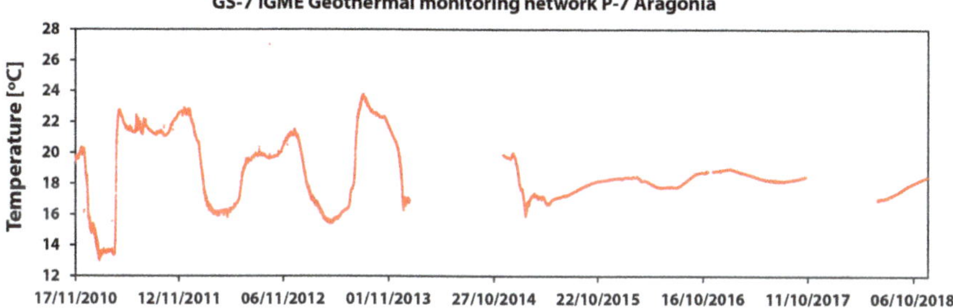

Fig. 11.12 Temporal oscillation of groundwater temperature at piezometer P-7 of the Zaragoza Geothermal Monitoring Network

of piezometer PZ-5 (380 m away from PZ-9), 90 m from the discharge of A1 towards A2 installation, the thermal interference process between A1 and A2 can be inferred.

In other patterns, such as the one followed by piezometer P-7 (Fig. 11.12), the interpretation resulting from the thermal records is almost immediate. In this case, the piezometer is affected by an amplitude of temperature of approximately 10 K, a high value if it is taken into account that the background temperature of the aquifer in that sector oscillates only a few tenths of a degree. In this example, the piezometer is located in a down gradient of an injection point of thermal discharge generated by the *Aragonia Shopping Centre* GWHP system. The thermogram also registers how the GWHP system in 2015 changed its system operation regimes due to the construction of a new injection geothermal well away from the piezometer P-7. After this major change in GWHP system operations, the temperature of the groundwater regularises, so that it once again reaches the background values for this sector of the aquifer.

11.5 Impact of Thermal Discharges from Shallow Geothermal Installations on the Aquifer

The operation of shallow geothermal installations with open-loop geothermal heat exchangers can be understood by the operators of shallow geothermal resources, and by society as a whole, as an environmentally friendly activity. It is generally assumed that the groundwater extracted is returned to the same aquifer with thermal modification that does not have any significant side effects. This is generally true; however, the groundwater circulating inside a plate heat exchanger of a shallow geothermal installation absorbs or gives up a quantity of heat that will subsequently be transferred to the aquifer by means of injection geothermal wells. Therefore, there is heated or cooled water flow with the capacity to generate more or less significant impacts of different nature into the aquifer and the environment.

The most important impacts that appear in the city of Zaragoza as a consequence of a thermal spill can be grouped into three categories:

11.5.1 Thermal Impact

In the urban alluvial aquifer of Zaragoza, the thermal impacts generated by GWHP systems are detected on the basis of groundwater temperature data obtained from the established shallow geothermal monitoring network and from the geothermal installation captation/injection wells. This is interpreted with the support of numerical models of groundwater flow and heat transport. The impacts are identified by the occurrence of two morphologies, depending on the extent of the thermally affected waterbody: thermal plumes and SUHI. Thermal plumes in Zaragoza occur to a greater or lesser extent in the vicinity of all the GWHP systems in Zaragoza (Fig. 11.13), but their extension is greater in the plumes linked to large hospital systems and shopping centres in the southwest and southeast of the city. Heat islands occur due to the coalescence of plumes and are observed in areas where there is a high concentration of small closely spaced systems, as appears to be the case in the historic quarter in Zaragoza city centre area.

Thermal impacts most frequently detected are self-interference, related to the alteration of the captation water temperature registered in the captation wells of

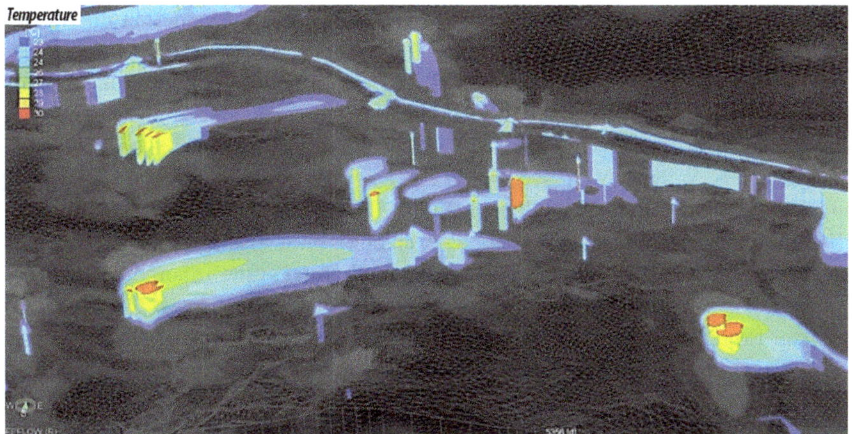

Fig. 11.13 Simulation results obtained with a heat transport model in the city of Zaragoza. Three-dimensional oblique image showing the discharge plumes generated by several shallow geothermal systems (IGME-CHE 2019)

GWHP systems sometime after their start-up. These self-interference impacts are important because they particularly affect the performance of heat pumps which, in extreme situations, reduce their performance significantly. The process of thermal self-interference or recirculation occurs when the thermal discharge feeds back into the captation well of the same geothermal installation. Another type of thermal interference occurs when the process of thermal interference is produced between different geothermal installations, in which the discharge from one installation is captured by a captation well of a different installation located downstream and in the same direction of the groundwater regional flow. The thermal interference between systems often goes unnoticed by the users, as it is detected at a lower intensity than the recirculation process and is likely to be found stationary (e.g., 2 K constant increase with respect to the background temperature of the aquifer throughout the year). It is masked if both processes coexist or is interpreted as a recirculation effect, even when the circumstances for this type of impact do not exist.

In order to minimise these risks and reduce the thermal impacts on the city of Zaragoza, it has been decided to control the groundwater exploitation rates and the injection-captation temperature change with which the heat pumps work. In addition, it is considered important to have precise knowledge of the hydrogeological functioning of the aquifer. Knowing the variables that govern the hydraulic and thermal regime of the aquifer is of great interest in order to properly manage the resources and design of the GWHP systems. Regional groundwater flow is considered a very important factor for the management of shallow geothermal resources, as it influences the arrival time of the heat front of a heat plumes to captation wells of other systems. In addition, the displacement of groundwater also takes into account the solid matrix of the aquifer as it must accommodate the new thermal state, which heats up much more slowly. All in all, it is recommended to consider heat advection mechanisms due to groundwater flow in addition to heat conduction of the solid matrix of the ground.

The first research conducted by IGME-CHE (2014), under conceptual assumptions and ideal aquifer functioning, offers a first assessment of the risk of self-interference and the arrival time of the thermal front in most of the geothermal well doublets of the GWHP systems in Zaragoza. With these assumptions, slightly more than 50% of the well doublets would be affected by recirculation of their own thermal discharge in less than 90 days. A calculation is also made of the minimum separation that the doublet of wells in each geothermal system should have in order to avoid the risk of recirculation. In addition, the impact of indirect thermal energy transfer from discharges on the unsaturated zone and the basement has been considered. This impact can be deduced from numerical heat transport simulation models built at the city scale (García-Gil et al. 2015d; IGME-CHE 2014, 2019; Rivas et al. 2013).

11.5.2 Chemical Impact

A second group of impacts generated by thermal discharges is identified through anomalous rises or drop downs in geothermal wells, excessive rises and even, in extreme cases, overflows in injection wells and excessive drop downs in captation wells. In other situations, it causes the collapse and subsidence of the catchment area, the appearance of gelatinous precipitates formed by microorganisms, carbonate precipitate scaling, and even corrosion of pipelines. Some real examples of these situations in shallow geothermal installations in Zaragoza are shown in the images in Fig. 11.14. All of these are simply manifestations of the processes of dissolution-precipitation or bacterial growth, which modify the concentrations of the physicochemical parameters and dissolved ions in the groundwater and which are triggered by alteration of the water temperature values or lack of isolation with respect to the atmospheric conditions of the groundwater pumped.

Geothermal installations with open-loop geothermal heat exchangers affect the physicochemical parameters of groundwater in very different ways. Theoretically, the increase in temperature increases the kinetics of reactions, shifts the equilibria of hydrolysis reactions or dissolution of solids towards the development of dissolved phases, decreases the solubility of gases such as carbon dioxide or oxygen and enhances the metabolic activity of micro-organisms involved in oxidation-reduction reactions.

As a result, the values of certain aquifer properties are altered, mainly porosity, permeability and transmissivity, due to the appearance of chemical precipitates, mainly calcium carbonate, which occlude pores, voids and filtering zones of well screens and pipes, reducing the permeability of the medium and, ultimately, the production capacity of the wells. In these situations, the pressure exerted by the water column in the wells forces a flow at the base of the aquifer that accelerates the dissolution of the aquifer basement made of evaporites, the generation of voids,

Fig. 11.14 **a** Carbonate precipitates in a submersible pump. **b** Ground subsidence with the collapse around an injection geothermal well. (C) Overflow in an injection geothermal well due to occlusion of the porous media and/or well screen in the intake zone

the loss of support and, finally, the collapse and sinking of the well. Figure 11.15 summarises schematically this sequence of processes induced by thermal discharges, corresponding to the examples in Fig. 11.14.

The thermal induced chemical impacts on the Zaragoza aquifer have been the subject of various experimental studies through the chemical analysis of groundwater samples and of the installations themselves, taken from the shallow geothermal control network. The evolution of physicochemical parameters has been monitored and the mineral speciation, dissolution or precipitation of gypsum, calcite and halite as characteristic minerals of the geological formations in the immediate vicinity of the Zaragoza aquifer have been investigated. Such studies have been carried out at three different scales of work: in captation-injection well doublets, thermally affected versus unaffected wells and also along regional groundwater flow lines.

Observing the behaviour of the physicochemical parameters in the well doublets (Garrido et al. 2016b; IGME-CHE 2014), it is found that the operation of these

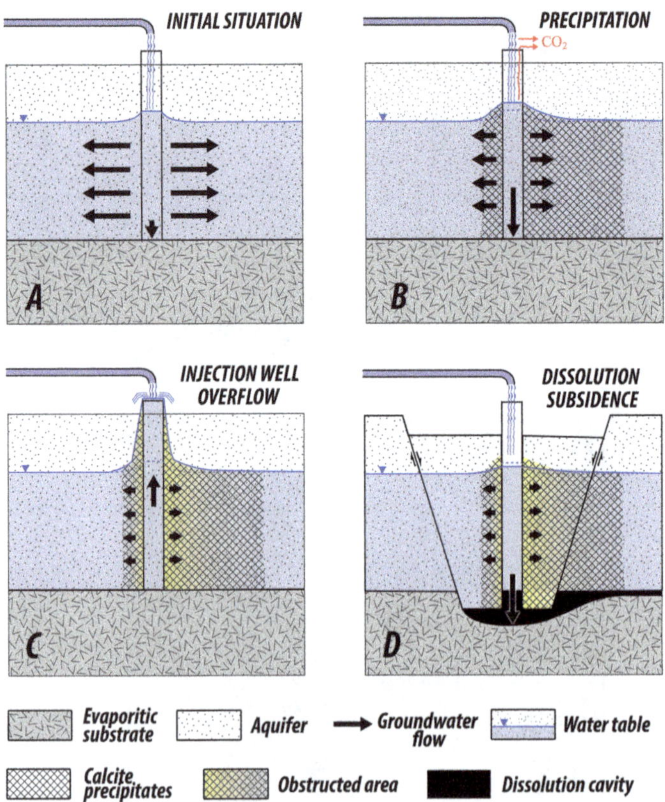

Fig. 11.15 Schematic conceptual model of the processes induced by a thermal discharge in the Zaragoza aquifer (García-Gil et al. 2016a). **a** Ideal discharge conditions in an initial stage. **b** Calcite precipitation in pores around the injection well due to dissolved CO_2 exolution. **c** Total clogging of pores and well screen, loss of intake capacity of the return well. **d** Dissolution of saline substrate due to downwards forced circulation of injected water

geothermal installations hardly modifies the pH of the discharge water, that the electrical conductivity remains stable and that the concentration of dissolved oxygen increases. This increase would be a consequence of the bubbling and aeration to which the water is subjected during the free-fall injection through unpressurised pipelines and wells.

Along the same groundwater flow line, what is observed is a progressive increase in oxygenation that occurs after the flow passes through different geothermal developments. Similar behaviour along the flow line is deduced for pH and alkalinity, while pH remains stable or increases slightly, and alkalinity decreases as the flow progresses, showing a decrease in the concentration of dissolved bicarbonate ions in the groundwater.

The study of geochemical speciation describes the mineral dissolution-precipitation processes currently occurring naturally in the aquifer and those resulting from the operation of geothermal systems. In the well doublets, there is naturally occurring dissolution of halite and more actively of gypsum, together with the equilibrium or saturated state of calcite. Among the effects of geothermal exploitation, there is some uncertainty about how it affects the stability of gypsum, but it is confirmed by the continuity and increase of the calcite precipitation process. When the groundwater flow passes through different thermal discharges of geothermal systems, the same processes are observed, perhaps detecting an increase in the subsaturation state of the gypsum in the water. In any case, the aforementioned processes are not unrelated to the existing uncertainty that mixtures with water from different sources may end up altering the composition of the water and, therefore, the final result of mineral speciation.

The hydrogeochemical studies conducted simulated the evolution of the saturation state of calcite and gypsum in the event of an increase in water temperature of a similar magnitude to that produced by thermal discharge. In this sense, the research would corroborate that there would be increasing states of calcite saturation, which could generate more precipitates, and it is more likely the greater the applied temperature increase with the heat pumps. As for gypsum, the high degree of undersaturation in the water would be maintained and, therefore, the dissolution capacity for this mineral species.

11.5.3 Microbiological Impact

In a first diagnosis of the open-loop geothermal heat exchangers in the city of Zaragoza, no gelatinous growths or significant damage were observed as a consequence of the growth of bacterial communities. With regard to the presence of microorganisms that are pathogenic for human health, and in the absence of specific background and studies, IGME-CHE (2019) analysed 12 parameters and organisms in 31 groundwater samples from different wells and piezometers in the city. Most of the samples provided higher or lower counts for some or several of the following microorganisms and microbiological parameters: total aerobic bacterial counts at 36 °C and at 22 °C, coliform bacteria, faecal streptococci, *Escherichia coli, Clostridium perfrigens, Staphylococcus aureus, Pseudomonas aeruginosa*, Salmonella, Legionella and free-living amoebae. Looking at the analytical results and their distribution in the aquifer, the organism count decreases significantly after the groundwater passes through the heat exchange loops and on injection into the discharge wells. Once the water is incorporated into the groundwater flow and the injected heat is dissipated by the aquifer, the count of microorganisms also undergoes an exponential recovery trend with respect to the initial values. In summary, the hypothesis that García-Gil et al. (2018b) propose in this research is that the plate heat exchangers of GWHP systems could act as pasteurisation devices, where groundwater is subjected to a sudden increase in temperature that affects and reduces the survival of microorganisms in the discharge water and, to some extent, in the thermal

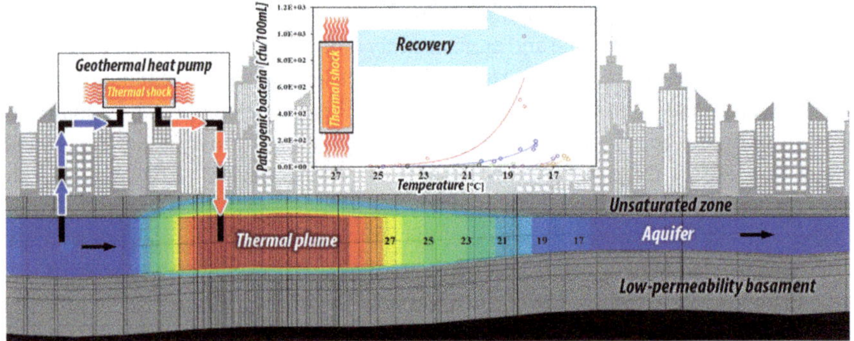

Fig. 11.16 Schematic diagram explaining the pseudopasteurisation process involving the passage of groundwater through plate heat exchanger of groundwater heat pump systems, which would affect the microbiological content of the groundwater. A decrease in the count of organisms would have been detected from the injection of the discharge water and a subsequent recovery with the advance of the thermal plume in the direction of groundwater flow (García-Gil et al. 2018b)

plumes (Fig. 11.16). To definitively confirm such a hypothesis would require additional effort, a thorough and detailed exploration of the processes and the subsurface environment, which would open the door to a promising field of research of interest for human health.

11.6 The 3D Numerical Model of Groundwater Flow and Heat Transport

In hydrogeology, the purpose of numerical modelling in the studies of this branch of Earth Sciences is to express a conceptual model of the functioning of an aquifer system in terms that are consistent with the main laws of fluid physics, thermodynamics, chemistry, etc. The best tool to achieve physical consistency between this conceptual model and reality is, therefore, the execution of numerical models. These models allow prediction of the state variables describing heat flow and transport processes of a given aquifer.

With all these studies, it has been proven that numerical modelling of the Zaragoza urban aquifer, together with high resolution monitoring of groundwater conditions, has been able to respond to different conceptual uncertainties and hydrogeological questions. It has enabled all the information on variables involved in the conceptual model of the aquifer (geometry, hydraulic parameters, potentials, boundary conditions, etc.) to be handled jointly, and it has served as a strategy for detecting information gaps that might require greater observation or research effort. In addition, the hydrogeological regime of the urban aquifer has been integrated with that of heat transport, whose behaviour depends on thermodynamic parameters and intrinsic properties of water and the subsurface media. Obtaining the value of these parameters experimentally requires a certain amount of effort, specialised techniques and investment, but with mathematical models a very approximate and contrasted calculation value is obtained when those are calibrated and validated.

Tackling numerical modelling of high predictive quality in the urban area of Zaragoza has proved viable and effective thanks to the impetus given by IGME and CHE since 2009. The quality of the data obtained from the shallow geothermal network and the monitoring of the exploitation regimes of the shallow geothermal installations in operation has made it possible to achieve a robust model by IGME-CHE in 2019. Other preliminary models had to be tentatively addressed, first at the local scale (Garrido et al. 2012a), using two-dimensional codes such as VS2DHI (Lappala et al. 1987) or TRANSIN-IV (Medina and Carrera 1996) by solving the general flow equation using the finite difference and finite element numerical methods, respectively. Subsequently, other models attempted to reproduce heat transport between various geothermal systems with the support of a commercial code FEFLOW 6.0 (Diersch 2013), also using finite elements, which is capable of discretising the space in high detail in three dimensions (García-Gil et al. 2015d; Rivas et al. 2014, 2013).

The numerical model of underground flow and heat transport developed by (IGME-CHE 2019), called GEOTERZ, is a complex model primarily designed to support the management of shallow geothermal energy resources in Zaragoza. Hundreds of hours of computation have been invested in its calibration. Through successive iterations, the aim is to improve the fit of these simulations and calibrate the result with the physical reality of the hydrogeological and thermal parameters.

The GEOTERZ model has been built with the FEFLOW 6.2 code, which solves the flow and heat transport equations using the finite element method, in this case by means of triangular prisms. The mesh, as seen in Fig. 11.17, consists of 13 layers of varying thickness according to the three-dimensional geometry of the urban aquifer, containing 34.6×10^6 elements and 18.6×10^6 calculation nodes. The size of the elements in distal areas is 100 m, 5 m in areas of higher hydraulic gradient, in the vicinity of the Ebro river and the exploitation wells. The maximum resolution of elements is 0.2 m. In order to reproduce the vertical heat transfer to the surface and to the basement, four layers have been assigned to the unsaturated zone, four to the saturated zone and five to the impermeable basement.

The delimitation of the domain has been based on the hydrodynamics of the regional aquifer of the Ebro alluvial, which includes part of the non-urban environment to avoid border effects in the boundary conditions. The boundary conditions selected are prescribed level (*Dirichlet*) for the south-eastern boundary; prescribed flow (*Neumann*) for the contact zones with impermeable materials (zero flow); and third-type boundary condition (*Cauchy*) or mixed condition for those boundaries where groundwater transfer is estimated due to the lateral continuity of the aquifer or the hydraulic connection to surface water bodies, i.e., in the northern, western and southern boundaries, and in the connection with rivers Ebro, Gállego and Huerva.

Regarding the zonal parameters, the hydraulic conductivity has been treated as spatially distributed, considering eight differentiated zones. The specific storage coefficient has been considered constant and of low value throughout the domain. Most

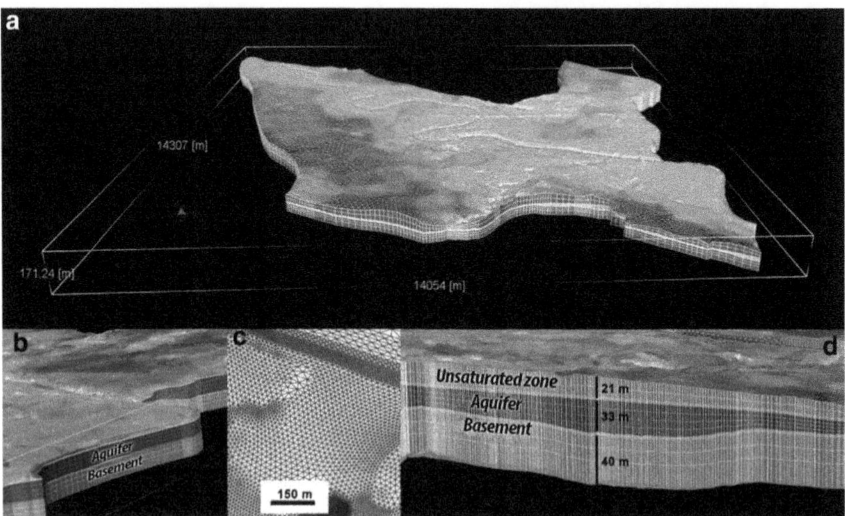

Fig. 11.17 **a** Three-dimensional view of the triangular finite element mesh of the GEOTERZ numerical model. **b** Eastern sector, **c** detail of triangular elements in plan, and **d** detail of the vertical discretisation, differentiating the unsaturated zone, aquifer and basement

of the domain to be modelled is over urbanised areas, which entails a high degree of uncertainty in assigning distributed values to the aquifer recharge. The values finally assigned were 500 mm year^{-1} in urban areas and 20 mm year^{-1} in non-urbanised areas. Since about 70% of the exploitation of the aquifer is destined to shallow geothermal installations, only the exploitation regime of these installations is taken into account for simulation of the transient regime of the aquifer. For this purpose, 122 abstraction and injection points have been considered and implemented in the model by imposing their corresponding daily equivalent flow rates measured as prescribed flow rates for each time in the model.

To solve the heat transport, boundary conditions consistent with the conditions and parameters defined to solve the subsurface flow have been used. For all boundary conditions, lateral borders and bottom surface, a value of 17 °C has been imposed. The surface temperatures in the river-aquifer relations are assigned according to the temperatures measured in the gauging stations controlled by the CHE. In points where geothermal wells are found, transient prescribed time functions are assigned from the equivalent daily operating temperatures of the injection wells.

The result of the calibration shows that the hydraulic conductivity values vary by several orders of magnitude, depending on the zoning of terraces and geological sectors considered in the discretisation of the model, showing a large spatial variability. The drip coefficients are high, indicating good hydraulic connection in the northern boundary, river Ebro (10^3 day^{-1}) and southern boundary (32 day^{-1}); they are low in the river Gállego (9.0×10^{-2} day^{-1}) and very low in the rest ($<1.5 \times 10^{-3}$ day^{-1}), where hydraulic connections are non-existent or minimal. Regarding dispersivities, it should be noted that the calibration results suggest low values in the low and current terraces of the Ebro, in contrast to the high values in the intermediate terrace zone. Porosity has been obtained as a spatially distributed parameter in the aforementioned sectors and as a constant in the rest. The thermal parameters have been considered constant throughout the area, although with different values between the unsaturated zone, aquifer and basement.

The simulation period adopted is from 1 January 2003 to 1 September 2018, which allows a relatively stationary condition to be reached around the year 2008, when rigorous and continuous data on groundwater levels and temperatures in the city are available. The time increment used is 1 day as a constant value throughout the modelled period.

The fit between measured and calculated piezometric levels has an absolute error of less than 0.4 m for 90% of the cases. In general, the hydrographs of temporal variability of levels and their oscillations generally reproduce the data measured at each piezometer and for almost the entire modelled area. The discrepancies are due to simplifications adopted in the city-scale model, such as disregarding the exploitation of irrigation wells, variations in storativity or more precise data from exploitations of geothermal systems. The head contour map obtained at the end of the simulation time shows that the Ebro conditions the general direction of the groundwater flow, although it is complex, with large changes in gradient in areas of lower permeability associated with the high terraces. The use of these models also allows the theoretical residence and transit times of groundwater to be studied (Fig. 11.18). Values are

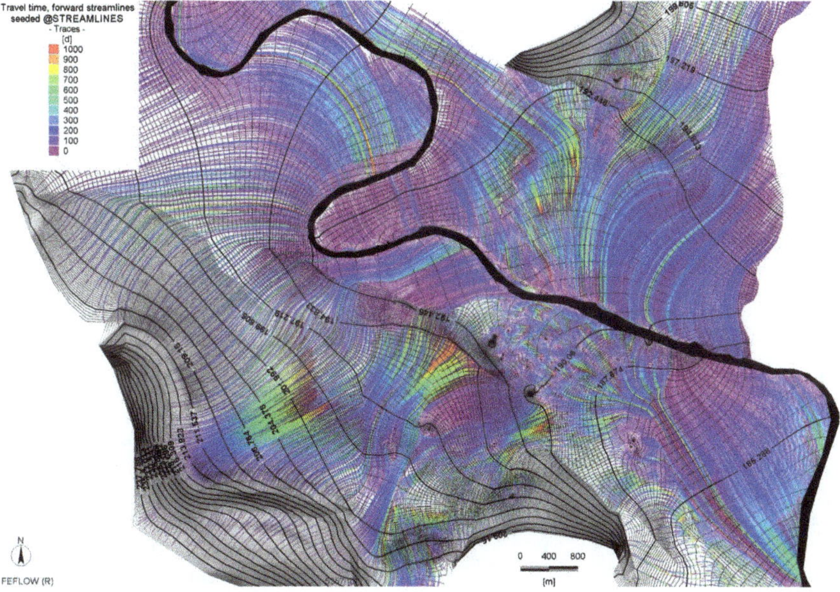

Fig. 11.18 Map of transit times [days], head contour map [m a.s.l.] and calculated groundwater flow lines in the city of Zaragoza

obtained for very permeable sediments, ranging from 300 to 600 days in the eastern zone from the Ebro weir, through 3000 days in the northern contour of the Gállego alluvial, to very slow flows of 20 years from the recharge areas in the high terraces of the Ebro.

The calculated groundwater temperatures are equally satisfactory. The model has reproduced both the general trends and ranges of thermal values at the assigned aquifer observation points, but it is very complex to achieve good fits, since small uncertainties in the conceptual model or initial errors subsequently translate into significant deviations in the results. In addition, factors such as the position of the control points with respect to the location of the discharges, exploitation and injection regimes, self-interference problems in geothermal exploitation systems, or values intrinsic to the aquifer such as the heat transport retard or the value of the aquifer's thermal parameters, significantly condition the results.

Regarding the areas detected with high temperatures, heat plumes associated with most of the GWHP systems in use are observed (Figs. 11.19 and 11.20). The largest plume observed in the underground flow direction is generated by a hospital building (8.5 MW), one of the main systems in the city, whose 20 °C isotherm reaches a distance of more than 1500 m downstream. Other buildings linked to museums, shopping centres and hospitals also generate significant plumes, usually more than 200 m long for 20 °C isotherms and up to 100 m for 25 °C isotherms. Among the thermal impacts reproduced by the model, the one that emerges in the city centre stands out, where the confluence of several plumes generates a subsurface urban heat

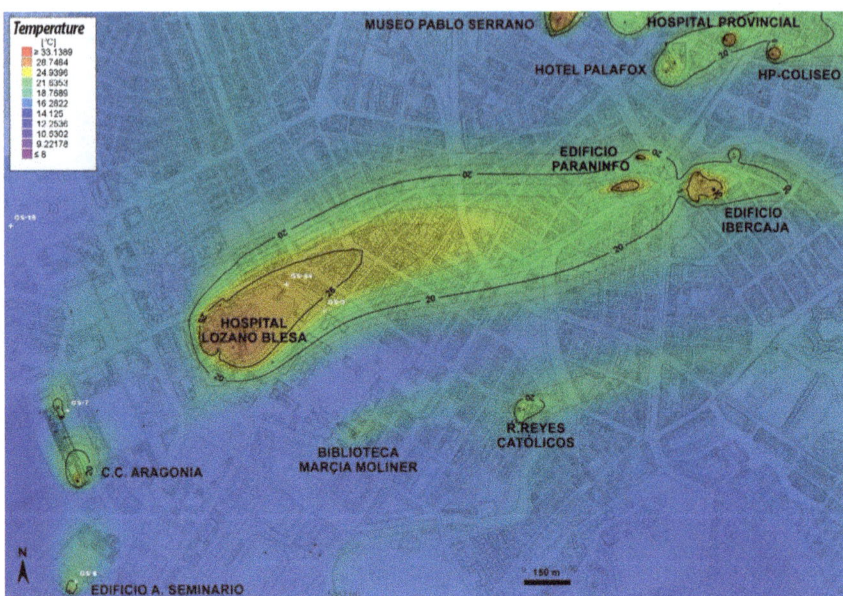

Fig. 11.19 Map of calculated groundwater temperature distribution in the southern sector of the city

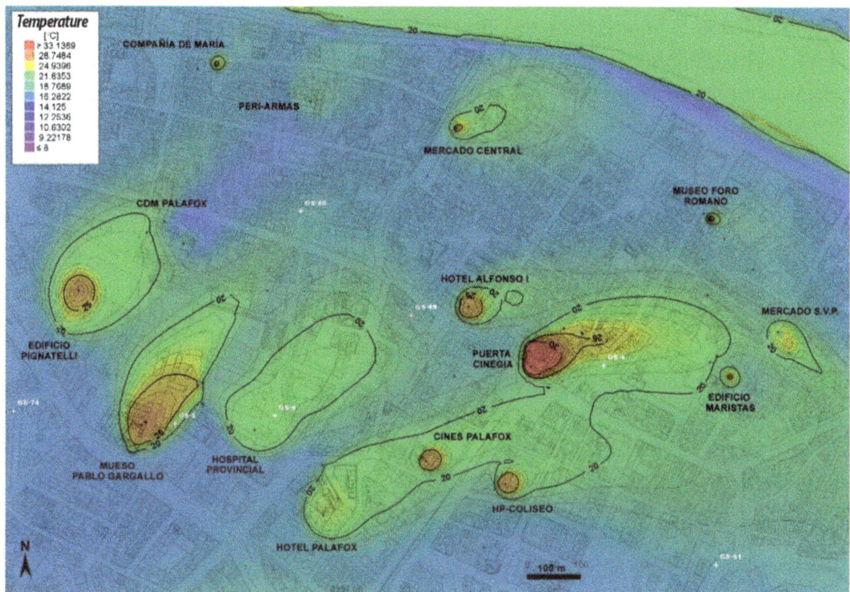

Fig. 11.20 Distribution map of calculated groundwater temperatures in the historic sector of the city

island of more than 20 °C around seven GWHP systems. The most intense thermal core is associated with a single system that generates an isotherm of 30 °C and is almost 100 m long.

In short, the quality of the simulations achieved by the GEOTERZ model with the data and the current level of existing knowledge of the Zaragoza urban aquifer represents an important advance in the understanding of the thermal regime of the subsurface in Zaragoza, providing a tool with predictive capacity to be used in the framework of an adequate management of shallow geothermal energy resources and in coordination with the use of groundwater resources.

11.7 Criteria and Policy for Adopted by Resource Managers

The demand for groundwater and the thermal discharges associated with the exploitation of shallow geothermal energy resources are, in both cases, additional elements of pressure on the qualitative status of groundwater bodies, and a challenge that must be faced by water management authorities, which are responsible for maintaining their general good status. The incipient implementation of technology for the use of shallow geothermal energy by means of open-loop geothermal exchangers, and the lack of knowledge regarding potential adverse effects on aquifers due to inadequate management of discharges has hardly had its counterpart in Spain, in the dissemination and implementation of methodological guides or criteria to regulate the sustainable use and management of shallow geothermal energy resources. There are a few guides that provide more or less detailed operational procedures for GWHP systems (ACA 2008; Llopis and Rodrigo 2008; Ruiz et al. 2010) and other technical standards, which are generally directed more towards BHEs or the exploitation of medium and high enthalpy geothermal energy. However, local, regional and national regulations governing the management of geothermal groundwater exploitation in Spain are still scarce and anecdotal (García-Gil and Mejías 2020). At present, only a small number of Hydrological Plans of the Duero, Ebro, Miño-Sil, Tagus, Western Cantabrian and Eastern Cantabrian basins (R.D. 1/2016) include, in their regulations, aspects related to shallow geothermal energy. When these are considered, they are generally unclear and with evident lack of coordination between one plan and another (Garrido et al. 2016a). The unequal treatment is reflected, for example, in the obligatory or non-obligatory nature of certain requirements; not all the inter-community plans even treat these systems as energy use requiring regulation, although most of them highlight the need to assign maximum relative temperature change values between captation and injection wells of between six and eight degrees Celsius and maximum temperature thresholds of up to 30 °C. A common consideration in the above-mentioned Hydrological Plans is the requirement that water abstractions for air-conditioning have a dossier for their concession and the requirement that pollution prevention measures must be adopted.

The enormous growth potential for shallow geothermal installations with open-loop geothermal exchangers suggests that, in a future revision of the Hydrological Plans, what are currently a series of criteria and recommendations aimed at achieving sustainable exploitation should be regulated as a standard. Zaragoza has established itself as a pioneering city of reference in the exploitation of shallow geothermal energy resources by means of open-loop geothermal heat exchangers, where the effects, consequences and results of applying procedures that regulate the exploitation and interaction of GWHP systems in operation can be observed, but also where innovative solutions that aim to improve the functioning and sustainability of the aquifer could be experimented with. As a result of previous research and experience, IGME proposes to implement different actions (García-Gil et al. 2015d; Garrido et al. 2016b; IGME-CHE 2014), which are grouped into five categories that include sustainable management criteria:

- ***Criteria for the improvement of administrative procedures***. These criteria are intended to improve resource management and to regulate the processing of administrative files in advance, with observance of the elements that form part of the projects, the control and monitoring of works for the construction of new geothermal heat exchangers, and the adaptation or reform of existing ones. The shallow geothermal energy installations should be classified according to their capacity to generate impacts, taking as a criterion the power supplied: small installations for power below 100 kW, large installations for power over

500 kW, the rest being an intermediate size. For installations of less than 50 kW, the construction of closed-loop systems would be recommended, simplifying the administrative procedure. In addition, the HVAC solutions adopted should be accompanied by appropriate environmental justification and other alternative solutions to geothermal heating and cooling; for pre-existing installations, a transitional procedure should be followed for their adaptation according to their size. All collected hydrogeological data plus other data on installations, energy demands and supplies could form a centralised geo-referenced database of shallow geothermal installations.

- ***Construction criteria for geothermal wells***. The aim is to ensure the monitoring and compliance with technical conditions for the construction of geothermal wells, which take into account the particularities and hydrogeological characteristics of the aquifer. These would include (1) the enforcement to provide hydrogeothermal studies during the authorisation procedure for exploitation, with a sufficiently detailed scale of at least one kilometre around the geothermal installation; (2) obtaining hydrogeological field parameters; (3) locating wells in accordance with hydrogeological and groundwater flow criteria; (4) specification of the technical characteristics of the works to be carried out, with an indication of depth, position of well screens, gravel packing, etc.; (5) ensuring that captation and injection wells are located in the same aquifer, suggesting a rule that water intake should take place in deeper layers or levels of the same aquifer and injection in shallower layers or levels; (6) distance of more than 100 m between the geothermal wells and other nearby installations, which should be more than 250 m if both installations are on the same groundwater flow line; and (7) recommended maximum injection flow rate of 50 $m^3\ h^{-1}$ per open-loop geothermal heat exchanger.
- ***Criteria for the sustainable exploitation of uses***. It contemplates measures for distributed and respectful use of shallow geothermal energy resources within the aquifer. These measures include, among others, (1) the recommendation and promotion of the reversible use of heat pumps with the aim of achieving a zero energy balance with the aquifer throughout the year (heat dissipated equal to heat absorbed); (2) limiting the number and flowrate of thermal discharges in areas affected by different heat plumes, even temporarily encouraging the exclusive use of installations for heating, which would allow the high temperature of the aquifer to be lowered; (3) the periodic review and control of the COP of geothermal heat pumps; (4) limiting the rate of water recirculation between geothermal wells of the same installation as much as possible; (5) limiting the admissible temperature change in geothermal heat exchangers from 6 to 11 °C; (6) taking into account in decision-making the criteria for the exploitation of other geothermal sources and their density in the area; (7) promoting the implementation of a *Relaxation Factor*, understood as a reserve of shallow geothermal energy resources available for use by third parties (García-Gil et al. 2015d); (8) Establishing for each installation the appropriate injection temperature, and defining in any case a maximum injection temperature limit, for example 30 °C; (9) Promoting and proposing alternatives that take into account the reduction in the volume of thermal discharge; (10) Recommending the use of several simultaneous wells for discharge, the

number of which should indicatively double the number of capture wells; (11) establishment of a moratorium on new installations if the average temperature of the aquifer at the proposed site exceeds 25 °C; and (12) preparation of thermal evolution forecasts of the aquifer for 10 and 20-year horizons, taking into account local hydrogeological and industrial design characteristics of the installations, a criterion which ultimately involves the use of numerical models of groundwater flow and heat transport, of varying degrees of complexity depending on the size of the installation.

- ***Groundwater quality protection criteria.*** The aim is to preserve the good qualitative status of the groundwater bodies under management, avoiding changes in chemical composition due to the addition of substances or alteration of physicochemical parameters. To this end, the following criteria and recommendations are proposed: (1) ensure pressurisation of the secondary circuit, avoiding aeration and degassing of the groundwater pumped from the captation wells; (2) control of the temperature change to avoid, among other aspects, changes in the solubility product, the appearance of incrustations or a growth of the bacterial mass in the presence of water with a high nutrient load; (3) avoid mixing of waters with different salinities and changes of the oxidation-reduction potential; (4) prevent the introduction of suspended particles and larger materials into open-loop geothermal heat exchangers, which facilitate the clogging of flow paths; (5) prohibit the use of chemical additives in the secondary circuit; (6) restrict or prohibit GWHP systems in areas of the aquifer with a significant presence of contaminated water, including areas close to septic tanks, areas of other direct discharges to groundwater, areas close to sewage treatment plants or contaminated soils; and (7) limit the implementation of these systems, especially those larger than 100 kW, in areas of maximum and moderate protection of groundwater abstraction for urban supply for human consumption.
- ***Criteria for the control of thermal discharges through open-loop geothermal heat exchangers.*** Its purpose is to address and detect in advance any circumstance that may affect the normal operation of the system, in order to reduce risks, ensure the maintenance of groundwater quality and minimise any impact arising from inadequate management of the geothermal installation. Particular attention should be paid to events such as (1) structural fractures in the ground that may be precursors of terrain collapse, (2) contamination by accidental discharges into wells, especially with nutrient input or unauthorised additives, (3) rising groundwater levels, which could have originated from clogging of the screens or reduction of porosity in the aquifer environment, and (4) excessive or higher than permitted temperature changes, etc. In order to maintain the compliance of installations with open-loop geothermal heat exchangers, the discharge authorisations should include a minimum protocol, control and monitoring of both the aquifer and infrastructural elements of geothermal developments, establishing (1) exploitation control points in the installations, (2) piezometers for monitoring the aquifer at different distances from the problematic installations and in the aquifer in general, (3) a methodology for the control, monitoring and surveillance of indicators for different parameters and frequencies, and (4) a panel of

physicochemical indicators, including at least the independent control in capture and injection wells of temperature, flowrate, pH, Eh, electrical conductivity, alkalinity, dissolved oxygen, Ca^{2+}, Mg^{2+}, Na^{+}, K^{+}, Cl^{-}, SO_4^{-2}, HCO^{3-}, Fe^{2+}, Mn^{2+}, organic matter and main pathogenic microorganisms.

11.8 Current Situation. The Procedure for Authorisation of Thermal Discharges and Thermal Impact Assessment Studies

As can be deduced from the previous sections and more specifically in Chap. 10, there is no specific standard or protocol in Spain that regulates from a legal point of view the aspects of licensing and operation of any type of shallow geothermal installation. Considering specifically the installations with open-loop heat exchangers in Zaragoza, the studies promoted by IGME and CHE were focused mainly on analysing the existing link between shallow geothermal energy exploitation with groundwater resources, the hydrogeological context and the environmental impact of thermal discharges. Zaragoza serves as a pioneering field of experimentation, as a management model from which procedures can be extracted and incorporated as standards in future revisions and improvements of the hydrological plans or the regulations that develop them.

In the regulatory development of the Hydrological Plan of the Ebro Basin, in force for the period 2015–2021, the thermal use of groundwater is not differentiated from industrial uses in general. Furthermore, it does not distinguish between installations with open-loop geothermal heat exchangers from closed-loop geothermal heat exchangers. However, reference is made to the discharge of water that has been used for geothermal purposes, which is treated in the same way as direct discharge of other pollutants into groundwater. Specifically, Article 57.7 of Annex XII of R.D.1/2016, referring to the regulations of the Ebro Hydrological Plan, states in section *(a)II* that the *injection of pollutants may be authorised in the case of injection into the same aquifer of water used for geothermal purposes or in the cases referred to in Article (11.j) of Directive 2000/60/EC, provided that: (1) the discharges do not compromise the achievement of the environmental objectives established for the waterbody to which it relates; (2) the best available techniques are applied to abate the polluting discharge body or in those bodies of water to which it relates; (3) specific monitoring mechanisms are established for the status of the affected bodies of water and periodic assessments are carried out of the effect of the discharges made.*

The work done by IGME and CHE reinforces the idea that it is necessary to regulate shallow geothermal energy exploitation, adopting the sustainability criteria that best suit each case and carrying out specific advisory studies that assess the thermal impact of the thermal discharge of each system individually and periodically. Thus, any progress towards improving the regulatory framework in this sense will contribute to more efficient management of the groundwater body. In this process, numerical modelling tools for groundwater flow and heat transport in combination with high resolution groundwater monitoring have been confirmed as a successful approach in the procedure followed, which involves carrying out simulations of the geothermal systems operations and proposing, when necessary, sustainable operating alternatives within the limits established by the competent authorities.

11.8.1 Authorisation Procedure for a Thermal Discharge

As has been discussed previously, the main impact of shallow geothermal energy systems has to do with the groundwater environment, especially with thermal discharges. In this sense, the regional-basin water authority is the competent authority for establishing the operating criteria and requesting the corresponding authorisations when thermal discharge is produced. In general, each owner of a shallow geothermal energy installation with open-loop heat exchangers needs to apply for a groundwater exploitation concession, which is managed according to standard procedures in a similar way to other groundwater exploitation. An additional authorisation is also required for the discharge of water from thermal exchange which, given its nature, is processed with some particularity. The minimum content of the discharge authorisations is established in article 245 and subsequent articles of R. D. 849/1986, of 11 April, which approves the Regulations of the Public Hydraulic Domain (RDPH).

In order to process the authorisation of the thermal discharge, the CHE's Water Quality Department opens an administrative file that involves several phases, as shown in Fig. 11.21. First, it is the owner of the installation who applies for the discharge authorisation, although if there is knowledge of existing discharges that do not have the aforementioned permit, the authority opens an ex officio file to require the relevant legalisations. Second, once the information in the application has been received and studied, a correction is required if its content is found to be insufficient, providing new technical information and/or requiring the installation of flow and temperature control devices at the intake and discharge points. With regard to the written documentation, it is necessary to have the discharge declaration, a technical explanatory report, a report on the thermal impact of the discharge on the aquifer and information on the flow and temperature control devices planned or, where appropriate, already installed.

Third, all the documentation and information received is reassessed, and the existence of the corresponding concession authorising the use of public water for HVAC is verified in the CHE database. If such a concession has not been granted, the owner is required to apply for it, together with the relevant administrative or technical documentation (path 1 in Fig. 11.21).

A preliminary technical report is then drawn up, the procedure for which is set out in articles 247.2 and 247.3 of the RDPH. This report assesses whether the requested discharge, a priori, does not cause non-compliance with the quality standards and environmental objectives and, therefore, would not cause a worsening of the status of the receiving waterbody. For urban discharges there are specific regulations on emission limits; for this specific type of thermal discharges there are no specific

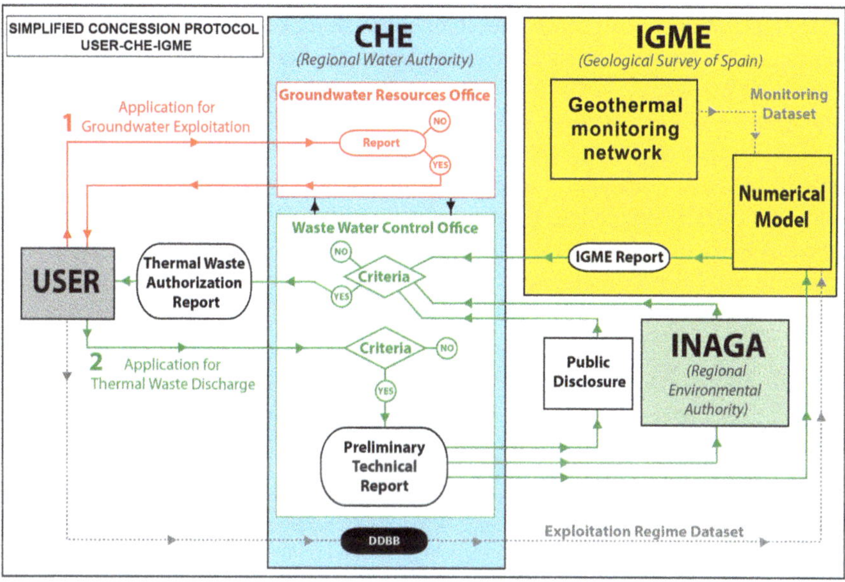

Fig. 11.21 Diagram of the procedure for the granting of the thermal discharge authorisation for groundwater heat pump systems

regulations on emission limits. Therefore, the CHE, based on the study (IGME-CHE 2014), applies the following limit values in Zaragoza: 30 °C for the maximum injection temperature, 8 °C for the average monthly thermal jump and a maximum instantaneous thermal jump (15-minutal) of 10 °C. There is also a limit on the annual volume discharged, the value of which must correspond to the granted water concession. The report finally proposes that the dossier be submitted to the obligatory public disclosure procedure and that a report be requested from the competent regional environmental body. As part of the procedure followed, the CHE also requests a report from the IGME. This report assesses the information available on geothermal exploitation, the data from the geothermal control network, the results of the simulations of the numerical model constructed on the impact of long-term discharges, the self-interference process, thermal interference with third parties and the extent of the heat plume in the aquifer, with a time horizon of 10 years.

With the assessments of the environmental body, the allegations, if any, and the IGME report, the thermal discharge (waste) authorisation report is drawn up, which establishes the conditions under which the discharge must be carried out, both in terms of discharge limit values (flow and temperature) and the control that must be carried out, the frequency with which information must be sent to the agency, compliance with the permit and the periodic analyses carried out on the physicochemical parameters representative of the discharge. Up to 2021, the IGME has reported on a score of facilities, providing timely and precise responses on some of the most important systems in operation or those with the greatest potential impact. In all cases, the response has been favourable, although on certain occasions it has been recommended that the temperature changes and discharge temperatures be adjusted to the criteria required by the CHE. The control established in the thermal discharge authorisation is a key element in the management of discharges. The CHE's Water Quality Department, which is responsible for monitoring the discharge authorisations granted, ensures compliance with the conditions established in the authorisation. The analysis of the evolution of the parameters established in the discharge authorisation will make it possible to detect situations in which it will be necessary to take measures to guarantee the sustainable use of the aquifer, for the discharge itself, as well as for all of them as a whole.

References

ACA (2008) Guía CLIMACA Climatizació d'edificis a partir de l'energia del subsòl. Departament de Medi Ambient i Habitatge. In: l'Aigua, ACd (ed). Generalitat de Catalunya, p 125

Banks D (2012) An introduction to thermogeology: ground source heating and cooling. Wiley

Benito G, Pérez-González A, Gutiérrez F, Machado MJ (1998) River response to Quaternary subsidence due to evaporite solution (Gállego River, Ebro Basin, Spain). Geomorphology 22:243–263

Benito G, Sancho C, Peña JL, Machado MJ (2010) Large-scale karst subsidence and accelerated fluvial aggradation during MIS6 in NE Spain: climatic and paleohydrological implications. Quatern Sci Rev 29:2694–2704

Benito G (1987) Karstificación y colapsos kársticos en los yesos del sector central de la depresión del Ebro (Aragón, España). Cuaternario y Geomorfología 1:61–76

CEDEX (2019) Anuario de aforos 2016–2017

Diersch HJ (2013) FEFLOW: finite element modeling of flow, mass and heat transport in porous and fractured media. Springer Berlin Heidelberg.

García-Gil A, Vázquez-Suñe E, Schneider EG, Sánchez-Navarro JÁ, Mateo-Lázaro J (2015) Relaxation factor for geothermal use development—Criteria for a more fair and sustainable geothermal use of shallow energy resources. Geothermics 56:128–137. https://doi.org/10.1016/j.geothermics.2015.04.003

Garrido EA, Arce MV, Van Ellen W (2006) Modelo matemático de flujo subterráneo del acuífero aluvial del Ebro en el entorno de Zaragoza. From data gathering and groundwater modelling to integrated management, Alicante, Spain, p 657

Garrido EA, Sánchez-Navarro JA, Coloma P (2010) Geothermic use of the urban alluvial aquifer of Zaragoza: first results. Geogaceta Geological Society of Spain 49:4

Garrido E, Moreno L, Azcón A, Padrino A (2003) Implementación de una red urbana para el control de calidad del agua subterránea en la ciudad de Zaragoza, Presente y futuro del agua subterránea en España y la Directiva Marco Europea. AIG-GE, Madrid, pp 279–284

Garrido E, Oroz M, Coloma P, Sánchez-Navarro J (2012b) Transformación de usos y demandas del acuífero urbano de Zaragoza derivado del aprovechamiento geotérmico de sus aguas subterráneas. In: AIH-GE (ed), Las aguas subterráneas: desafíos de la gestión para el siglo XXI. International Association of Hydrogeologists, Zaragoza, p 7

Garrido E, García-Gil A, Sánchez-Navarro JA, Coloma P, Delgado F (2012a) Evaluación del impacto térmico de los aprovechamientos geotérmicos someros del acuífero aluvial urbano de Zaragoza, Las Aguas Subterráneas: Desafíos de la Gestión para el siglo XXI. AIH-GE, Zaragoza

García-Gil A et al (2020a) Defining the exploitation patterns of groundwater heat pump systems. Sci Total Environ 710:136425. https://doi.org/10.1016/j.scitotenv.2019.136425

Garrido E, Sánchez Navarro JA (2009) Aprovechamiento geotérmico de las aguas subterráneas de Zaragoza Hacia Un Crecimiento Sostenible. Obras Urbanas 14:64–69
García-Gil A, Vázquez-Suñe E, Garrido E, Sánchez-Navarro JA, Mateo-Lázaro J (2014) The thermal consequences of river-level variations in an urban groundwater body highly affected by groundwater heat pumps. Sci Total Environ. https://doi.org/10.1016/j.scitotenv.2014.03.123
Garrido E, García-Gil A, Arrazola Martínez C, Escayola Calvo O, Sánchez Navarro JA (2017) Usefulness of a groundwater temperature baseline monitoring network for the identification of thermal interferences between shallow geothermal exploitation systems in urban environments. In: Calvache ML, Duque C, Pulido-Velazquez D (Eds), Congress on Groundwater and Global Change in the Western Mediterranean International Association of Hydrogeologists, Granada
Garrido EA, García-Gil A, Vázquez-Suñè E, Sánchez-Navarro JÁ (2016) Geochemical impacts of groundwater heat pump systems in an urban alluvial aquifer with evaporitic bedrock. Sci Total Environ 544:354–368. https://doi.org/10.1016/j.scitotenv.2015.11.096
García-Gil A et al (2016) A reactive transport model for the quantification of risks induced by groundwater heat pump systems in urban aquifers. J Hydrol 542:719–730. https://doi.org/10.1016/j.jhydrol.2016.09.042
García-Gil A et al (2018) Decreased waterborne pathogenic bacteria in an urban aquifer related to intense shallow geothermal exploitation. Sci Total Environ 633:765–775. https://doi.org/10.1016/j.scitotenv.2018.03.245
García-Gil A, Mejías M (2020) Current legal framework on shallow geothermal energy use in Spain. J Sustain Res 2(1):e200005. https://doi.org/10.20900/jsr20200005
Garrido E et al (2016a) Aspectos normativos de los aprovechamientos geotérmicos someros. La experiencia del acuífero urbano de Zaragoza, Las aguas subterráneas y la planificación hidrológica. Congreso hispano-luso. AIH-GE, Madrid. ISBN: 978-84-938046-5-7
IGME (2017) Continuous digital geological map of Spain (GEODE), Cuenca del Ebro (Zona-2700)
IGME-DGA (2005) Trabajos técnicos para la aplicación de la Directiva Marco del Agua en materia de aguas subterráneas. Caracterización adicional Aluvial del Ebro-Zaragoza. IGME-Dirección General del Agua, Madrid
IGME-CHE (2014) Análisis del impacto térmico generado por los pozos de climatización en las aguas subterráneas de la ciudad de Zaragoza
IGME-CHE (2019) Aplicación de un modelo de transporte de calor en aguas subterráneas para la simulación y alternativas de gestión de aprovechamientos geotérmicos en la ciudad de Zaragoza. Informe interno. In: Geológicos, D.d.I.e.R. (Ed.)
Lappala EG, Healy RW, Weeks ER (1987) Documentation of computer program VS2D to solve the equations of fluid flow in variably saturated porous media. In: Investigations WR (ed), Report 83–4099. USGS, p 184
Llopis G, Rodrigo V (2008) Guía de la Energía Geotérmica. Fundación de la Energía de la Comunidad de Madrid, p 185
Muela Maya S et al (2018) An upscaling procedure for the optimal implementation of open-loop geothermal energy systems into hydrogeological models. J Hydrol 563:155–166. https://doi.org/10.1016/j.jhydrol.2018.05.057
Medina A, Carrera J (1996) Coupled estimation of flow and solute transport parameters. Water Resour Res 32(10):3063–3076. https://doi.org/10.1029/96wr00754
Moreno L, Garrido EA, Azcón A, Durán JJ (2008) Hidrogeología urbana de Zaragoza. IGME, Instituto Geológico y Minero de España.
Rivas E, Garrido E, de la Orden J, Elorza FJ, Azcón A (2013) Simulación mediante elementos finitos de un aprovechamiento geotérmico en el acuífero aluvial urbano de Zaragoza, pp 317–324. AIH-GE, Barcelona., Actas Congreso sobre Aspectos Tecnológicos e Hidrogeológicos de la Geotermia
Ruiz E, Hendriks M, Toimil D (2010) Guía Técnica de Sistemas Geotérmicos Abiertos. Fundación de la Energía de la Comunidad de Madrid, pp 129

Rivas E, Elorza FJ, Garrido E (2014) Simulación numérica de la interferencia termo-hidrodinámica entre aprovechamientos geotérmicos urbanos de muy baja entalpía, II Congreso Ibérico de las Aguas Subterráneas, pp 759–778, Universitat Politècnica de València
Soriano MA, Simón JL, Gracia J, Salvador T (1994) Alluvial sinkholes ower gypsum in the Ebro Basin (Spain): genesis and environmental impact. Hydrol Sci J 39:257–268

Chapter 12
Example of Application (II): The Exploitation of Shallow Geothermal Energy Resources in the Canary Islands

The energy system of the Canary Islands has important singularities with respect to mainland systems: remoteness from the continent, geographical fragmentation, as there are several islands, difficulty in developing an economy of scale in order to obtain competitive energy production costs, etc. Apart from these difficulties, the Canary Islands present a promising outlook with regard to renewable energies. In addition to those already installed, to a greater or lesser extent, such as wind, solar and reversible hydroelectric energy, in the medium term energy can be harnessed through the use of shallow geothermal energy, mainly for use in buildings, small industries and the restoration industry. This chapter introduces the energy singularity of the Canary Islands, the interest in promoting the study and use of shallow geothermal energy in the archipelago and, finally, the analysis of nine cases of application.

12.1 Renewable Energy in the Canary Islands

The use of energy in the Canary Islands has gone through some unique stages: the first is the aboriginal stage, with the use of biomass, which the mountains supplied in the form of firewood, and later, after the conquest, it evolved towards hydraulic energy, such as the mills for kneading gofio (flour) or the hydraulic sugar mills. Wind energy was initially introduced in the Canary Islands in the form of water elevation systems, which gradually fell into disuse when fossil fuels such as oil came into play, a situation that has lasted until the present day.

According to the government of the Canary Islands (2012), the islands produce approximately 9400 GWh annually (626 GWh are of renewable origin) and their installed electrical power is 3044 MW, of which the renewable energy generation park contributes 327.3 MW. Gran Canaria and Tenerife provide 80% of the energy of the archipelago. The Canary Islands is an isolated system, which causes certain singularities in the electricity supply, which is largely supplied by fossil fuels. This isolated system in general is saturated on the islands and presents certain problems

A. García Gil et al., *Shallow Geothermal Energy*, Springer Hydrogeology,
https://doi.org/10.1007/978-3-030-92258-0_12

that can be summarised as follows: (1) an exponential increase in electricity demand until the 2008 crisis; (2) the pressure to integrate renewable energies into the generation system; (3) the lack of interconnection between the islands' electricity systems; (4) the higher costs of construction and operation; (5) the atomisation of the islands' electricity systems; and (6) the lack of interconnection between the islands' electricity systems; (7) the atomisation of production centres; (8) the high dependence on fossil fuels; (9) the administrative and environmental difficulties for the development of new power lines and energy projects; (10) the complexity for the location of new power plants; as well as (11) the complexity and increased bureaucracy for the authorisation of new generation equipment.

One of the most important conditions for promoting renewable energies in the Canary Islands, including geothermal energy, is insularity. There is a total energy dependence on the external sources, and, therefore, a great vulnerability to energy crises.

For this reason, the Canary Islands government, society and other autonomous bodies are committed to the use of renewable energies in all areas, including buildings. The energy contribution of renewable energies implemented in buildings can lead to an energy balance in the use of the building, so that the energy consumed in the building is equal to the energy generated by its active production systems. This was the European Union's target in 2019: the construction of zero-energy buildings, new buildings that produce as much energy as they consume. This energy production will have to be carried out through the use of renewable energies in the environment (solar, wind, geothermal, etc.). The islands remain far behind the levels of renewable energy participation recorded in other energy systems in the European Union.

As mentioned above, renewable energies currently account for a very small percentage of energy demand in the Canary Islands. The combination of the significant growth in energy consumption and associated CO_2 emissions (well above what Spain has assumed within the framework of the Kyoto Protocol and the subsequent distribution within the EU) will require a very active policy of efficient energy use and favouring energies with low or zero CO_2 production, and all of Spain's autonomous communities will have to show solidarity in this effort.

Although the use of shallow geothermal is still at a juvenile market stage compared to other renewable energy technologies, its concept is encouraging within energy systems that rely heavily on fossil fuels (Chen et al. 2007; Quick et al. 2013). This fact is especially relevant for many small islands and archipelagos that rely heavily on energy imports. Indeed, the European Union recognises the special and very vulnerable role of EU islands in the context of clean energy transition. The impacts of climate change are strongest on small islands, which are highly dependent on local ecological systems. As well, the energy transition faces greater challenges on islands, as they are conditioned by smaller economies and the lack of geographical connection to neighbouring regions for joint smart energy grids. In addition, energy demand profiles may differ from mainland regions when islands are used for tourism purposes. For that reason, in 2017, the European Commission, together with 14 EU member states, signed the policy statement on *Clean Energy for EU islands* (EC 2017). This declaration emphasises the integration of local renewable energy sources, together with the empowerment of island communities, to become part of their energy transition. The declaration also recognises the opportunities offered by a transition to clean energy supply, as it can reduce dependence on imported fossil fuels, which often leads to high energy costs. This transition also introduces green jobs, thus adding the tourism sector as a way to diversify the employment situation. In cooperation with the European Parliament, the EU Commission set up a secretariat to support the Clean Energy for EU Islands initiative. In 2019, the secretariat published a handbook on *Clean Energy Transition Agendas* to support regional communities in their transition procedures, from planning to citizen participation. It also proposes transition indicators for monitoring the process (De-Clercq et al. 2019). By 2020, six EU islands had already published a Clean Energy Transition Agenda.

There are different reference thermal installations with shallow geothermal energy in the Canary Islands; they are designed to provide air conditioning, swimming pool heating and DHW supply. The use of these systems not only saves energy costs but also reduces CO_2 emissions. In total, it is estimated that there are more than 7200 kW of total installed power.

12.2 Shallow Geothermal Installations in the Canary Islands

In the Canary Islands, there is growing emerging development of efficient heating and cooling of tourist infrastructures using low enthalpy geothermal energy, sometimes combined with solar thermal energy. This chapter shows the research results (Santamarta et al. 2021) of nine thermal installations that switched from conventional *heating* and *cooling* technologies to systems based on geothermal heat pump technology. The energy, economic and environmental benefits of this transition were calculated on the basis of real operating data, which makes it possible to argue for this strategy of decarbonisation of the thermal sector, even in volcanic environments.

12.2.1 Geological and Hydrogeological Framework

The Canary Islands Archipelago consists of eight islands and five islets, covering an area of approximately 7500 km^2 (Fig. 12.1a). They are located approximately 1400 km from the nearest coasts of the European continent (Iberian Peninsula) and 100 km from the west coast of the African continent (Western Sahara). In geological terms, the Canary archipelago emerged from *hot* zones associated with a residual heat plume, which has been active since the beginning of the opening of the Atlantic 200 million years ago (Anguita and Hernán 2000).

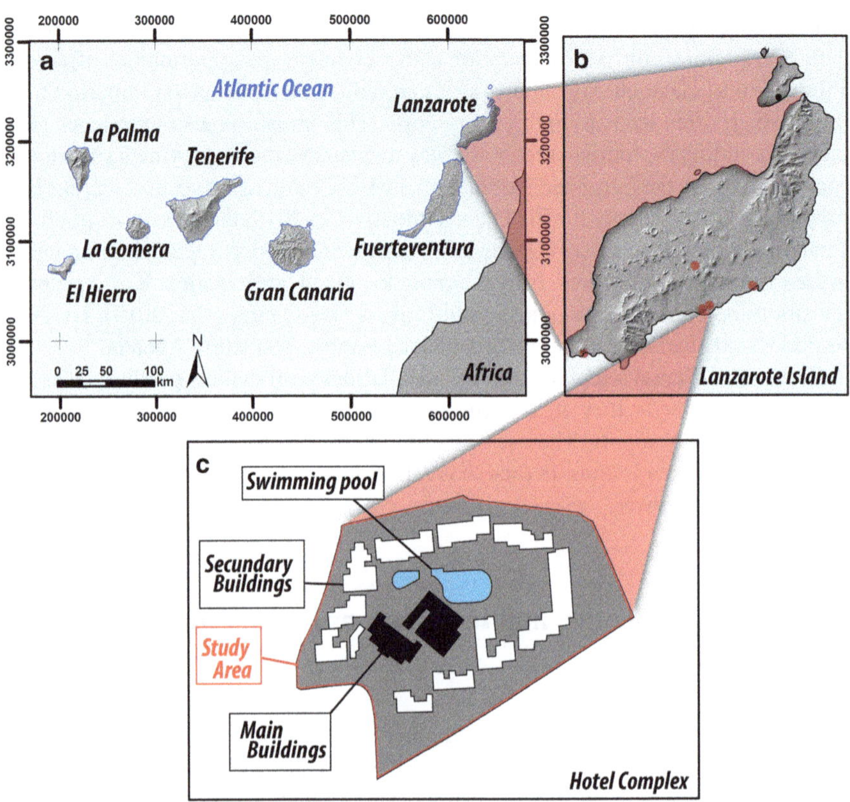

Fig. 12.1 Location of the shallow geothermal installations (**a**) investigated in the Canary Islands (**b**). Projection: WGS 1984 Complex UTM Zone 28 N, Datum DWGS1984. The typical hotel complex installation is shown (**c**)

The rock types found in the Canary Islands cover the full range of the typical oceanic alkaline suite (Anguita and Hernán 2000; Gómez-Ulla et al. 2018), including melilitites, nephelinites, basanites, toleiitic and alkaline olivine basalts, tephrites, rhyodacites, rhyolites, pantellerites and cornendites, trachytes, phonolites and carbonatites. This book section describes nine low-enthalpy geothermal installations on the islands of Lanzarote and Fuerteventura. These are the northernmost and easternmost of all the Canary Islands. The geological history of these islands is relatively straightforward, as they consist almost entirely of basaltic rocks generated from magma emissions from the Miocene to the present day. The location of the case studies and a detailed description of the petrological characteristics are provided in Fig. 12.2.

Shallow geothermal resources are strongly related to groundwater flow due to heat advection (Alcaraz et al. 2016a, b). Therefore, the groundwater regime in shallow aquifers must be well understood before starting the design of a shallow geothermal energy system. This is often the main challenge for the use of shallow geothermal

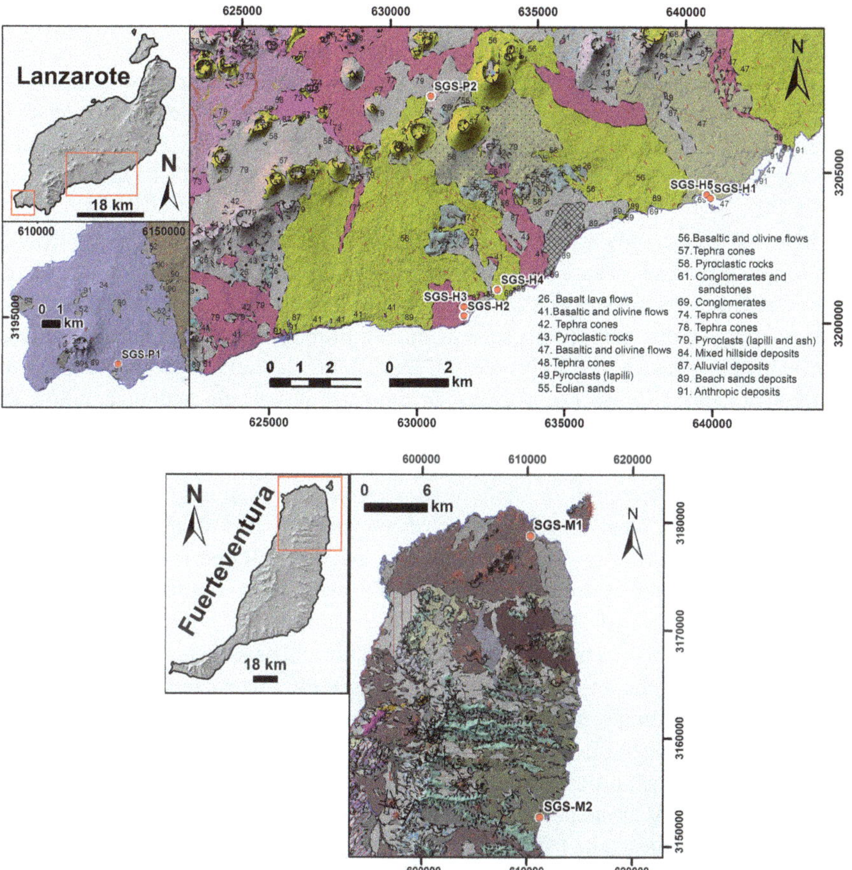

Fig. 12.2 Geological map of the investigated shallow geothermal installations on the islands of Lanzarote and Fuerteventura, Canary Islands (Spain). The detailed geological legend can be found in IGME (IGME 2017). Projection: WGS 1984 Complex UTM Zone 28 N, Datum DWGS1984

energy resources in volcanic areas, as groundwater resources are often scarce (Fetter 2018). Most of the existing groundwater wells in Lanzarote and Fuerteventura have low production (~1 $m^3 \cdot day^{-1}$), and documented hydraulic conductivities are in the range of 10^{-3} to 0.5 $m \cdot day^{-1}$ (Custodio 1989; Herrera Lameli and Custodio Gimena 2011). This low permeability explains why the groundwater level is found between 2 and 10 m depth (Herrera Lameli and Custodio Gimena 2011). Higher permeabilities can be found on the coast, where groundwater is used for industrial purposes, including geothermal utilisation. The use of groundwater for heating and cooling indoor spaces is increasingly common in tourist recreational infrastructures.

According to the Spanish Geological Survey (IGME 1987), the Canary Islands are considered *geothermal regions* of Spain, with considerable deep geothermal potential due to their volcanic origin and the presence of very recent historical volcanic eruptions. Although there are some estimates of deep geothermal potential (IGME 1987), shallow geothermal potential is still unexplored.

In this research, a total of nine case studies representative of tourism infrastructures using geothermal heat pump technology have been selected. The selected systems represent the following tourism facilities: hotel complexes (SGS-H1, 2, 3, 4 and 5, see Fig. 12.1c), shopping centres (SGS-1 and 2), water parks (SGS-P1) and wineries (SGS-P2).

Most of the shallow geothermal installations identified in the Canary Islands are hotel complexes. These installations usually require the heating of swimming pool water throughout the year. Although hotels can be single block buildings, they are generally tourist complexes, several small buildings, sports and leisure facilities surrounded by green areas. This typical Canary Island layout has given rise to micro-grids for district heating and cooling. In the most renewable and energy-efficient cases, the networks are powered by geothermal heat pumps. In addition, the existing installations of this type use groundwater in most of the cases studied. Therefore, successful implementation of shallow geothermal energy relies, to a large extent, on the hydrogeological characteristics of the ground. In volcanic environments, groundwater resources are limited. However, as demonstrated in Santamarta et al. (2021), groundwater resources, when available, provide enormous energy potential for HVAC of tourism infrastructures. Figure 12.3 summarises typical HVAC systems based on heat pump systems with open-loop geothermal heat exchangers (geothermal wells) used in hotel complexes in the Canary Islands. The systems use geothermal well doublets to pump groundwater from shallow aquifers, with aquifer background temperatures of about 22 °C under standard conditions. Using water-to-water plate heat exchangers, heat is transferred to the geothermal heat pump that provides chilled water for the HVAC systems in the hotel facilities. In this process, the heat pumps require approximately one fifth of the thermal energy provided as mechanical (electrical) energy. The required electrical energy is obtained from the electricity grid and, in some cases, from photovoltaic modules, further reducing CO_2 emissions. By recovering waste heat from the geothermal heat pump, tap water is preheated to produce DHW and to heat swimming pool water. In addition to this waste heat recovery, solar thermal panels are sometimes also used to produce DHW and heat the pool water. The remaining heat produced during the operation of the ground source heat pumps is redirected to the plate heat exchangers, and the water is then injected back into the coastal aquifers as heated water, usually at 24–30 °C. Hotel complexes using geothermal heat pumps are usually already constructed buildings with centralised heating and cooling systems that became obsolete years ago. Due to the global financial crisis in the last decade, existing conventional systems have become obsolete due to their low performance. The obsolescence, together with the rising prices of fossil resources, has resulted in an unsustainable situation.

In this category of hotel complexes, five facilities were investigated as representative and best practice examples of this type of facility. The first system (SGS-H1) is a 15-storey building with a total of 164 rooms, commercial area, spa, swimming pool and common areas, located in the centre-east of the island of Lanzarote (Fig. 12.2). It is an open-loop system using a well doublet and feeding two geothermal heat pumps,

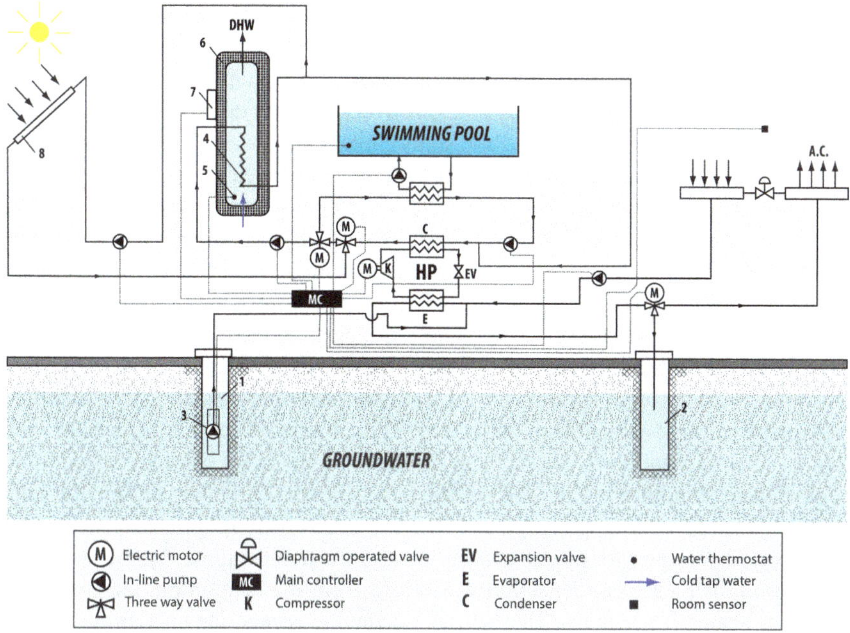

Fig. 12.3 Schematic diagram of the general design principle of shallow geothermal installations for hotel complexes in the Canary Islands using open-loop geothermal heat exchangers. (1) Geothermal production well; (2) geothermal injection well; (3) submersible pump; (4) DHW coil; (5) DHW thermostat; (6) DHW storage; (7) additional heat source; (8) solar collector

with a total installed thermal capacity of 876 kW and 1076 kW for cooling and heating, respectively. The wells used by this system are sufficiently productive (>80 $m^3 \cdot day^{-1}$) to guarantee the necessary groundwater flow to support the geothermal heat pump and, thus, the thermal demand of the resort. Open-loop systems are the most efficient technology for geothermal energy recovery (Saeid et al. 2015) and, in this case, two wells were used, one for chilled water production and one for *hot* water injection. In summer, the installation supplies chill water for air-conditioning of common areas and rooms. Part of the waste heat is recovered for the spa areas and DHW preheating. During the winter, the geothermal heat pump switches to heating mode and extracts heat from the groundwater to heat the pool water and produce DHW. Before the cooled water is rejected and injected back into the aquifer, the thermal energy of the *cold* fluid is recycled for air conditioning of the hotel's common areas. The second case study (SGS-H2) is a hotel complex located in Puerto del Carmen, the main tourist resort on the island of Lanzarote (Fig. 12.2). The hotel complex has 187 rooms distributed in 14 single-storey buildings and is equipped with a geothermal heat pump of 691 and 849 kW for cooling and heating, respectively. The cooling and heating system uses an open-loop geothermal heat exchanger with similar characteristics to the SGS-H1. In this case, pool water heating and DHW production are supported by solar thermal panels. The third case studied (SGS-H3) is also located in Puerto del Carmen (Fig. 12.2). It is also a hotel complex consisting of 211 rooms distributed in small block buildings between swimming pools and common areas. Its open-loop geothermal installation has an installed capacity of 506 and 622 kW for cooling and heating, respectively, and the design and operation are essentially the same as SGS-H1. Case study SGS-H4 is located in the same municipality. It is another hotel resort with 242 rooms distributed in 16 blocks, showing an installed capacity of 253 and 311 kW for cooling and heating, respectively. This complex uses groundwater wells near the sea. The geothermal wells are connected to a liquid-liquid plate heat exchanger, and the latter is connected to the heat pump which is, finally, connected to a system of fan coils that provide air conditioning for the room spaces, as well as producing DHW and heating the pool water. The heating is provided by three sub-systems: (1) 300 solar thermal panels intended to generate DHW and heat for the swimming pool, (2) two gas boilers, as a legal requirement for water treatment against legionella, and (3) geothermal heat pumps, whose waste heat from condensation is used to heat the water in the hotel complex's swimming pool. For air-conditioning, a GHP is used to generate chilled water for the hotel complex's air-conditioning equipment. The installed water chiller is used in summer to generate cold water for the air-conditioning of the hotel complex's interior spaces. In addition, using the condensation waste heat from the heat pump, the swimming pools are heated if necessary. In addition, during the winter, the heat pump runs for approximately two and a half hours every morning to generate chilled water for air conditioning of the dining room. During this mode of operation, the heat pump dissipates the heat, which is then used to heat the pool water or is otherwise injected into the aquifer. The production of DHW is achieved by existing solar thermal panels and gas boilers. Thus, the heat pump simultaneously generates cold water for air conditioning and *hot* water for heating the pool water. Finally, the fifth case study (SGS-H5) is a 115-room hotel building in the centre-east of the island of Lanzarote (Fig. 12.2). The system has an installed thermal capacity of 275 and 325 kW for cooling and heating, respectively, which is used by this hotel in a similar way to the SGS-H1 system, which is located very close to it.

A second type of shallow geothermal installation in the Canary Islands is found in shopping centres. Two (SGS-M1 and M2) using shallow geothermal energy have been investigated on the island of Fuerteventura (Fig. 12.2). The SGS-M1 system is an open-loop system with an installed cooling capacity of 1498 kW for cooling and 1,882 kW for heating. This system serves a three-storey building with a total floor area of 5500 m^2. The system uses groundwater for air conditioning, space heating and space cooling. The SGS-M2 system is also an open-loop system serving a three-storey shopping centre building equipped with a 1285 kW chiller. The current open-loop geothermal system replaced one based on cooling tower technology, which generated excessive costs due to the chemicals required for its use and the corrosion that occurred in the installation. The current geothermal system consists of a low-temperature (20–23 °C) district cooling system with a ring network, where each shop in the shopping centre has its distribution system, and the heated water is then injected through a return well into the coastal aquifer from which it was extracted.

Another type of tourism infrastructure using shallow geothermal energy is water parks. SGS-P1 is an open-loop system in the southwest of the island of Lanzarote (Fig. 12.2) that uses groundwater to heat the water used in the park. This installation, with an installed heating capacity of 450 kW, was selected as a reference for World Tourism Day 2012. Finally, the SGS-P2 system represents installations in wineries, which use geothermal heat pumps for cooling demands in the winemaking process. The system is equipped with closed-loop BHEs with an installed cooling and heating thermal capacity of 35 kW and 15 kW, respectively.

Prior to the installation of geothermal heat exchangers in the above-mentioned installations, the installations used conventional systems consisting of water chillers with air-cooled condensers and gas boilers. In most cases, the installations were originally designed as two groups of installations: (1) heat generating and heat receiving installations for the production and consumption of DHW and for heating of swimming pool water and, (2) installations pertaining to the production of air-conditioning. In general, a centralised heating and *cooling* system was used to meet the thermal demand of the facilities.

12.2.2 *Impact of the Energy Transition Through Shallow Geothermal Energy*

Table 12.1 presents the main results obtained from the heat balance and comparative analyses performed as a decarbonisation process. The methodology followed can be consulted in Santamarta et al. (2021). It can be seen that the total annual cooling demand of the nine DHC systems is 4.35 GWh (GWh thermal), while the heating demand is 5.21 GWh, which shows a heating bias of 9% from the balanced demand. The average cooling and heating demand of the studied systems is 483 and 652 MWh·year^{-1}, respectively, a situation that is well represented by SGS-H5. Shopping centres have the highest thermal energy demand, more than three times higher than hotels and almost two orders of magnitude higher than the domestic small industry systems represented by SGS-P2. The energy consumption of conventional systems allows the annual performance factor (APF) of these systems to be assessed. Prior to installation of the shallow geothermal energy technology, the APF for conventional systems (Table 12.1) had an average value of 1.73–3.39. These are low values, indicating obsolete conventional installations. The nine conventional installations investigated required 2.12 and 3.94 GWh·year^{-1} to meet cooling and heating demands, respectively. It should be noted that, although solar collectors were already used for heating before the transition to shallow geothermal energy technology, heating with gas boilers accounted for 3.67 GWh·year^{-1} of primary energy consumption, equivalent to 303,731 kg·year^{-1} of commercial propane. When the systems investigated transitioned to geothermal technology, the final energy requirements were significantly reduced, from 6.07 to 2.59 GWh·year^{-1}, which was a 40% reduction in final energy consumption each year. This achievement can also be seen

Table 12.1 Performance of shallow geothermal installations studied in the Canary Islands where Q_{GE} is the average heat produced by the generation equipment, Q_{SC} is the thermal energy provided by the solar thermal panels, Q_{RC} is the heat recovered from the generation equipment, Q_{DL} is the heat losses through the distribution network pipes connecting the substations, W_{PR} is the primary energy work required for heat generation, W_{EE} is the electricity consumption required for heat generation, W_{GHE} is the electricity consumption used by the pumps of the geothermal heat exchangers, W_{FC} is the electricity consumption of fan coils, and W_{DP} is the electricity consumption used by the distribution pumps through the grid

DHC system				SGS-H1	SGS-H2	SGS-H3	SGS-H4	SGS-H5	SGS-M1	SGS-M2	SGS-P1	SGS-P2
Conventional technology	Thermal energy (kWh-year^{-1})	Cooling (Q_{DHC_C})	Q_{GE}	370,832	260,657	168,495	100,804	443,475	629,171	2,345,125	0	32,621
			Q_{DL}	18,542	13,033	8425	5040	22,174	31,459	117,256	0	1631
		Heating (Q_{DHC_H})	Q_{GE}	518,025	366,039	238,795	142,862	362,389	1,535,936	0	269,518	9,356
			Q_{SC}	0	520,121	620,569	629,583	0	0	0	0	0
			Q_{DL}	25,901	44,308	42,968	38,622	18,119	76,797	0	13,476	468
	Energy consumption (Wh-year^{-1}) (kWh-year^{-1})	Cooling (W_{DHC_C})	W_{EE}	105,952	81,455	73,259	69,044	129,671	171,905	1,202,628	0	14,828
			W_{FC}	7,521	7,520	6,333	3,294	11,888	10,006	189,636	0	1388
			W_{DP}	2,124	1,675	1,489	1,423	2,632	3,487	24,145	0	311
		Heating (W_{DHC_H})	W_{PR}*	558,111	377,360	262,413	155,285	414,159	1,610,123	0	280,011	10,754
			W_{FC}	34,126	31,724	18,252	12,190	25,728	61,868	0	16,982	918
			W_{DP}	11,204	7,618	5,251	3,128	8,321	32,244	0	5,651	232
	$SEER_{DHC}[-]$			3.05	2.73	1.97	1.30	2.92	3.22	1.57	–	1.88
	$SCOP_{DHC}[-]$			0.82	2.02	2.86	4.30	0.77	0.86	–	0.85	0.75
	$\eta s_{DHC}[\%]$			92.8	97.0	91.0	92.0	87.5	95.4	–	96.3	87.0
	$APF_{DHC}[-]$			1.17	2.15	2.66	3.39	1.29	1.09	1.57	0.85	1.40
Shallow geothermal technology	Thermal energy (kWh-year^{-1})	Cooling (Q_{DHC_C})	Q_{GE}	371,560	261,413	168,759	101,224	443,863	630,005	2,345,484	0	32,850
			Q_{DL}	19,359	13,702	9186	5392	23,100	31,697	118,189	0	2194

(continued)

Table 12.1 (continued)

		Heating (Q_{DHC_H})	Q_{GE}	388,895	274,856	180,718	107,725	363,143	1,536,352	0	269,937	9536
			Q_{SC}	0	521,105	621,366	629,597	0	0	0	0	0
			Q_{RC}	130,167	91,539	59,300	36,101	0	0	0	0	0
			Q_{DL}	11,204	17,773	17,727	15,844	7880	31,004	0	6269	1141
	Energy consumption (kWh-year^{-1}) (kWh-year^{-1})	Cooling (W_{DHC_C})	W_{EE}	23,952	27,535	20,870	12,349	44,180	34,581	683,763	0	4703
			W_{GHE}	5376	6200	4731	2797	10,026	7822	153,907	0	1109
			W_{FC}	6662	7646	5882	3423	12,357	9660	190,320	0	1327
			W_{DP}	381	521	343		803	602	13,659	0	113
		Heating (W_{DHC_H})	W_{PR}[a]	22,251	41,729	38,575	29,680	45,351	0	0	0	0
			W_{EE}	122,520	114,931	65,227	45,171	93,677	220,503	0	63,049	3180
			W_{GHE}	27,645	25,875	14,546	10,115	21,025	49,521	0	14,031	717
			W_{FC}	34,168	32,009	17,979	12,488	25,938	61,292	0	17,294	903
			W_{DP}	2965	3211	2072	1517	2793	4412	0	1336	
	$SEER_{DHC}[-]$			9.68	5.91	5.01	5.12	6.25	11.36	2.14	–	4.23
	$SCOP_{DHC}[-]$			2.42	3.99	6.10	7.65	1.88	4.48	–	2.75	1.70
	$\eta s_{DHC}[\%]$			92.8	97.0	91.0	92.0	87.5				
	$APF_{DHC}[-]$			3.50	4.30	5.89	7.25	3.03	5.42	2.14	2.75	3.21
Savings	Energy (kWh-year–^{1}) (kWh-year^{-1})			461,914	229,919	179,044	110,816	328,367	1,470,235	374,760	200,666	15,109
	Monetary (€-year-1)			85,387	42,502	33,097	20,485	60,700	271,780	69,276	37,094	2793
	Emissions (tCO_2-year^{-1})			345.8	172.1	134.1	83.0	245.9	1100.8	280.6	150.2	11.3

[a] A primary energy conversion factor of 0.0715 kWh·Kg^{-1} was used for electricity production for the energy system of the Canary Islands

in the improvement of the APF, which increased positively, on average, by 41%, from 1.73 to 4.17 (Table 12.1). With reference to the seasonal performance factors SEER and SCOP (also shown in Table 12.1), they showed an average positive increase of 45% and 40%, respectively, from 2.33 to 6.21 and from 1.65 to 3.87, respectively. The transition from conventional to geothermal systems in the investigated cases resulted in a final energy saving of 3.48 GWh·year^{-1} or a saving of 40%, leading to savings of 623,115 € and 2524 tCO_2. These savings clearly demonstrate the economically attractive decarbonisation of the heating and cooling energy sector. These results are relevant to justify the promotion of shallow geothermal energy on energy-dependent islands in volcanic areas, always in the context of a sustainable development framework under proper resource governance (García-Gil et al. 2020).

Distribution losses showed that the distance from heat production facilities to individual consumers does not result in significant pumping energy consumption or heat losses, and accounts for approximately 2% and 5% of the energy required in DHC micro-grids and heat generated, respectively. These observations are in line with ultra-low temperature (5G) district heating and cooling (DHC) networks, which have shown relatively low power consumption during pumping of heat transfer fluid to substations, which is below 1.6–3.0%, as well as heat losses during distribution, presenting values below 5–15% of the heat supplied (Buffa et al. 2019). This means that, in the DHC micro-grids studied, both energy consumption and heat loss values are lower than those found in DHC grids.

The environmental analysis of shallow geothermal installations assesses the environmental impact, understood as their potential to reduce GHG emissions (Bayer et al. 2012; Saner et al. 2010). In the cases studied in the Canary Islands, the energy transition from conventional systems to shallow geothermal energy produces, each year, average reductions of 280 tCO_2 in each thermal installation, and up to 1100 tCO_2·year^{-1} in the most successful case studied (SGS-M1).

The nine case studies help to justify investment in shallow geothermal technology in general, and specifically justify its use by the energy-dependent island tourism sector in a volcanic environment. More specifically, the results of this study highlight the fact that open-loop geothermal systems dominate over closed-loop systems, due to the hydrogeological conditions on the coast, where these systems are concentrated. In such places, fresh groundwater discharges occur naturally, mixing with salt water, thus generating the so-called saline interface zone where brackish and saline groundwater can be found. Open-loop systems operate effectively under such conditions, without affecting possible seawater intrusion due to groundwater pumping, as the pumped groundwater is injected back into the aquifer without affecting the magnitude of fresh groundwater discharge, i.e., a non-consumptive use of groundwater resources.

The results of the economic analysis are shown in Table 12.2. On average, based on the economic savings of 52,189 €·year^{-1}, it takes 3.72 years to recover the initial investment (0.16 M€) to achieve the energy transition target by using shallow geothermal energy in the cases studied. The minimum payback period is 2.06 years for the SGS-M1 shopping centre and three times longer for the SGS-P1 closed loop system. The NPV values are 0.63 and 0.93 M€ on average for the project lifetime of

Table 12.2 Economic variables of the economic analysis, including net present value (NPV), internal rate of return (IRR) and payback period indicators calculated for the shallow geothermal energy systems studied in the Canary Islands

System	SGS-H1	SGS-H2	SGS-H3	SGS-H4	SGS-H5	SGS-M1	SGS-M2	SGS-P1	SGS-P2
Initial investment (€)	176,511	137,469	193,738	100,506	150,000	281,289	160,257	148,526	116,344
Savings (€·year^{-1})	65,112	51,375	37,639	18,823	53,386	113,885	77,759	24,500	27,230
NPV - 15 year (€)	618,463	487,422	261,082	121,351	499,790	1,117,562	791,304	143,615	209,598
NPV - 25 year (€)	80,666	780,537	471,191	217,758	805,058	1,788,425	1,243,858	274,325	356,806
IRR-15 year (%)	36.31	36.73	17.38	16.19	34.91	40.09	48.18	13.70	21.72
IRR-25 year (%)	36.64	37.05	18.80	17.66	35.28	40.33	48.30	15.50	22.74
Payback (year)	2.71	2.68	5.15	5.23	2.81	2.47	2.06	6.06	4.27

15 and 25 years, respectively, and up to 2.07 M€ in the case of the SGS-M1 system for an NPV of 25 years. The IRR values at 15 and 25 years showed an investment rate above 13, 70, 30% on average and up to 48.30%, showing very profitable investment rates.

12.2.3 Environmental and Economic Benefits

Open-loop geothermal systems using groundwater are also economically viable in volcanic environments. Indeed, recent cost increases and uncertainty about the future of conventional energy supplies for power generation are boosting the attractiveness of very low temperature geothermal resources (Boschetti 2013). The exploration and exploitation of geothermal energy has increased worldwide in recent decades, in the search for sustainable energy resources with low carbon emissions (Poulsen et al. 2015). In the Canary Islands, it is essential to promote the use of renewable energies, due to their dependence on external energy, based on the characteristics conferred by their insularity. For this reason, the Canary Islands government is promoting the use of renewable energies in all sectors, including the construction sector, where it has been demonstrated that shallow geothermal energy can play a crucial role in the process of decarbonisation and energy security.

The global energy crisis caused by high dependence on fossil fuels, their rising prices, the need to reduce CO_2 emissions to avoid further aggression to the environment, and technological advances are the key points to promote the use of geothermal energy, as it does not depend on direct external factors. Its energy source is the earth, which has constant and optimal conditions for energy production. It has been technically concluded that the low enthalpy geothermal systems used are feasible in a volcanic environment, where groundwater resources might be scarce, and these systems can significantly reduce costs compared to conventional fossil fuel sources.

The heat balance and comparative analyses demonstrate that the energy transition to shallow geothermal technology in the nine facilities studied, which are representative of the infrastructures of the tourist islands, is cost-effective. The results of the comparative study carried out showed a 40% reduction in final energy consumption in the investigated heating and cooling systems. The nine systems investigated significantly reduced their electricity needs from 6.07 GWh to 2.59 GWh, saving 623,115 € and 2524 tCO_2. Evaluation of the efficiency of the systems (COP) showed a performance proportional to the installed capacity of the systems, which was four times more efficient than conventional systems using air-to-water heat pumps, air-condensed liquid chillers, cooling towers and gas boilers. The financial value of the investment was evaluated using three indicators, NPV and IRR, showing, on average, a payback period of 3.72 years and investment rates above 13%. The payback is clearly offset by the performance advantages of shallow geothermal systems and therefore financial incentives are required to overcome this well-known development barrier (Jaudin 2013).

Despite the successful implementation of ultra-low temperature DHC micro-grids, performance may decrease due to poor hydrogeological conditions. Further studies are needed to determine whether groundwater used near the coast may cause long-term problems, such as corrosion of geothermal wells and heat exchangers or changes in hydraulic conductivity due to groundwater mixing processes during operations. Low permeability zones of exploited volcanic aquifers are not observed in this study as only successful case studies were covered. However, this does not mean that there have been unsuccessful attempts to realise the energy transition using shallow geothermal energy resources. The case studies are fairly spatially dispersed, but low permeability zones combined with high operating flow rates could potentially create thermal interference between production and injection wells, which was not assessed in this study. However, our data provide additional evidence of successful operation for the cases studied and are likely to be representative of situations where groundwater resources are available. This fact shows the enormous energy potential for heating and cooling of tourism infrastructures. The cases investigated in this work have shown that the operation of geothermal systems in the coastal aquifers of volcanic islands has, so far, not been affected by brackish and saline groundwater found in the natural saltwater mixing zone of the coast. It can be concluded that the non-consumptive use of groundwater resources observed in the investigated geothermal systems tends to ensure the sustainable use of groundwater in the coastal aquifers, as thermal injection seems to prevent seawater intrusion phenomena.

This work provides evidence of the economic, energy and environmental advantages of using geothermal heat pump technology on volcanic islands and serves as the basis for a roadmap for energy planning on volcanic islands. It also demonstrates the feasibility of such thermal installations and their ability to meet the enormous heating and cooling demands of year-round tourism. Shallow geothermal technology is available for consideration in the energy plans required in the current energy decarbonisation scenario, thus helping to meet the Clean Energy Transition Agenda that the EU is demanding.

References

Alcaraz M, García-Gil A, Vázquez-Suñé E, Velasco V (2016a) Advection and dispersion heat transport mechanisms in the quantification of shallow geothermal resources and associated environmental impacts. Sci Total Environ 543(Part A):536–546

Alcaraz M, García-Gil A, Vázquez-Suñé E, Velasco, V (2016b) Use rights markets for shallow geothermal energy management. Applied Energy 172:34–46. https://doi.org/10.1016/j.apenergy.2016.03.071

Anguita F, Hernán F (2000) The Canary Islands origin: a unifying model. J Volcanol Geoth Res. 103(1):1–26. https://doi.org/10.1016/S0377-0273(00)00195-5

Bayer P, Saner D, Bolay S, Rybach L, Blum P (2012) Greenhouse gas emission savings of ground source heat pump systems in Europe: a review. Renew Sustain Energy Rev 16(2):1256–1267. https://doi.org/10.1016/j.rser.2011.09.027

Boschetti T (2013) Oxygen isotope equilibrium in sulphate-water systems: a revision of geothermometric applications in low-enthalpy systems. J Geochem Explor 124:92–100. https://doi.org/10.1016/j.gexplo.2012.08.011

Buffa S, Cozzini M, D'Antoni M, Baratieri M, Fedrizzi R (2019) 5th generation district heating and cooling systems: a review of existing cases in Europe. Renew Sustain Energy Rev 104:504–522. https://doi.org/10.1016/j.rser.2018.12.059

Chen F, Duic N, Manuel Alves L, da Graça Carvalho M (2007) Renew islands—Renewable energy solutions for islands. Renew Sustain Energy Rev 11(8):1888–1902. https://doi.org/10.1016/j.rser.2005.12.009

Custodio E (1989) Groundwater characteristics and problems in volcanic rock terrains. IAEA, International Atomic Energy Agency (IAEA)

De-Clercq S, Proka A, Jensen J, Carrero MM (2019) Clean Energy for EU Islands Secretariat. In: Secretariat, C.E.f.E.I. (Ed.), Brussels

Fetter CW (2018) Applied hydrogeology, 4th edn. Waveland Press

García-Gil A. et al., (2020) Governance of shallow geothermal energy resources. Energy Policy 138:111283. https://doi.org/10.1016/j.enpol.2020.111283

Gómez-Ulla A, Sigmarsson O, Huertas MJ, Devidal J-L, Ancochea E (2018) The historical basanite – alkali basalt - tholeiite suite at Lanzarote, Canary Islands: Carbonated melts of heterogeneous mantle source? Chem Geol 494:56–68. https://doi.org/10.1016/j.chemgeo.2018.07.015

Herrera Lameli C, Custodio Gimena E (2011) Flujo de agua subterránea en la parte central de la isla de Fuerteventura, archipiélago de Canarias, España. In: cabrera MC, Jiménez J, Custodio E. (Eds.), El conocimiento de los recursos hídricos en Canarias cuatro décadas después del proyecto SPA-15. Asociación Internacional de Hidrogeólogos - Grupo Español, Las Palmas de Gran Canaria, pp 93–107

IGME (1987) Geothermal research in the Canary Islands and geothermal resources and reserves assessment of Spain (in Spanish). Instituto Geológico y Minero de España

IGME (2017) Continuous digital geological map of Spain (GEODE), Cuenca del Ebro (Zona-2700)

Jaudin F (2013) D2.2: general report of the current situation of the regulative framework for the SGE systems. Regeocities 50

Poulsen SE, Balling N, Nielsen SB (2015) A parametric study of the thermal recharge of low enthalpy geothermal reservoirs. Geothermics 53:464–478. https://doi.org/10.1016/j.geothermics.2014.08.003

Quick H, Michael J, Arslan U, Huber H (2013) Geothermal application in low-enthalpy regions. Renewable Energy 49:133–136. https://doi.org/10.1016/j.renene.2012.01.047

Saeid S, Al-Khoury R, Nick HM, Hicks MA (2015) A prototype design model for deep low-enthalpy hydrothermal systems. Renewable Energy 77:408–422. https://doi.org/10.1016/j.renene.2014.12.018

Saner D et al (2010) Is it only CO2 that matters? A life cycle perspective on shallow geothermal systems. Renew Sustain Energy Rev 14(7):1798–1813. https://doi.org/10.1016/j.rser.2010.04.002

Santamarta JC, et al., (2021) The clean energy transition of heating and cooling in touristic infrastructures using shallow geothermal energy in the Canary Islands. Renewable Energy 165(1):712–718

GPSR Compliance
The European Union's (EU) General Product Safety Regulation (GPSR) is a set of rules that requires consumer products to be safe and our obligations to ensure this.

If you have any concerns about our products, you can contact us on

ProductSafety@springernature.com

In case Publisher is established outside the EU, the EU authorized representative is:

Springer Nature Customer Service Center GmbH
Europaplatz 3
69115 Heidelberg, Germany

www.ingramcontent.com/pod-product-compliance
Ingram Content Group UK Ltd.
Pitfield, Milton Keynes, MK11 3LW, UK
UKHW021009290726
14059UKWH00001BA/35

* 9 7 8 3 0 3 0 9 2 2 5 7 3 *